Frontiers in Mathematics

Francesco Catoni
Dino Boccaletti
Roberto Cannata
Vincenzo Catoni
Enrico Nichelatti
Paolo Zampetti

The Mathematics
of
Minkowski
Space-Time

With an Introduction to
Commutative Hypercomplex Numbers

Birkhäuser Verlag
Basel · Boston · Berlin

Authors:
Francesco Catoni
Via Veglia 10
00141 Roma
Italy

Vincenzo Catoni
Via Veglia 10
00141 Roma
Italy
e-mail: vjncenzo@yahoo.it

Dino Boccaletti
Dipartimento di Matematica
Università di Roma "La Sapienza"
Piazzale Aldo Moro 2
00185 Roma
Italy
e-mail: boccaletti@uniroma1.it

Enrico Nichelatti
ENEA, C.R. Casaccia
Via Anguillarese 301
00123 Roma
Italy
e-mail: nichelatti@casaccia.enea.it

Roberto Cannata
ENEA, C.R. Casaccia
Via Anguillarese 301
00123 Roma
Italy
e-mail: cannata@casaccia.enea.it

Paolo Zampetti
ENEA, C.R. Casaccia
Via Anguillarese 301
00123 Roma
Italy
e-mail: zampetti@casaccia.enea.it

2000 Mathematical Subject Classification: 20G20, 22E43, 30G35, 32Q35, 52A55, 53A05, 53B30

Library of Congress Control Number: 2008923068

Bibliographic information published by Die Deutsche Bibliothek
Die Deutsche Bibliothek lists this publication in the Deutsche Nationalbibliografie;
detailed bibliographic data is available in the Internet at <http://dnb.ddb.de>.

ISBN 978-3-7643-8613-9 Birkhäuser Verlag AG, Basel · Boston · Berlin

© 2008 Birkhäuser Verlag AG
Basel · Boston · Berlin
P.O. Box 133, CH-4010 Basel, Switzerland
Part of Springer Science+Business Media
Printed on acid-free paper produced from chlorine-free pulp. TCF ∞

ISBN 978-3-7643-8613-9   ISBN 978-3-7643-8614-6 (eBook)
9 8 7 6 5 4 3 2 1   www.birkhauser.ch

Non omnes arbusta iuvant humilesque myricae

To our friend and colleague Mario Lizzi,
who always strove for a non-humdrum life,
for his fundamental help in all the steps of our research.

Preface

This book arises from original research of the authors on hypercomplex numbers and their applications ([8] and [15]–[23]). Their research concerns extensions to more general number systems of both well-established applications of complex numbers and of functions of a complex variable.

Before introducing the contents of the book, we briefly recall the epistemological relevance of *Number* in the development of Western Science. In his "Metaphysics of number", Pythagoras considered reality, at its deepest level, as mathematical in nature. Following Pythagoras, Plato (Timaeus) explained the world by the regular polygons and solids of Euclidean geometry, laying a link between Number, Geometry and Physical World that represents the foundation of Modern Science. Accordingly, Galileo (Il Saggiatore § 6) took geometry as the language of Nature. These ideas that may appear trivial to modern rationalism, still have their own validity. For example, imaginary numbers make sense of algebraic equations which, from a geometrical point of view, could represent problems that admit no solutions. Despite such an introduction, complex numbers are strictly related to Euclidean geometry (Chap. 3) and allow formalizing Euclidean trigonometry (Chap. 4). Moreover, their functions are the means of representing the surface of the Earth on a plane (Chap. 8). In more recent times, another astonishing coincidence has been added to the previous ones: the space-time symmetry of two-dimensional Special Relativity, which, after Minkowski, is called Minkowski geometry, has been formalized [18] by means of *hyperbolic numbers*, a number system which represents the simplest extension of complex numbers [81].

Finally, N-dimensional Euclidean geometries and number theory have found a unified language by means of "Clifford algebra" [14], [42] and [45], which has allowed a unified formalization of many physical theories.

In this book, we expose first the same thread of association between *numbers* and geometries; secondly we show how the applications of the functions of a complex variable can be extended. In particular, after providing the basics of the classical theory of hypercomplex numbers, we show that with the commutative systems of hypercomplex numbers, a geometry can be associated. All these geometries except the one associated with complex numbers, are different from the Euclidean ones. Moreover the geometry associated with hyperbolic numbers is as distinctive as the Euclidean one since it matches the two-dimensional space-time geometry. This correspondence allows us to formalize space-time geometry and trigonometry with the same rigor as the Euclidean ones. As a simple application, we obtain an exhaustive solution of the "twin paradox". We suggest that, together with the introductory Sect. 2.2, these topics could be used as background for a university course in two-dimensional hyperbolic numbers and their application to space-time geometry and physics, such as *the mathematics of two-dimensional Special Relativity*.

After such algebraic applications of hyperbolic numbers, we broaden the study of space-time symmetry by introducing the functions of a hyperbolic variable. These functions allow us to extend the studies usually performed in Euclidean space, by means of functions of a complex variable, to two-dimensional space-time varieties. In addition, to offer to a larger audience the opportunity of appreciating these topics, we provide a brief discussion of both the introductory elements of Gauss' differential geometry and the classical treatment of constant curvature surfaces in Euclidean space. Nevertheless, for a better understanding, the reader should have a good knowledge of advanced mathematics, such as the theory of functions of a complex variable and elements of differential geometry.

The applications of hyperbolic numbers to Special Relativity may increase interest in multidimensional commutative hypercomplex systems. For these systems, functions can be introduced in the same way as for complex and hyperbolic variables. Therefore, we introduce in three appendices an outline of a research field that should be further developed for both a more complete mathematical formalization and an examination of physical applications.

In Appendices A and B, we begin the study of commutative hypercomplex systems with the four-unit system that has two relevant properties.

– Four unities closely recall the four-dimensional space-time.

– Their two-dimensional subsystems are given by complex and hyperbolic numbers whose applicative relevance is shown in the book.

Coming back to hypercomplex number systems, their algebraic theory was completed at the beginning of the XXth century [76] and concluded, in our view, with the article *Théorie des nombres* written by E. Cartan for the French edition of the *Enciclopédie des sciences mathématiques* [13]. This article is an extensive revision of E. Study's article for the German edition of the Encyclopedia (*Enzyklopädie der Mathematischen Wissenschaften*). Both these authors made contributions to the development of the theory of hypercomplex numbers. Today these numbers are included as a part of abstract algebra [46], and only a few uncorrelated papers introduce their functions. Therefore, to give new insights and inspiration to scientists interested in other fields (not abstract algebra), in Appendix C we give a rigorous and self-consistent exposition of algebra and function theory for commutative hypercomplex numbers by means of matrix formalism, a mathematical apparatus well known to the scientific community.

As a final observation, we remark that in this book many different mathematical fields converge as a confirmation of David Hilbert's assertion: *Mathematics is an organism that keeps its vital energy from the indissoluble ties between its various parts*, and — we shall add following Klein, who refers to Riemann's ideas — *from the indissoluble ties of Mathematics with Physics and, more generally, with Applied Sciences.*

Since the content involves different fields, this book is addressed to a larger audience than the community of mathematicians. As a consequence, also the language employed is aimed at this larger audience.

It is a pleasant task for us to thank Prof. Stefano Marchiafava of Rome University "La Sapienza", for useful discussions and encouragement in many steps of our work.

Contents

List of Figures

List of Tables

Chapter 1

Introduction

Complex numbers represent one of the most intriguing and emblematic discoveries in the history of science. Even if they were introduced for an important but restricted mathematical purpose, they came into prominence in many branches of mathematics and applied sciences. This association with applied sciences generated a synergistic effect: applied sciences gave relevance to complex numbers and complex numbers allowed formalizing practical problems. A similar effect can be found today in the "system of hyperbolic numbers", which has acquired meaning and importance as the *Mathematics of Special Relativity*, as shown in this book. Let us proceed step by step and begin with the history of complex numbers and their generalization.

Complex numbers are today introduced in relation with square roots of negative numbers and are considered as an extension of the real numbers. However, they were introduced in the XVIIth century for solving a mathematical paradox: to give a sense to the real solutions of cubic equations that appear as the sum of square roots of negative quantities. Their introduction was thorny and the square roots of negative quantities are still called *imaginary numbers* and contain the symbol "i" which satisfies the relation $i^2 = -1$. *Complex numbers* are those given by the symbolic sum of one real and one imaginary number. This sum is a symbolic one because it does not represent the usual sum of "homogeneous quantities", rather a "two component quantity" written as $z = x + iy$. Today we know another two-component quantity: the plane vector, which we write $\mathbf{v} = \mathbf{i}x + \mathbf{j}y$, where $\mathbf{i}$ and $\mathbf{j}$ represent two unit vectors indicating the coordinate axes in a Cartesian representation. Despite there being no a priori indication that a complex number could represent a vector on a Cartesian plane, complex numbers were the first representation of two-component quantities on a Cartesian (or Gauss–Argand) plane. Klein explained this correspondence by associating geometries with groups and showing ([30], p. 451) that both additive and unimodular multiplicative groups of complex numbers are the same groups characterizing Euclidean geometry (translations and rotations, respectively). This association was largely used in the XIXth century also in the extensions of Euclidean geometry to differential and non-Euclidean geometries (see Chapters 8 and 9). The mathematical formalization of group theory by Sophus Lie is one of the milestones in the development of science in the XIXth century. Actually, also from a philosophical point of view, the concept of group recalls an idea as old as the beginning of scientific research: to find the common properties in the changes of Nature. This connection generated a strong link between the mathematics of group theory and applied sciences [9].

Coming back to complex numbers, their association with groups allowed Lie to generalize complex numbers to hypercomplex numbers [50]. However, Gauss had observed, half a century before, that the generalization of real and complex numbers cannot be done without losing some peculiarities of algebraic operations. In particular, one must either renounce the commutative property of the product or acknowledge that the product between some non-zero numbers is zero. These numbers have been called *divisors of zero.*

Following the first possibility, Hamilton introduced the non-commutative quaternions to represent vectors in space. This representation was the first step for the introduction of vectors as they are used today.

Acknowledgement of the second property took longer, and only in recent years has it been recognized that as complex numbers are associated with Euclidean geometry, so two other systems of two-dimensional hypercomplex numbers (called parabolic and hyperbolic) can be associated with groups (or geometries) of physical relevance. Specifically, parabolic numbers can be associated with Galileo's group of classical mechanics and hyperbolic numbers with Lorentz's group of special relativity [80] and [81].

It is easy to infer the synergy that can grow from the link between hyperbolic numbers and Special Relativity. Actually the scientific relevance of Special Relativity stimulated the studies and developments of hyperbolic numbers and, vice versa, these numbers were recognized as the most suitable mathematical tool for studying problems in space-time. In this book, we show their potential beginning with an exhaustive formalization of space-time trigonometry. This formalization is obtained by looking for analogies and differences between complex and hyperbolic numbers. In particular, we start from the relations:

- Euclidean geometry ⇔ complex numbers,

- complex numbers ⇔ hyperbolic numbers,

- hyperbolic numbers ⇔ Minkowski space-time geometry,

in the following way: it is well known that all theorems of Euclidean geometry and trigonometry are obtained through elementary geometry observations. Otherwise complex numbers allow one to formalize, in a Cartesian plane, the trigonometric functions as a direct consequence of Euclid's rotation group (Sect. 4.2.1), and allow one to obtain all the trigonometry theorems by an analytical method as mathematical identities (Chapter 4). These considerations within Euclidean geometry are extended to the space-time geometry associated with hyperbolic numbers. Actually, the definition of hyperbolic trigonometric functions directly from the Lorentz group of Special Relativity allows us to demonstrate theorems through an algebraic approach. In this way, we obtain a mathematical tool that, from an axiomatic and practical point of view, is equivalent to Euclidean geometry (see the introduction to Chapter 9).

A similar parallel approach, between complex and hyperbolic numbers, is used in Chapters 7–10 to introduce and apply the functions of a hyperbolic variable. Actually the applicative relevance of functions of a complex variable was shown for the first time by Euler ([25], p. 37), who used these functions for representing the motion of a fluid, and later by Gauss, who showed that they give the best way for mapping an arbitrary surface onto another and, in particular, the Earth's surface onto a plane. The Gauss' paper is summarized in Chapter 8, where we extend his results to space-time geometry.

These obtained results can be considered as the starting points for further developments, since the properties of complex numbers can represent a guideline for studying multidimensional number systems, as they have been for the applications of both hyperbolic numbers and functions of a hyperbolic variable. Actually, the rediscovery of hyperbolic numbers and their applications to Special Relativity show that also numbers with "zero divisors" can have applications. This insight increases the interest in commutative multidimensional systems of numbers that have two notable properties in common with complex and hyperbolic numbers:

1. Every hypercomplex number generates its own geometry.

2. As M. G. Scheffers demonstrated [64], for the aforesaid number systems, the differential calculus can be introduced and functions can be defined, as they are for a complex variable.

As far as the first point is concerned, the "invariant quantity" of these geometries is a form of degree N (Chapter 3), instead of the quadratic Euclidean distance as it is the case for Hamilton quaternions and for complex and hyperbolic numbers. This implies that, if from one point of view this observation could limit the interest in these geometries, from another one their investigation could open the way to new applications.

As for the second point, the possibility of so introducing these functions opens two ways for applications. The first one comes from the fact that these functions satisfy some particular partial differential equations (Generalized Cauchy–Riemann conditions), which may represent physical fields as well as the functions of complex variables do[1]. The second application is related to the fact that, as the functions of complex and hyperbolic numbers can be used for studying non-flat surfaces (Chapter 9), so the functions of a hypercomplex variable can be used for studying (Appendix B) non-flat spaces associated with the geometries introduced by the multidimensional commutative hypercomplex numbers.

---

[1]In particular L. Sobrero applied the functions of a particular four-dimensional hypercomplex number to the theory of elasticity [71].

Chapter 2

N-Dimensional Commutative Hypercomplex Numbers

As summarized in the preface, hypercomplex numbers were introduced before the linear algebra of matrices and vectors. In this chapter, in which we follow a classical approach, their theory is developed mainly by means of elementary algebra, and their reference to a representation with matrices, vectors or tensors is just for the practical convenience of referring to a widely known language. In particular, the down or up position of the indexes, which in tensor calculus are named *covariance and contravariance*, respectively, indicates if the corresponding quantities (vectors) are transformed by a direct or inverse matrix (see Section 2.1.4). We also use Einstein's convention for tensor calculus and omit the sum symbol on the same covariant and contravariant indexes; in particular, we indicate with Roman letters the indexes running from 1 to $N - 1$, and with Greek letters the indexes running from 0 to $N - 1$.

2.1 N-Dimensional Hypercomplex Numbers

2.1.1 Equality and Sum

Hypercomplex numbers are defined by the expression ([13], Vol. **II**, p. 107) and [50]

$$x = \sum_{\alpha=0}^{N-1} e_\alpha x^\alpha \Rightarrow e_\alpha x^\alpha , \qquad (2.1.1)$$

where $x^\alpha \in \mathbf{R}$ are called *components* and $e_\alpha \notin \mathbf{R}$ are called *units*, *versors* or *bases* as in vector algebra. Also the elements zero, equals and sum are defined as in vector or complex algebra.

- The null (or zero) element is the element with all components equal to zero.

- Two hypercomplex numbers, x and y, are equal if $x^\alpha = y^\alpha$, $\forall\, \alpha$.

- z represents the sum $x + y$ if $z^\alpha = x^\alpha + y^\alpha$, $\forall\, \alpha$.

- As in vector algebra, two hypercomplex numbers are mutually proportional if $x^\alpha = \lambda\, y^\alpha \,\forall\, \alpha$, with $\lambda \in \mathbf{R}$.

From these definitions it follows that the algebraic sum satisfies commutative and associative properties and the null element exists.

Given $K \leq N$ hypercomplex numbers, these numbers are linearly independent if the characteristic of the matrix of the components is K. As in vector
algebra, N also represents the maximum number of independent hypercomplex
numbers that can be taken as *unities or basis* of the hypercomplex system.

Two systems of hypercomplex numbers are equivalent systems if one can be
obtained from the other by means of a non-singular linear transformation of the
versors and the corresponding inverse transformation of the variables, as is usual
for quadratic forms in tensor calculus (see Section 2.1.4).

2.1.2   The Product Operation

An essential feature of hypercomplex numbers is the definition of the product.
We know from tensor calculus that for vectors in an N-dimensional space the
scalar product is a scalar quantity, whereas the cross-product is a tensor of rank
$N - 2$, thus it is a vector just for $N = 3$. Conversely, the product of hypercomplex
numbers is defined so that the result is still a hypercomplex number. This property
is also true for the "inverse operation", i.e., division, which does not exist for
common vectors, whereas it generally exists for hypercomplex numbers. For real
and complex numbers the product operation satisfies the following properties:

1. distributive with respect to the sum,

2. associative,

3. commutative,

4. does not have *divisors of zero*.

We note that property 1 is taken as an axiom in linear algebra both for
addition of numbers $((\alpha+\beta)\mathbf{v} = \alpha\,\mathbf{v}+\beta\,\mathbf{v})$ and for addition of vectors $(\alpha(\mathbf{v_1}+\mathbf{v_2}) =
\alpha\,\mathbf{v_1}+\alpha\,\mathbf{v_2})$ ([81], p. 243). About property 3, it is shown in Appendix C, p. 217 that
commutative hypercomplex systems with unity are associative, while associative
systems can be non-commutative, such as the Hamilton quaternions.

The first three properties are well known for real and complex numbers. As
far as the fourth one is concerned, we shall explain below the meaning of divisors
of zero.

It can be shown [65] that, only the real and complex numbers can satisfy
these four properties all together. As far as other systems are concerned, *if they
satisfy the first two properties they can satisfy just one of the last two.*

The most famous and most studied system of hypercomplex numbers is the
system of non-commutative Hamilton quaternions. In this book, we consider commutative systems for which the following *fundamental theorem*, due to Scheffers,
is valid [64].

Theorem 2.1. *For distributive systems with the unity, the differential and integral
calculus does exist only if the systems are commutative.*

In Appendix C, p. 237 we demonstrate, by using matrix formalism, that for commutative systems the differential calculus does exist. The *differential calculus* allows us to define the functions of a hypercomplex variable and to associate them with the infinite-dimensional Lie group of functional mappings, as we shall see in Section 7.2.3. Thanks to these important properties, commutative hypercomplex numbers can be considered as an extension of real and complex numbers.

As in vector algebra, the product of two hypercomplex numbers is defined if the product between the versors is defined. This product, for the considerations discussed above, must be given by a linear combination of versors. Equation (2.1.1) defines a hypercomplex number if the versor multiplication rule is given by

$$e_\alpha e_\beta = e_\gamma C_{\alpha\beta}^\gamma, \tag{2.1.2}$$

where the constants $C_{\alpha\beta}^\gamma \in \mathbf{R}$ are called *structure constants* and define the characteristics of the system. If $C_{\alpha\beta}^\gamma = C_{\beta\alpha}^\gamma$, $\forall\, (\alpha, \beta, \gamma)$, the system is commutative.

As stated by Scheffers' theorem, we consider systems which have a *unity* versor e_0 such that $e_0^2 = e_0$, $e_0\, e_\alpha = e_\alpha$. This unity versor can be replaced by "1" and then omitted.

Let us consider the product, $z = e_\gamma z^\gamma$, of two hypercomplex numbers, $x = e_\alpha x^\alpha$ and $y = e_\beta y^\beta$, which, as a function of the components, is given by

$$e_\gamma z^\gamma = e_\alpha x^\alpha e_\beta y^\beta \equiv e_\gamma C_{\alpha\beta}^\gamma x^\alpha y^\beta \quad \Rightarrow \quad z^\gamma = C_{\alpha\beta}^\gamma x^\alpha y^\beta \equiv X_\beta^\gamma y^\beta, \tag{2.1.3}$$

where we have set

$$X_\beta^\gamma \equiv C_{\alpha\beta}^\gamma x^\alpha. \tag{2.1.4}$$

The expressions (2.1.3) are symmetric with respect to x and y, therefore we could also set $Y_\alpha^\gamma \equiv C_{\alpha\beta}^\gamma y^\beta$. We shall see that these two substitutions are equivalent.

2.1.3 Characteristic Matrix and Characteristic Determinant

The two-index elements X_β^γ can be considered as the elements of a matrix in which the upper index indicates the rows and the lower one the columns. We note that the N^2 elements of this matrix depend on the structure constants and on the N components of x. Therefore, even if introduced in a formal way, this matrix can be used as a representation of the hypercomplex number x as shown in Section 2.1.3 and more diffusely in Appendix C. For this reason, it is called the *characteristic matrix* of the hypercomplex number x. The determinant

$$\|X\| \tag{2.1.5}$$

is called the *characteristic determinant* and is an *invariant quantity* of the number systems (Section 2.1.4). In general, we indicate the characteristic matrix by capital letters with two indexes and the characteristic determinant by capital letters.

The elements of the characteristic matrix can be obtained by considering the N hypercomplex numbers $e_\beta x$ for $\beta = 0, \ldots, N-1$. We have:

Theorem 2.2. *The N^2 coefficients of the versors give the elements of the characteristic matrix.*

Proof.
$$e_\beta\, x \equiv e_\beta\, e_\alpha\, x^\alpha \equiv e_\gamma\, C^\gamma_{\alpha\beta} x^\alpha = e_\gamma\, X^\gamma_\beta. \qquad (2.1.6)$$

Then β determines the columns and the coefficients of the versors e_γ give the rows.   □

The system of hypercomplex numbers is associative for multiplication if $(x\,y)\,z = x\,(y\,z)$. If this condition is expressed as a function of the components, we obtain

$$(xy)z = (e_\gamma C^\gamma_{\alpha\beta} x^\alpha y^\beta)\, e_\delta\, z^\delta \equiv e_\epsilon\, C^\epsilon_{\gamma\delta}\, C^\gamma_{\alpha\beta} x^\alpha y^\beta z^\delta,$$

$$x(yz) = e_\alpha x^\alpha (e_\gamma C^\gamma_{\beta\delta} y^\beta\, z^\delta) \equiv e_\epsilon\, C^\epsilon_{\alpha\gamma}\, C^\gamma_{\beta\delta} x^\alpha y^\beta z^\delta.$$

Then, a system of hypercomplex numbers is associative if the structure constants satisfy the relations

$$C^\epsilon_{\gamma\delta} C^\gamma_{\alpha\beta} = C^\epsilon_{\alpha\gamma} C^\gamma_{\beta\delta}, \ \forall\, (\, \alpha,\, \beta,\, \delta,\, \epsilon). \qquad (2.1.7)$$

Some Properties

The product of two hypercomplex numbers is something new if compared with the product of vectors. Nevertheless, (2.1.3) allows us to establish a link between hypercomplex numbers and the linear algebra of matrices and vectors. The hypercomplex numbers y and z can be represented as row vectors, whereas the combination of the structure constants and the hypercomplex number x, represented by (2.1.4), is a matrix. In this way, the product of two hypercomplex numbers is equivalent to a vector-matrix product. Indeed, we have:

Theorem 2.3. *All the associative hypercomplex numbers can be represented by a characteristic matrix and their product by a matrix-matrix product.*

Proof. From the definition of the characteristic matrix (2.1.4) and from (2.1.3), (2.1.7), we have
$$Z^\gamma_\beta \equiv C^\gamma_{\alpha\,\beta} z^\alpha = C^\gamma_{\alpha\,\beta} C^\alpha_{\mu\,\nu} x^\mu\, y^\nu \overset{(2.1.7)}{\longrightarrow}$$
$$C^\gamma_{\mu\,\alpha} C^\alpha_{\nu\,\beta} x^\mu\, y^\nu \equiv \left(C^\alpha_{\nu\,\beta}\, y^\nu\right)\left(C^\gamma_{\mu\,\alpha}\, x^\mu\right) \equiv Y^\alpha_\beta\, X^\gamma_\alpha. \qquad (2.1.8)$$
□

It follows, from (2.1.8) ([65], p. 291), that

Theorem 2.4. *The product of two characteristic matrices is a characteristic matrix.*

All the products in (2.1.8) are commutative; then, the same property holds for the final product of matrices. For commutative hypercomplex numbers, the matrix representation (instead of the vector representation) is more suitable. Indeed for these systems the form of the characteristic matrices is such that their product is commutative, in agreement with the properties of the numbers. From the characteristic matrix, one can also obtain the matrix representation of the basis e_α. Actually the matrix that represents the versor e_α is obtained by setting

$$x^\beta = \delta_\alpha^\beta \text{ where } \begin{cases} \delta_\alpha^\beta = 1 & \text{for } \alpha = \beta, \\ \delta_\alpha^\beta = 0 & \text{for } \alpha \neq \beta \end{cases} \tag{2.1.9}$$

in the characteristic matrix of x^1.

As a consequence of (2.1.8), we can extend to the characteristic determinant a property that is well known for complex numbers. We have

Theorem 2.5. *The product of two characteristic determinants is a characteristic determinant.*

Then, given the hypercomplex numbers x and y, from a property of determinants it follows that

$$\|X \cdot Y\| = \|X\| \cdot \|Y\|. \tag{2.1.10}$$

2.1.4 Invariant Quantities for Hypercomplex Numbers

Let us briefly recall the algebraic invariant quantities for equivalent hypercomplex systems. Let us consider a linear transformation from the versors e_α to the versors $\bar{e}_\beta$ by means of a non-singular matrix with elements (a_α^β), and let us call $(a_\beta^\alpha)^{-1}$ the elements of the inverse matrix; then

$$e_\alpha = a_\alpha^\beta \bar{e}_\beta \text{ and for the components } x^\alpha = (a_\gamma^\alpha)^{-1} \bar{x}^\gamma \text{ with } a_\alpha^\beta (a_\gamma^\alpha)^{-1} = \delta_\gamma^\beta. \tag{2.1.11}$$

As it happens for quadratic forms, (2.1.1) is invariant,

$$e_\alpha x^\alpha = a_\alpha^\beta \bar{e}_\beta (a_\gamma^\alpha)^{-1} \bar{x}^\gamma \equiv \bar{e}_\beta \bar{x}^\beta. \tag{2.1.12}$$

This property justifies the position of indexes in agreement with the convention of tensor calculus for covariant and contravariant indexes.

By means of transformation (2.1.11), the structure constants become

$$C_{\alpha\beta}^\gamma = \overline{C}_{\epsilon\mu}^\nu (a_\nu^\gamma)^{-1} a_\alpha^\epsilon a_\beta^\mu. \tag{2.1.13}$$

Proof. Putting $\bar{e}_\alpha \bar{e}_\beta = \bar{e}_\gamma \overline{C}_{\alpha\beta}^\gamma$ and substituting the first of (2.1.11) into (2.1.1), we get

$$e_\alpha e_\beta \equiv \begin{cases} a_\alpha^\epsilon \bar{e}_\epsilon a_\beta^\mu \bar{e}_\mu \equiv a_\alpha^\epsilon a_\beta^\mu \overline{C}_{\epsilon\mu}^\nu \bar{e}_\nu \\ e_\gamma C_{\alpha\beta}^\gamma \equiv \bar{e}_\nu a_\gamma^\nu C_{\alpha\beta}^\gamma \end{cases} \Rightarrow a_\alpha^\epsilon a_\beta^\mu \overline{C}_{\epsilon\mu}^\nu \bar{e}_\nu = \bar{e}_\nu a_\gamma^\nu C_{\alpha\beta}^\gamma \tag{2.1.14}$$

and, by multiplying the last expression by $(a_\nu^\gamma)^{-1}$, (2.1.13) follows. □

---

[1]The characteristic matrix is composed of real elements. Therefore, also this representation of versors is given by real numbers. This fact is not in contradiction with the statement $e_k \notin \mathbf{R}$ because one must not confuse e_k for its representation.

We have also

Theorem 2.6. *The characteristic determinant is invariant for equivalent systems of hypercomplex numbers.*

Proof. The elements of the characteristic matrix are transformed as

$$X_\alpha^\gamma \equiv C_{\alpha\beta}^\gamma \, x^\beta = \overline{C}_{\epsilon\mu}^\nu \, (a_\nu^\gamma)^{-1} \, a_\alpha^\epsilon \, a_\beta^\mu \, (a_\lambda^\beta)^{-1} \, \bar{x}^\lambda$$

$$\equiv \overline{C}_{\epsilon\mu}^\nu \, (a_\nu^\gamma)^{-1} \, a_\alpha^\epsilon \, \delta_\lambda^\mu \, \bar{x}^\lambda \equiv \overline{C}_{\epsilon\mu}^\nu \, \bar{x}^\mu \, (a_\nu^\gamma)^{-1} \, a_\alpha^\epsilon \equiv \overline{X}_\epsilon^\nu \, (a_\nu^\gamma)^{-1} \, a_\alpha^\epsilon \qquad (2.1.15)$$

and, by taking the determinants of the first and last terms, we obtain the statement. □

Theorem 2.7. *The trace of the characteristic matrix is invariant.*

Proof. From the first and last terms of (2.1.15), we get

$$X_\alpha^\alpha = \overline{X}_\epsilon^\nu \, (a_\nu^\alpha)^{-1} \, a_\alpha^\epsilon \equiv \overline{X}_\epsilon^\nu \, \delta_\nu^\epsilon \equiv \overline{X}_\nu^\nu. \qquad (2.1.16)$$

□

2.1.5   The Division Operation

Let us consider the hypercomplex numbers z and x. The division z/x is defined if a hypercomplex number y exists such that $z = x\,y$. Thus the components y^β must satisfy the linear system (2.1.3), and this means that the determinant (the characteristic determinant) of the system (2.1.3) must be $\neq 0$. Since the elements of this determinant depend on the structure constants as well as on the components of the hypercomplex number, the determinant (2.1.5) can be equal to zero for particular numbers, i.e., some numbers $x \neq 0$ may exist for which $\|X\| = 0$. For these numbers, called *divisors of zero* [13] and [81], the division is not possible, as shown in an example in Section 2.2.2. For these numbers also *the product is not univocally determined.*

Proof. Let us consider a divisor of zero, $\bar{y} = e_\beta \bar{y}^\beta$. The system (2.1.3) in the unknown x^α has its determinant equal to zero. Thus, for the homogeneous system ($z = 0$), infinite solutions with $\bar{x} \neq 0$ exist. Actually if $x\bar{y} = z$ also $(x + \bar{x})\bar{y} = z$, i.e., the multiplication of y by different numbers gives the same result. □

For commutative numbers, also $\bar{x}$ are divisors of zero, generally different from $\bar{y}$. The divisors of zero are divided into groups; products of numbers of one group times numbers of another one are zero (see Appendix C).

2.1.6   Characteristic Equation and Principal Conjugations

Hypercomplex numbers can be related to matrix algebra [65] and, as in matrix algebra, a characteristic equation exists. This equation can be obtained by considering for $n = 1, \ldots, N$ the nth power of the hypercomplex number x. From $N - 1$

equations of this set, we can determine the $N-1$ versors[2] as rational functions of the hypercomplex number and of the components. By substituting these versors in the Nth equation, we obtain an equation of degree N for the hypercomplex number x. This equation is called the *characteristic equation* or the *minimal equation*. We have ([13], Section 25) and ([65], p. 309)

Theorem 2.8. *The characteristic equation can be obtained from the eigenvalue equation for the characteristic matrix*

$$\|X_\beta^\gamma - \delta_\beta^\gamma x\| = 0, \quad where \quad \begin{cases} \delta_\beta^\gamma = 1 & for\ \gamma = \beta, \\ \delta_\beta^\gamma = 0 & for\ \gamma \neq \beta. \end{cases} \quad (2.1.17)$$

Proof. Equation (2.1.17) can be obtained from (2.1.6) by setting $e_\beta = e_\gamma\,\delta_\beta^\gamma$;

$$e_\beta\,x \equiv e_\gamma\,\delta_\beta^\gamma\,x = e_\gamma\,X_\beta^\gamma \quad \Rightarrow \quad e_\gamma\,(X_\beta^\gamma - \delta_\beta^\gamma\,x) = 0. \quad (2.1.18)$$

Equation (2.1.18) is equivalent to a linear system in the unknown e_γ. Then, since $e_\gamma \neq 0$, (2.1.17) follows. $\qquad \square$

The hypercomplex number x is a solution of (2.1.17); the other $N-1$ solutions are named *principal conjugations*. Let us now anticipate some assertions that are more thoroughly discussed in Appendix A for commutative quaternions and in Appendix C for general hypercomplex numbers. Let us indicate with $^k\bar{x}$ the principal conjugations of x; they have the following properties that are the same as the conjugate of a complex number:

1. the transformation $x^\alpha \leftrightarrow {}^k\bar{x}$ is bijective;

2. the principal conjugations are hypercomplex numbers with the same structure constants;

3. the product of the principal conjugations is the characteristic determinant of the hypercomplex number.

Let us demonstrate this last assertion, leaving the other two to the reader.

Proof. We can write the characteristic equation as the product of linear terms given by the differences between the unknown x and the roots $^k\bar{x}$: $\displaystyle\prod_{k=1}^{N-1} (x - {}^k\bar{x})$.
From this expression we see that the product of the roots is equal, but for the sign, to the known term of the characteristic equation (2.1.17), thus it is equal to the characteristic determinant. $\qquad \square$

Extending the concept of modulus of a complex number, the Nth root of the absolute value of the characteristic determinant is called the *modulus* of the

---

[2]The number of versors is $N-1$ since we are considering numbers with the unity versor. Otherwise for systems without the unity versor (decomposable systems, see Section 2.1.7) we have to consider the 0th power.

system and is indicated by $|x|$ [81]. It is well known that the modulus of a complex number coincides with the Euclidean distance, i.e., it is the *invariant* quantity of the Euclidean geometry (roto-translations). We shall see in Chapter 3 that a geometry in which $|x|$ is the invariant quantity can be defined and associated with all the systems of commutative hypercomplex numbers.

2.1.7 Decomposable Systems

There are hypercomplex numbers for which the versors can be classified into groups, if necessary by means of a suitable non-singular linear transformation, so that the product of two versors of a group belongs to the group and the product of versors of different groups is zero [13]. These systems are called *decomposable systems*, and an example will be shown in Section 2.2.2. As is known from group theory, each of these sub-systems must have its own unity. These unities are called *sub-unities*. We can also "compose" systems; in this case the composed system has as *main unity* the sum of the sub-unities and the other versors are obtained by a linear transformation of the versors of the component sub-systems. For these decomposable systems, the characteristic determinant is given by the product of the characteristic determinants of the sub-systems ([65], p. 310).

2.2   The General Two-Dimensional System

Taking into account that two-dimensional hypercomplex numbers are the basic elements of Chapters 4–10 of this book, we introduce them in the present section, giving a self-consistent exposition.

Let us consider the general two-dimensional system [47] for which we indicate the versors and the components with specific letters

$$\{z = x + u\,y;\ u^2 = \alpha + u\,\beta;\ x,\,y,\,\alpha,\,\beta \in \mathbf{R},\,u \notin \mathbf{R}\}, \tag{2.2.1}$$

where the second relation assures that the product of two hypercomplex numbers is a hypercomplex number itself. The structure constants are $C_{00}^1 = C_{01}^0 = 0$; $C_{00}^0 = C_{01}^1 = 1$; $C_{11}^0 = \alpha$; $C_{11}^1 = \beta$. From (2.1.6) and (2.2.1) one has

$$u\,z = \alpha y + u\,(x + \beta y), \tag{2.2.2}$$

and the characteristic matrix and the characteristic determinant are

$$\begin{pmatrix} x & \alpha\,y \\ y & x + \beta\,y \end{pmatrix};\qquad \begin{vmatrix} x & \alpha\,y \\ y & x + \beta\,y \end{vmatrix} \equiv x^2 - \alpha\,y^2 + \beta\,x\,y. \tag{2.2.3}$$

Regarding the product, if $z_1 = x_1 + u\,y_1$ and $z_2 = x_2 + u\,y_2$ are two hypercomplex numbers, on the strength of (2.2.1), we have

$$z_1\,z_2 = (x_1 + u\,y_1)(x_2 + u\,y_2) \equiv x_1\,x_2 + \alpha\,y_1\,y_2 + u\,(x_1\,y_2 + x_2\,y_1 + \beta\,y_1\,y_2). \tag{2.2.4}$$

For the inverse operation (division), given a number $a + u\,b$, one has to look for a number $z = x + u\,y$ such that

$$(a + u\,b)(x + u\,y) = 1. \tag{2.2.5}$$

If (2.2.5) is satisfied, the inverse z of $a + u\,b$ exists and we can divide any number by $a + u\,b$, by multiplying it by z.

The hypercomplex equation (2.2.5) is equivalent to the real system

$$a\,x + \alpha\,b\,y = 1,$$
$$b\,x + (a + \beta\,b)\,y = 0, \tag{2.2.6}$$

whose determinant is

$$D = a^2 + a\,\beta\,b - \alpha\,b^2 \equiv \left(a + \frac{\beta}{2}\,b\right)^2 - \left(\alpha + \frac{\beta^2}{4}\right) b^2.$$

If we set

$$\Delta = \beta^2 + 4\alpha, \tag{2.2.7}$$

then the parabola $\alpha = -\beta^2/4$ $(\Delta = 0)$ divides the (β, α) plane into three regions (see Fig. 2.1).

- Inside the region I $(\Delta < 0)$, D is always > 0 except for $a = b = 0$ (in this case $D = 0$). Therefore the division is always possible for every non-null number.

- In region II, the parabola $\Delta = 0$, D only vanishes on $a + \beta/2\,b = 0$.

- In region III $(\Delta > 0)$, D vanishes on the two lines $a + (\beta/2 \pm \sqrt{\Delta}/2)\,b = 0$. If $D = 0$, the system (2.2.6) has solutions for certain non-null a and b; the associated homogeneous system $a\,x + \alpha\,b\,y = 0$, $b\,x + (a + \beta\,b)\,y = 0$ admits non-null solutions. This is related to the existence of *divisors of zero*, numbers for which the product of two non-null hypercomplex numbers $a + u\,b$ and $x + u\,y$, is zero.

Therefore the sign of the real quantity $\Delta = \beta^2 + 4\alpha$ determines the possibility of executing the division between two hypercomplex numbers. The sign of Δ also determines the subdivision of the general two-dimensional hypercomplex number $z = x + u\,y$ into three different types. We start from the characteristic equation for z obtained by means of (2.1.17)

$$\begin{vmatrix} x - z & \alpha\,y \\ y & x + \beta\,y - z \end{vmatrix} = 0 \Rightarrow z^2 - z(2x + \beta y) + x^2 - \alpha y^2 + \beta xy = 0, \quad (2.2.8)$$

whose solutions are

$$z = \frac{2\,x + \beta\,y \pm y\,\sqrt{\beta^2 + 4\alpha}}{2}. \tag{2.2.9}$$

Equation (2.2.8), as mentioned in Section 2.1.6, can be also obtained by calculating the powers of $z = x + u\,y$; $z^2 = x^2 + \alpha\,y^2 + u\,y\,(2\,x + \beta\,y)$ and by substituting

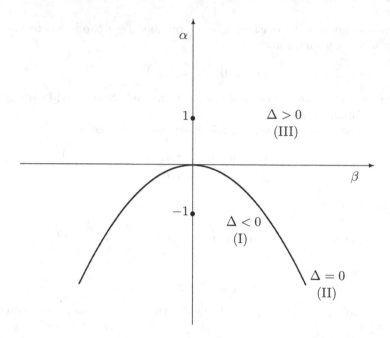

Figure 2.1: In this figure we represent in the $(\beta,\ \alpha)$ plane the parabola $\alpha = -\frac{\beta^2}{4}$ obtained by setting $\Delta \equiv \beta^2 + 4\alpha = 0$. The position of the point $P \equiv (\beta,\ \alpha)$ determines the three types of two-dimensional hypercomplex numbers.
(I) Inside the parabola ($\Delta < 0$) we call these systems *elliptic hypercomplex numbers*. (II) On the parabola ($\Delta = 0$), we call these systems *parabolic hypercomplex numbers*. (III) Outside the parabola ($\Delta > 0$), we call these systems *hyperbolic hypercomplex numbers*.

in the second equation the value of $u\,y$ obtained from the first one. Therefore Δ determines the nature of the solutions of the characteristic equation. On the other hand, we know that $u^2 = \alpha + u\,\beta$ is fundamental in the definition of the product of two hypercomplex numbers, according to (2.2.1) and (2.2.10); therefore, we can classify the hypercomplex numbers into three classes according to the position of the point $P \equiv (\beta,\ \alpha)$ in the $(\beta,\ \alpha)$ plane

1. If P is inside the parabola ($\Delta < 0$), any non-null element has an inverse and, as a consequence, division is possible for any non-null hypercomplex number. We call these systems *elliptic hypercomplex numbers*.

2. If P is on the parabola ($\Delta = 0$), we call these systems *parabolic hypercomplex numbers*. Each of them admits divisors of zero satisfying $x + (\beta/2)\,y = 0$. Division is possible for all the other numbers.

3. If P is outside the parabola ($\Delta > 0$), we call these systems *hyperbolic hypercomplex numbers*. Each of them admits divisors of zero satisfying $x + (\beta/2 \pm \sqrt{\Delta})\,y = 0$. Division is possible for all the other numbers.

Let us consider the conic with equation

$$x^2 + \beta\,x\,y - \alpha\,y^2 = 0,$$

where the left-hand side is the characteristic determinant (2.2.3). According to whether $\Delta \equiv \beta^2 + 4\alpha$ is < 0, $= 0$, > 0, the curve is an ellipse, a parabola or a hyperbola. This is the reason for the names used for the three types of hypercomplex numbers. For the three cases, we define our systems as *canonical systems* if

1. $u^2 = -1$ (i.e., $\alpha = -1$, $\beta = 0$). This is the case of the ordinary complex numbers. One can verify that any elliptic system is isomorphic to the canonical system.

2. $u^2 = 0$ (i.e., $\alpha = 0$, $\beta = 0$). One can verify that any parabolic system is isomorphic to the canonical system.

3. $u^2 = 1$ (i.e., $\alpha = 1$, $\beta = 0$). This system is related to the pseudo-Euclidean (space-time) geometry, as we shall see later in this book. One can verify that any hyperbolic system is isomorphic to the canonical system. In this case the divisors of zero satisfy $y = \pm x$.

Coming back to the characteristic equation, we have

Theorem 2.9. *The square root in* (2.2.9) *can be replaced with* $2\,u - \beta$.

Proof. Let us look for a complex number $a + u\,b$ so that $(a + u\,b)^2 \equiv \beta^2 + 4\alpha$. Taking into account the multiplication rule for u^2, we can write for a and b ($b \neq 0$), a system of degree 2,

$$2\,a + \beta\,b = 0; \qquad a^2 + \alpha\,b^2 = \beta^2 + 4\alpha.$$

The solutions are $a = \mp\beta$, $b = \pm 2$. Selecting the upper signs, we obtain the relation

$$\Delta \equiv \beta^2 + 4\alpha = (2\,u - \beta)^2. \tag{2.2.10}$$

$\square$

We can write the solutions of (2.2.9) as

$$\text{a)} \quad z = x + u\,y; \qquad \text{b)} \quad \bar{z} = x + \beta y - u\,y. \tag{2.2.11}$$

We call $\bar{z}$ *the conjugate* of z. In the case of canonical systems, $\beta = 0$ and then $\bar{z} = x - u\,y$ ($u = \mathrm{i}$ in the particular case of complex numbers). The product $z\,\bar{z} = x^2 - \alpha\,y^2 + \beta\,x\,y \in \mathbf{R}$ is the square modulus of z ($z\,\bar{z} = x^2 + y^2$ in the particular case of complex numbers).

We end this section with a matrix representation (that we write in *sans serif*) of the versors and of z. From (2.1.9) and (2.2.3), we have for the unit versor, u and z

$$1 = \begin{pmatrix} 1 & 0 \\ 0 & 1 \end{pmatrix}; \qquad u = \begin{pmatrix} 0 & \alpha \\ 1 & \beta \end{pmatrix}; \qquad z = x \cdot 1 + y \cdot u \qquad (2.2.12)$$

and also

$$u^2 = \begin{pmatrix} \alpha & \alpha\beta \\ \beta & \alpha + \beta^2 \end{pmatrix} \equiv \begin{pmatrix} \alpha & 0 \\ 0 & \alpha \end{pmatrix} + \begin{pmatrix} 0 & \alpha\beta \\ \beta & \beta^2 \end{pmatrix} \equiv \alpha \cdot 1 + \beta \cdot u. \quad (2.2.13)$$

2.2.1   Canonical Two-Dimensional Systems

Let us go back to (2.2.10). For canonical systems ($\beta = 0$) we have $\Delta = (2\,u)^2$ and the solutions (2.2.9) become

$$z = \frac{2x + 2\,u\,y}{2} \equiv x + u\,y; \qquad \bar{z} = \frac{2x - 2\,u\,y}{2} \equiv x - u\,y. \qquad (2.2.14)$$

These solutions are the ones we consider for complex numbers ($\Delta < 0$), whereas for hyperbolic numbers, because $u^2 = 1$, it is usual to omit the versor. Now, as we stressed in the preface, even if complex numbers were introduced as square roots of negative numbers, they have acquired other important meanings, which can be usefully applied to other two-dimensional number systems. Then, in agreement with (2.2.14), also for a canonical hyperbolic system we write $z = x \pm u\,y$. These solutions are obtained by setting the appropriate versors before the square root. In Chapters 8 and 9 we shall see how these considerations are fruitfully used for extending the applications of hyperbolic numbers.

2.2.2   The Two-Dimensional Hyperbolic System

Let us introduce some properties of hyperbolic numbers, considered as belonging to a decomposable system.

Let us consider the canonical hyperbolic system [47], [80] and [81], defined as

$$\{z = x + h\,y;\ h^2 = 1\ \ x, y \in \mathbf{R}, h \notin \mathbf{R}\}, \qquad (2.2.15)$$

where we have set $u = h$, as we shall do in the following, when we refer to a canonical hyperbolic system. Let us apply the substitution of versors

$$e_1 = \frac{1}{2}(1 + h);\ \ e_2 = \frac{1}{2}(1 - h)\ \ \Rightarrow\ \ e_1^2 = e_1;\ \ e_2^2 = e_2;\ \ e_1\,e_2 = 0, \quad (2.2.16)$$

with inverse transformations

$$1 = e_1 + e_2, \quad h = e_1 - e_2, \qquad (2.2.17)$$

and of variables

$$x + y = \xi, \qquad x - y = \eta, \tag{2.2.18}$$

with inverse transformations

$$x = \frac{\xi + \eta}{2}, \qquad y = \frac{\xi - \eta}{2}. \tag{2.2.19}$$

The basis with the versors satisfying (2.2.16), i.e., where the powers of the versors are equal to the versors themselves and the product of different versors is zero, is called the *idempotent basis* and the numbers represented in this basis are said to be in *decomposed form* [43], [60] and [70]. By setting $\zeta = e_1\xi + e_2\eta$, it is easy to check that the conjugate (which we indicate by ˜ if we refer just to hyperbolic numbers) is given by

$$\tilde{\zeta} = e_2\xi + e_1\eta \tag{2.2.20}$$

and that the quantity

$$\zeta\tilde{\zeta} = (e_1\xi + e_2\eta)(e_1\eta + e_2\xi) \equiv (e_1 + e_2)\xi\eta \equiv \xi\eta \tag{2.2.21}$$

is real. $\xi\eta \equiv x^2 - y^2$ represents the invariant of hyperbolic numbers in their decomposed form. It coincides with the product of the invariants (the distances [81]) of the component systems. Now, let us consider the numbers $\zeta_1 = e_1\xi_1 + e_2\eta_1$ and $\zeta_2 = e_1\xi_2 + e_2\eta_2$. From (2.2.16), it follows [43] that

$$\zeta_1\zeta_2 = e_1\xi_1\xi_2 + e_2\eta_1\eta_2, \tag{2.2.22}$$

$$\begin{aligned}
\frac{\zeta_1}{\zeta_2} &\equiv \frac{e_1\xi_1 + e_2\eta_1}{e_1\xi_2 + e_2\eta_2} \equiv \frac{(e_1\xi_1 + e_2\eta_1)(e_1\eta_2 + e_2\xi_2)}{(e_1\xi_2 + e_2\eta_2)(e_1\eta_2 + e_2\xi_2)} \\
&\equiv \frac{e_1\xi_1\eta_2 + e_2\eta_1\xi_2}{(e_1 + e_2)\xi_2\eta_2} = e_1\frac{\xi_1}{\xi_2} + e_2\frac{\eta_1}{\eta_2}.
\end{aligned} \tag{2.2.23}$$

We see from (2.2.23) that division is not possible if $\xi_2 = 0$ or $\eta_2 = 0$; the coordinate axes are the divisors of zero for this system in decomposed form. It is immediate to verify that $(e_1\,0 + e_2\,\eta)(e_1\,\xi + e_2\,0) \equiv 0$, i.e., the product of any number lying on the $\xi = 0$ axis by any number lying on the $\eta = 0$ axis is zero although the factors are $\neq 0$. For the decomposed systems, this general rule holds: *The divisors of zero are the zeros of the component systems.*

Chapter 3

The Geometries Generated by Hypercomplex Numbers

3.1 Linear Transformations and Geometries

3.1.1 The Continuous Lie Groups

The groups of transformations were introduced and formalized by S. Lie in the second half of the XIXth century. The concept of group is well known; here we only recall the definition of *transformations or Lie groups* that is used in this section.

Let us consider N variables x^m and N variables y^i that are functions of x^m and of K parameters a^l. Now, let us consider a second transformation from y^i to the variables z^n, given by the same functions but with other values of the parameters (b^l). We say that these transformations are a group if, by considering them one after the other (*product of two transformations*), we have the same functional dependence between z and x with other parameters, whose values depend on the parameters of the previous transformations. To express these concepts in formulas, we indicate by

$$y^i = f^i(x^1 \cdots x^N, a^1 \cdots a^K) \equiv f^i(x, a) \quad \text{with} \ \ i = 1, \ldots, N$$

the first transformation and by $z^n = f^n(y, b)$ a second transformation with other parameters. The relation between z and x (composite transformation) is $z^n = f^n[f(x, a), b]$. If these last transformations can be written as $z^n = f^n(x, c)$ with $c = g(a, b)$, they represent a group. The number of parameters (K) is the *order of the group*. If the parameters can assume continuous values, these groups are called *continuous groups* (see Section 7.2.3). A fundamental result obtained by Lie is that these groups can be related to systems of differential equations and the number of parameters is linked to the initial conditions.

3.1.2 Klein's Erlanger Programm

In a talk that introduced his famous "Erlanger Programm", Felix Klein in 1872 associated the continuous groups with geometries. The guideline is the following: associated with all the geometries (elementary, projective, etc.) is the notion of equivalence. It can be shown that the transformations that give rise to equivalent forms have the properties of groups. Then, the geometries are the theories related to the invariants of the corresponding groups. This concept can be inverted and in this way is applied to hypercomplex numbers.

We begin by recalling the importance of linear transformations as related to important geometries. An arbitrary linear transformation can be written as $y^\gamma = c_\beta^\gamma x^\beta$, with the condition $\|c_\beta^\gamma\| \neq 0$; therefore it depends on N^2 parameters. By identifying y^γ and x^β as vector components, we can write, following the notation of linear algebra,

$$\begin{pmatrix} y^1 \\ \vdots \\ y^N \end{pmatrix} = \begin{pmatrix} c_1^1 & \cdots & c_N^1 \\ \vdots & \ddots & \vdots \\ c_1^N & \cdots & c_N^N \end{pmatrix} \begin{pmatrix} x^1 \\ \vdots \\ x^N \end{pmatrix}. \tag{3.1.1}$$

These transformations are known as *homographies* and are generally non-commutative. From a geometrical point of view, they represent a change of reference frame through an arbitrary rotation of its axes and a change of length measures on the axes. The geometry that considers equivalence of geometric figures with respect to these transformations is called *affine geometry*, and the corresponding group is called an *affine group*. An invariant quantity for such a group is the distance between two points, expressed by a positive definite quadratic form. *Euclidean geometry*, for which the distance is given by Pythagoras' theorem, is a subgroup of this group. For the Euclidean group, the constants c_β^γ must satisfy $N(N + 1)/2$ conditions, then the remaining parameters are $N^2 - N(N + 1)/2 = N(N - 1)/2$. From a geometrical point of view, the allowed transformations of Euclidean geometry are given by rotations of the Cartesian axes.

If we consider also the *translations group*, we must add N parameters and we get a group with $N(N + 1)/2$ parameters, which corresponds to the allowed motions of geometrical figures in a Euclidean space. These motions also represent the allowed motions in a homogeneous space (*a constant curvature space*). In Chapter 9 we shall recall that Riemann's and Lobachevsky's plane geometries are represented on *positive* and *negative constant curvature surfaces*, respectively. Now let us see how the above notions are applied to hypercomplex systems.

3.2   Groups Associated with Hypercomplex Numbers

A hypercomplex number x can be represented in an N-dimensional space by identifying its components x^α with the Cartesian coordinates of a point $P \equiv (x^\alpha)$. In this space we introduce the metric (distance of a point from the coordinate origin), given by the Nth root of the characteristic determinant. More generally, the distance between two points shall be given by substituting in the characteristic determinant the differences of the points' coordinates. In this space we have

Theorem 3.1. *Every commutative system of hypercomplex numbers generates a geometry. For such geometry, the metric is obtained from a form of degree N given by the characteristic determinant. Motions are characterized by $2N - 1$ parameters.*

Proof. As in Euclidean geometry, we define as *motions* the transformations of coordinates that leave the metric unchanged. There are two kinds of such transformations.

1. We begin by considering the transformations corresponding to Euclidean translations. Let us consider the hypercomplex constant $a = e_\alpha a^\alpha$ and the hypercomplex variables $y \equiv e_\alpha y^\alpha$ and $x \equiv e_\alpha x^\alpha$. The transformation $y = a + x$, whose components are $y^\gamma = a^\gamma + x^\gamma$, represents the translation group, which has the same number of parameters as the equivalent Euclidean one.

2. The second group of motions in Euclidean geometry is given by the rotations that, like the previous one, leave unchanged the Pythagorean distance between two points. Let us see what this group becomes for hypercomplex numbers. Let us consider the transformations $y = ax$, whose components are

$$e_\gamma y^\gamma = e_\alpha a^\alpha e_\beta x^\beta \equiv e_\gamma C^\gamma_{\alpha\beta} a^\alpha x^\beta.$$

$C^\gamma_{\alpha\beta} a^\alpha$ are the elements of the characteristic matrix of the hypercomplex constant a. Setting $C^\gamma_{\alpha\beta} a^\alpha = A^\gamma_\beta$, we have

$$y^\gamma = A^\gamma_\beta x^\beta, \tag{3.2.1}$$

therefore y is a linear mapping of x. Let us consider another constant, $b = e_\alpha b^\alpha$, and the corresponding transformation

$$\begin{aligned} z & \equiv e_\gamma z^\gamma = b\,y \equiv b\,a\,x \equiv e_\delta\,b^\delta\,e_\alpha\,a^\alpha\,e_\beta\,x^\beta \\ & \equiv e_\gamma\,(C^\gamma_{\alpha\,\epsilon}\,a^\alpha)\,(C^\epsilon_{\delta\,\beta}\,b^\delta)\,x^\beta \equiv e_\gamma\,A^\gamma_\epsilon\,B^\epsilon_\beta\,x^\beta. \end{aligned} \tag{3.2.2}$$

Since $A^\gamma_\epsilon = C^\gamma_{\alpha\,\epsilon}\,a^\alpha$ and $B^\epsilon_\beta = C^\epsilon_{\delta\,\beta}\,b^\delta$ are the elements of characteristic matrices, then, from Theorem 2.4, their product $D^\gamma_\beta = A^\gamma_\epsilon\,B^\epsilon_\beta$ is a characteristic matrix too and $z = e_\gamma\,D^\gamma_\beta\,x^\beta$ is a linear mapping of x via a characteristic matrix. This mapping is a group called the *multiplicative group*. Following Klein's program, we can associate the multiplicative group of every hypercomplex number with a geometry. From (2.1.10) and (3.2.1), it follows that the characteristic determinants $\|X\|$, $\|Y\|$ of the hypercomplex numbers x, y are the same if

$$\|A\| = 1. \tag{3.2.3}$$

Therefore the *characteristic determinant of the hypercomplex number is an invariant quantity for the unimodular multiplicative group.* Since the unimodular characteristic matrix takes the place of the orthogonal matrix of the N-dimensional Euclidean rotation group, we name this group the *hypercomplex rotation group* (HRG). The unimodularity condition for the characteristic determinant reduces the parameters of this group to $N - 1$ real constants: the components a^α that are mutually linked by the condition (3.2.3). $\qquad\square$

In the following table (3.2.4) we compare the numbers of parameters of these geometries with the numbers of parameters $(N(N-1)/2)$ of the Euclidean rotation group (ERG).

Space dimensions$[N]$	2	3	4	5	6	
HRG$[N-1]$	1	2	3	4	5	(3.2.4)
ERG$[N(N-1)/2]$	1	3	6	10	15	

Then, the number of parameters is the same for the two geometries only for $N = 2$.

The unimodular multiplicative group of the complex numbers is the same as the Euclidean rotation group, as we briefly recall in Section 3.2.1. The geometries generated by the other two-dimensional systems of hypercomplex numbers are known too. In particular, the geometry generated by parabolic numbers is extensively studied in [81], where also the geometry associated with hyperbolic numbers is introduced. A complete formalization of this last geometry and wide-ranging applications of hyperbolic numbers are the subject of this book.

None of the geometries generated by hypercomplex numbers for $N > 2$ corresponds to a Euclidean geometry. This fact is a straightforward consequence of the differences existing between the invariant quantities. Actually the characteristic determinant of a multidimensional ($N > 2$) hypercomplex number is an algebraic form of degree N, whereas the Euclidean invariant is the Pythagorean distance of degree 2. In addition, it follows that the invariable hypersurfaces for the motions (compared to the Euclidean hyperspheres) are obtained by equating the characteristic determinant to a constant.

In the studies of hyperbolic geometry by means of hyperbolic numbers, it has been pointed out ([81], p. 251) that the most suitable geometrical representation of these numbers is over a plane with the metric (scalar product) that derives from the structure of the hyperbolic numbers, i.e., from the characteristic determinant. If we extend this consideration to the above introduced geometries, we can say that their most suitable representations are obtained in spaces with the metric fixed by the algebraic form of degree N given by the characteristic determinant. In these spaces, *generated by commutative hypercomplex numbers*, we introduce the functions of a hypercomplex variable, and see (Chapter 7 and Appendix A) that the hypercomplex conformal mappings have the same geometrical properties of the conformal mappings corresponding to the functions of a complex variable.

Let us say some words about the geometries just introduced and the ones associated with the more general mappings of (2.1.11). The differences among them are the same existing for the following geometries:

1. The *Euclidean geometry*, associated with orthogonal groups;

2. The *affine geometry*, associated with affine groups of all non-singular linear transformations.

If we consider two-dimensional geometries, the first case corresponds to the geometry generated by complex numbers. In other words, the matrix of a linear mapping has the same structure as the characteristic matrix of complex numbers. Similar considerations hold for all the systems of hypercomplex numbers, for which a specific geometry can be associated with a linear mapping that has the same structure as the characteristic matrix. For complex and hypercomplex numbers, the geometries associated with all non-singular linear mappings are equivalent to the affine geometry of the second point.

3.2.1 Geometries Generated by Complex and Hyperbolic Numbers

The Geometry Generated by Complex Numbers

Complex numbers can represent plane vectors and the related linear algebra ([54], p.73). Actually $z = x + i y$ can be interpreted as a vector of components x and y, along the versors 1 and i, where x and y are the coordinates on the plane. Considering the product by a constant,

$$z_1 = az \equiv (a_r + i a_i)(x + i y), \qquad (3.2.5)$$

the complex numbers play the role of both a vector and an operator (matrix) ([54], p.73). Actually, (3.2.5) is equivalent to the familiar expression of linear algebra

$$\begin{pmatrix} x_1 \\ y_1 \end{pmatrix} = \begin{pmatrix} a_r & -a_i \\ a_i & a_r \end{pmatrix} \begin{pmatrix} x \\ y \end{pmatrix}.$$

In this expression, the double representation of a complex number, introduced in Section 2.1.3, has been used. In particular, the complex constant a is written in matrix form (like the operators in linear algebra), while z is represented as a column vector. If we write the constant a in its exponential form,

$$a \equiv (a_r + i a_i) = \exp \rho \, (\cos \phi + i \sin \phi),$$

with the inverse transformation

$$\rho = \ln \sqrt{a_r^2 + a_i^2} \, ; \quad \phi = \tan^{-1} a_i / a_r,$$

we see that the constant a plays the role of an operator representing an orthogonal-axis rotation with a homogeneous dilatation (homothety). If $\rho = 0$, and if we add another constant $b = b_r + i b_i$, then $z_1 = az + b$ gives the permissible vector transformations in a Euclidean plane. Then, the additive and unitary multiplicative groups of complex numbers are equivalent to the Euclidean group of roto-translations, which depends on the three parameters ϕ, b_r, b_i. Vice versa, we can use complex numbers for describing plane-vector algebra, because the vectors are usually represented in an orthogonal coordinate system and the additive and multiplicative groups of complex numbers are the same as Euclidean groups.

The Geometry Generated by Hyperbolic Numbers

In Chapter 4 we shall apply to hyperbolic numbers the same considerations expounded for the geometry generated by complex numbers and, in Section 4.1.2, we show that the multiplicative group of hyperbolic numbers (*hyperbolic geometry*) represents the space-time geometry. Now, we consider this geometry as the simplest one associated with decomposable systems of numbers.

The multiplicative group of hyperbolic numbers, expressed in vector-matrix form, is given by

$$\begin{pmatrix} y^1 \\ y^2 \end{pmatrix} = \begin{pmatrix} A_1^1 & A_2^1 \\ A_1^2 & A_2^2 \end{pmatrix} \begin{pmatrix} x^1 \\ x^2 \end{pmatrix} \quad \text{with } A_1^1 = A_2^2, \ A_2^1 = A_1^2. \tag{3.2.6}$$

The unimodularity condition requires that $(A_1^1)^2 - (A_1^2)^2 = 1$ and this allows introducing a hyperbolic angle θ so that $A_1^1 = \cosh\theta$, $A_1^2 = \sinh\theta$. This position is equivalent to writing, in complex analysis, the constant of the multiplicative group in its polar form. For hyperbolic numbers too, the polar hyperbolic coordinates are obtained from the exponential transformation $x = \rho\cosh\theta$, $y = \rho\sinh\theta$ [37], [47] (see Section 4.1.1). In this way, the multiplicative group becomes an additive group for θ, and the unimodularity condition becomes $\rho = 1$.

Let us consider the hyperbolic numbers and the hyperbolic constants in the decomposed form of Section 2.2.2. Equation (3.2.6) becomes

$$\begin{pmatrix} \eta^1 \\ \eta^2 \end{pmatrix} = \begin{pmatrix} \alpha_1^1 & 0 \\ 0 & \alpha_2^2 \end{pmatrix} \begin{pmatrix} \xi^1 \\ \xi^2 \end{pmatrix}. \tag{3.2.7}$$

If we express the constants and the variables in exponential form $\alpha_1^1 = \exp[\lambda]$, $\alpha_2^2 = \exp[\mu]$ and $\xi^i = \exp[\omega^i]$, the transformation (3.2.7) corresponds to translations along the ω^1, ω^2 axes.

If we consider one real variable, the multiplicative group is $y = a\,x$, but the unimodularity condition requires that $a = 1$; then this mapping is the trivial identity. When we compose two real variables and obtain the hyperbolic numbers, the unimodularity condition gives $\lambda = -\mu$ which, as we shall better see in Section 4.1.2, represents the Lorentz transformation of Special Relativity [70] and [81]. The above discussed example shows that the geometry associated with a composite system adds something new with respect to the geometries associated with the component systems.

3.3   Conclusions

Hypercomplex numbers, considered as an extension of real and complex numbers, can not satisfy the four properties of multiplication; if these systems are commutative, they must have divisors of zero. On the basis of these properties, associative hypercomplex numbers can be grouped in two classes: to the first class belong the non-commutative systems (Hamilton, hyperbolic quaternions etc.), while to the second class the commutative systems. The peculiar differences between non-commutative and commutative systems are their invariants and the existence of differential calculus. As far as the invariant is concerned, for the non-commutative systems it can be an algebraic quadratic form that can be related with Euclidean geometry so that, by means of Hamilton's quaternions, we can represent vectors in the three-dimensional Euclidean space.

For commutative systems, the invariant is represented by an N-form that, for $N > 2$, generates new geometries. Since for these systems the differential calculus exists, the above-mentioned geometries can be extended for studying non-flat spaces as it has been done with the complex variable for Euclidean geometry and, as it will be shown in the following chapters, with the hyperbolic variable for studying pseudo-Euclidean geometries [81]. Such multidimensional geometries have not been completely investigated and this allows us to assert the following consideration: the kind of two-dimensional numbers derives from the solutions of an equation of degree 2. We find the same classification in other mathematical fields. We have

- *imaginary solutions* → *complex numbers* → *Euclidean geometry* → *Gauss differential geometry (definite quadratic differential forms)* → *elliptic partial differential equations;*

- *real solutions* → *hyperbolic numbers* → *Minkowski (space-time) geometry* → *differential geometry on Lorentz surfaces (non-definite quadratic differential forms)* → *hyperbolic partial differential equations.*

Moreover, in more than two dimensions we suggest the following general links:

- *the kind of solutions of an algebraic equation of degree N* → *systems of hypercomplex numbers* → *multiplicative group* → *geometries* → *differential geometries.*

In this way, the differential geometry in an N-dimensional space would derive from a differential form of degree N instead of the quadratic Euclidean or pseudo-Euclidean differential forms. These peculiar properties could open new ways for applications in field theories.

Chapter 4

Trigonometry in the Minkowski Plane

We have seen in Section 3.2 how commutative hypercomplex numbers can be associated with a geometry, in particular the two-dimensional numbers can represent the Euclidean plane geometry and the space-time (Minkowski) plane geometry. In this chapter, by means of algebraic properties of hyperbolic numbers, we formalize the space-time geometry and trigonometry. This formalization allows us to work in Minkowski space-time as we usually do in the Euclidean plane, i.e., to give a Euclidean description that can be considered similar to Euclidean representations of non-Euclidean geometries obtained in the XIXth century by E. Beltrami [2] on constant curvature surfaces, as we recall in Chapter 9.

Let us consider the two-dimensional system of hyperbolic numbers defined as

$$\{z = x + h\,y;\ h^2 = 1;\ x, y \in \mathbf{R}\},$$

and see how it is strictly related to space-time geometry [81]. Actually, by calling $\tilde{z} = x - h\,y$ the *hyperbolic conjugate* of z, we have:

- The square modulus given by $|z|^2 = z\tilde{z} \equiv x^2 - y^2$ represents the Lorentz invariant of the two-dimensional special relativity.

- The unimodular multiplicative group is the special relativity Lorentz group [81].

These relations make hyperbolic numbers relevant for physics and stimulate us to find their application in the same way complex numbers are applied to Euclidean plane geometry. In this chapter we present a formalization of space-time trigonometry which we derive by first remarking that hyperbolic (complex) numbers allow us to introduce two invariant quantities with respect to Lorentz's (Euclid's) group. The first invariant is the scalar product, the second one is equivalent to the modulus of the vector product (an area). These two invariant quantities allow us to define, in a Cartesian representation, the hyperbolic trigonometric functions that hold in the whole hyperbolic plane, by which we can solve triangles with sides in any directions, except parallel to axes bisectors. More precisely we start from the experimental axiom that the Lorentz transformations represent a "symmetry of nature" and look for their geometrical "deductions". In this way the following formalization is an axiomatic-deductive theory, starting from experimental axioms, equivalent to Euclid's geometry construction.

4.1   Geometrical Representation of Hyperbolic Numbers

Now let us introduce a hyperbolic plane by analogy with the Gauss–Argand plane of a complex variable. In this plane we associate points $P \equiv (x, y)$ to hyperbolic numbers $z = x + h y$. If we represent these numbers on a Cartesian plane, in this plane the square distance of the point P from the origin of coordinates is defined as

$$D = z\tilde{z} \equiv x^2 - y^2. \tag{4.1.1}$$

The definition of distance (metric element) is equivalent to introducing the bilinear form of the *scalar product*. The scalar product and the properties of hypercomplex numbers allow one to state suitable axioms ([81], p. 245) and to give to a pseudo-Euclidean plane the structure of a vector space.

Now let us consider the multiplicative inverse of z that, if it exists, is given by $1/z \equiv \tilde{z}/z\tilde{z}$. This implies that z does not have an inverse when $z\tilde{z} \equiv x^2 - y^2 = 0$, i.e., when $y = \pm x$, or alternatively when $z = x \pm h\,x$, that are the "divisors of zero" defined in Section 2.1. These two straight lines, whose elements have no inverses, divide the hyperbolic plane in four sectors that we shall call *Right sector (Rs), Up sector (Us), Left sector (Ls),* and *Down sector (Ds)*. This property is the same as that of the special relativity representative plane and this correspondence gives a physical meaning (space-time interval) to the definition of distance. Now let us consider the quantity $x^2 - y^2$, which is positive in Rs, Ls ($|x| > |y|$) sectors, and negative in Us, Ds ($|x| < |y|$) sectors. As it is known from special relativity and, as we shall better see in this chapter, this quantity must have its sign and appear in this quadratic form.

When we must use the linear form (the modulus of hyperbolic numbers or the side length), we follow the definition of B. Chabat ([47], p. 51), ([25], p. 72) and I. M. Yaglom ([81], p. 180)

$$\rho = \sqrt{|z\,\tilde{z}|} \equiv \sqrt{|D|} \tag{4.1.2}$$

where $|D|$ is the absolute value of the square distance.

4.1.1   Hyperbolic Exponential Function and Hyperbolic Polar Transformation

The hyperbolic exponential function in pseudo-Euclidean geometry plays the same important role as the complex exponential function in Euclidean geometry. In Chapter 7 we shall introduce the functions of a hyperbolic variable and point out analogies and differences with respect to functions of a complex variable and see that functions in decomposable systems can be obtained as a continuation from the real field. Here we define the exponential function of a hyperbolic number following Euler's introduction of the complex exponential. Actually, in Euler's time the theory of power series was not sufficiently developed. Then it was not known that displacement of terms is possible only for absolutely convergent series. Since the following series have this quality, we apply to the hyperbolic exponential the properties of the exponential of a real and complex variable and have

$$\exp\left(\rho' + h\,\theta\right) = \exp\rho' \sum_{l=0}^{\infty} \frac{(h\,\theta)^l}{l!} = \exp\rho' \left(\sum_{l=0}^{\infty} \frac{(h\,\theta)^{2l}}{(2l)!} + \sum_{l=0}^{\infty} \frac{(h\,\theta)^{2l+1}}{(2l+1)!}\right)$$

$$= \exp\rho' \left(\sum_{l=0}^{\infty} \frac{(\theta)^{2l}}{(2l)!} + h \sum_{l=0}^{\infty} \frac{(\theta)^{2l+1}}{(2l+1)!}\right) = \exp\rho'(\cosh\theta + h\,\sinh\theta). \qquad (4.1.3)$$

By means of an exponential function, we introduce the exponential transformation and extend it to all the sectors; thus we have

$$\text{if } |x| > |y|; \quad x + h\,y = \mathrm{sign}(x)\exp[\rho' + h\,\theta] \equiv \mathrm{sign}(x)\exp[\rho'](\cosh\theta + h\,\sinh\theta);$$
$$(4.1.4)$$

$$\text{if } |x| < |y|; \quad x + h\,y = \mathrm{sign}(y)\exp[\rho' + h\,\theta] \equiv \mathrm{sign}(y)\exp[\rho'](\sinh\theta + h\,\cosh\theta).$$
$$(4.1.5)$$

Following [47] and [81] we define the *radial coordinate* as

$$\exp[\rho'] = \rho \equiv \sqrt{|x^2 - y^2|}$$

and the *angular coordinate* as

$$\text{for } |x| > |y|, \quad \theta = \tanh^{-1}(y/x); \quad \text{for } |x| < |y|, \quad \theta = \tanh^{-1}(x/y).$$

For $|x| > |y|$, $x > 0$ (i.e., $x, y \in Rs$), we introduce the *hyperbolic polar transformation* as

$$x + h\,y = \rho\exp[h\,\theta] \equiv \rho(\cosh\theta + h\,\sinh\theta). \qquad (4.1.6)$$

Equation (4.1.6) represents the map for $x, y \in Rs$; the extension of the map to the complete (x, y) plane is reported in Tab. 4.1.

Right sector (Rs)	Left sector (Ls)	Up sector (Us)	Down sector (Ds)
$|x| > |y|$		$|x| < |y|$	
$z = \rho\exp[h\,\theta]$	$z = -\rho\exp[h\,\theta]$	$z = h\,\rho\exp[h\,\theta]$	$z = -h\,\rho\exp[h\,\theta]$
$x = \rho\cosh\theta$	$x = -\rho\cosh\theta$	$x = \rho\sinh\theta$	$x = -\rho\sinh\theta$
$y = \rho\sinh\theta$	$y = -\rho\sinh\theta$	$y = \rho\cosh\theta$	$y = -\rho\cosh\theta$

Table 4.1: Map of the complete (x, y) plane by hyperbolic polar transformation

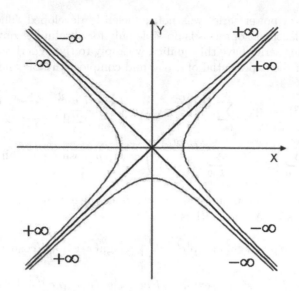

Figure 4.1: For $\rho = 1$ the x, y in Tab. 4.1 represent the four arms of equilateral hyperbolas $|x^2 - y^2| = 1$. Here we indicate how each arm is traversed as the parameter θ goes from $-\infty$ to $+\infty$. In particular there is a symmetry with respect to axis bisectors.

4.1.2 Hyperbolic Rotations as Lorentz Transformations of Special Relativity

Let us write a space-time vector as a hyperbolic variable[1], $w = t + h\,x$ and a hyperbolic constant $a = a_r + h\,a_h$ with $a_r > a_h$ in the exponential form

$$a_r + h\,a_h \equiv \rho_a \exp[\,h\theta_a\,] \equiv \rho_a\,(\cosh\theta_a + h\sinh\theta_a)$$

$$\text{where} \qquad \rho_a = \sqrt{(a_r^2 - a_h^2)}; \quad \theta_a = \tanh^{-1}(a_h/a_r).$$

Then the multiplicative group, $w' \equiv t' + h\,x' = a\,w$ becomes

$$t' + h\,x' = \sqrt{(a_r^2 - a_h^2)}\,[t\,\cosh\theta_a + x\,\sinh\theta_a + h(t\,\sinh\theta_a + x\,\cosh\theta_a)]. \quad (4.1.7)$$

In this equation, by letting $(a_r^2 - a_h^2) = 1$ and considering as equal the coefficients of versors "1" and "h", as we do in complex analysis, we get the Lorentz transformation of two-dimensional special relativity [55] and [62]. It is interesting to

---

[1]In all the problems which refer to Special Relativity we change the symbols by indicating the variables with letters reflecting their physical meaning x, y $\Rightarrow$ t, x, i.e., t is a normalized time variable (light velocity $c = 1$) and x a space variable.

note that the same result is normally achieved by following a number of "formal" steps ([54], p. 94) and ([52], p. 50) i.e., by introducing an "imaginary" time $t' = it$ which makes the Lorentz invariant $(x^2 - t^2)$ equivalent to the Euclidean invariant $(x^2 + y^2)$, and by introducing the hyperbolic trigonometric functions through their equivalence with circular functions of an imaginary angle. Let us stress that this procedure is essentially formal, while the approach based on hyperbolic numbers leads to *a direct description of Lorentz transformation of special relativity explainable as a result of symmetry (or invariants) preservation*: the Lorentz invariant (space-time "distance") is the invariant of hyperbolic numbers. Therefore we can say that the hyperbolic numbers have *space-time symmetry*, while the complex numbers have the symmetry of two spatial variables represented in a Euclidean plane. Within the limits of our knowledge, the first description of Special Relativity directly by these numbers was introduced by I.M. Yaglom [81].

With this formalism we have

Theorem 4.1. *The Lorentz transformation is equivalent to a "hyperbolic rotation".*

Proof. Let us write in the Lorentz transformation (4.1.7), the hyperbolic variable $t + h\,x$ in exponential form (4.1.6)

$$t + h\,x = \rho \exp[h\,\theta].$$

Then

$$w' = aw = \rho \exp[\,h\,(\theta_a + \theta)\,]. \tag{4.1.8}$$

From this expression we see that the Lorentz transformation is equivalent to a "hyperbolic rotation" of the $t + h\,x$ variable. $\qquad\square$

Then the invariance under Lorentz transformation can also be expressed as independence of the hyperbolic angle θ.

4.2 Basics of Hyperbolic Trigonometry

4.2.1 Complex Numbers and Euclidean Trigonometry

Euclidean geometry studies the properties of figures which do not depend on their position in a plane. If these figures are represented in a Cartesian plane we can say, in group theory language, that Euclidean geometry studies the invariant properties under coordinate axes roto-translations and these properties can be expressed by complex numbers.

Let us consider the Gauss–Argand complex plane where a vector is represented by $v = x + i\,y$. As is known, an axes rotation of an angle α transforms this vector into a new vector $v' \equiv v \exp[i\,\alpha]$. Therefore we can promptly verify that the quantity

$$|v'|^2 \equiv v'\bar{v}' = v \exp[i\,\alpha]\bar{v} \exp[-i\,\alpha] = v\,\bar{v} \equiv |v|^2 \tag{4.2.1}$$

is invariant under axes rotation. In a similar way we find two more invariants related to any pair of vectors. Let us consider two vectors $v_1 = x_1 + i\, y_1 \equiv \rho_1 \exp[\phi_1]$, $v_2 = x_2 + i\, y_2 \equiv \rho_2 \exp[\phi_2]$; we have

Theorem 4.2. *The real and imaginary parts of the product $v_2 \bar{v}_1$ are invariant under axes rotations and these two invariants allow us an operative definition of trigonometric functions by means of the components of the vectors:*

$$\cos(\phi_2 - \phi_1) = \frac{x_1\, x_2 + y_1\, y_2}{\rho_1\, \rho_2}; \qquad \sin(\phi_2 - \phi_1) = \frac{x_1\, y_2 - x_2\, y_1}{\rho_1\, \rho_2}. \qquad (4.2.2)$$

Proof. Actually

$$v_2' \bar{v}_1' = v_2 \exp[i\,\alpha]\bar{v}_1 \exp[-i\,\alpha] \equiv v_2\, \bar{v}_1, \qquad (4.2.3)$$

and let us represent the two vectors in polar coordinates $v_1 \equiv \rho_1 \exp[i\,\phi_1]$, $v_2 \equiv \rho_2 \exp[i\,\phi_2]$. Consequently we have

$$v_2 \bar{v}_1 = \rho_1 \rho_2 \exp[i\,(\phi_2 - \phi_1)] \equiv \rho_1 \rho_2 [\cos(\phi_2 - \phi_1) + i\,\sin(\phi_2 - \phi_1)]. \qquad (4.2.4)$$

As is well known, the resulting real part of this product represents the scalar product, while the imaginary part represents the modulus of a cross product, i.e., the area of the parallelogram defined by v_1 and v_2. In Cartesian coordinates we have

$$v_2\, \bar{v}_1 = (x_2 + i\, y_2)(x_1 - i\, y_1) \equiv x_1\, x_2 + y_1\, y_2 + i\,(x_1\, y_2 - x_2\, y_1), \qquad (4.2.5)$$

and, by comparing (4.2.4) with (4.2.5), we obtain (4.2.2). $\qquad \square$

We know that the theorems of Euclidean trigonometry are usually obtained following a geometric approach; now we have

Theorem 4.3. *Using the Cartesian representation of trigonometric functions, given by (4.2.2), the trigonometry theorems are simple identities.*

Proof. We know that the trigonometry theorems represent relations between angles and side lengths of a triangle. If we represent a triangle in a Cartesian plane it is defined by the coordinates of its vertexes P_1, P_2, P_3. From the coordinates of these points we obtain the side lengths by Pythagoras' theorem and the trigonometric functions from (4.2.2). By these definitions we can verify that the trigonometry theorems are identities. $\qquad \square$

The extension and applications of this procedure to a hyperbolic plane is the subject of this chapter.

4.2.2   Hyperbolic Rotation Invariants in Pseudo-Euclidean Plane Geometry

By analogy with the Euclidean trigonometry approach, just summarized, we can say that *pseudo-Euclidean plane geometry* studies the properties that are invariant under two-dimensional Lorentz transformations (Lorentz–Poincaré group of special relativity) corresponding to hyperbolic rotation (Section 4.1.2). We show afterwards, how these properties allow us to formalize hyperbolic trigonometry.

Let us define in the hyperbolic plane a hyperbolic vector, from the origin to point $P \equiv (x,\, y)$, as $v = x + h\, y$. A hyperbolic rotation of an angle θ transforms this vector into a new vector $v' \equiv v \exp[h\, \theta]$. Therefore the quantity

$$|v'|^2 \equiv v' \tilde{v}' = v \exp[h\, \theta] \tilde{v} \exp[-h\, \theta] \equiv v \tilde{v} = |v|^2 \qquad (4.2.6)$$

is invariant with respect to hyperbolic rotation.

In a similar way we can find two more invariants related to any pair of vectors. Let us consider two vectors $v_1 = x_1 + h\, y_1$ and $v_2 = x_2 + h\, y_2$; we have

Theorem 4.4. *The real and the hyperbolic parts of the product $v_2 \tilde{v}_1$ are invariant under hyperbolic rotation, and these two invariants allow us an operative definition of hyperbolic trigonometric functions by means of the components of the vectors:*

$$\cosh(\theta_2 - \theta_1) = \frac{x_1 x_2 - y_1 y_2}{\rho_1 \rho_2} \equiv \frac{x_1 x_2 - y_1 y_2}{\sqrt{|(x_2^2 - y_2^2)||(x_1^2 - y_1^2)|}}, \qquad (4.2.7)$$

$$\sinh(\theta_2 - \theta_1) = \frac{x_1 y_2 - x_2 y_1}{\rho_1 \rho_2} \equiv \frac{x_1 y_2 - x_2 y_1}{\sqrt{|(x_2^2 - y_2^2)||(x_1^2 - y_1^2)|}}. \qquad (4.2.8)$$

Proof. We have

$$v_2' \tilde{v}_1' = v_2 \exp[h\alpha] \tilde{v}_1 \exp[-h\alpha] \equiv v_2 \tilde{v}_1. \qquad (4.2.9)$$

Let us suppose $(x,\, y) \in R_s$, and represent the two vectors in hyperbolic polar form $v_1 = \rho_1 \exp[\,h\, \theta_1\,]$, $v_2 = \rho_2 \exp[\,h\, \theta_2\,]$. Consequently we have

$$v_2 \tilde{v}_1 \equiv \rho_1 \rho_2 \exp[h(\theta_2 - \theta_1)] \equiv \rho_1 \rho_2 [\cosh(\theta_2 - \theta_1) + h \sinh(\theta_2 - \theta_1)]. \qquad (4.2.10)$$

As we know from differential geometry [34], the real part of this product represents the scalar product; as far as the hyperbolic part is concerned, we shall see in Section 4.4.1 that, as for the Euclidean plane, it represents a *pseudo-Euclidean area*. In Cartesian coordinates, we have

$$v_2 \tilde{v}_1 = (x_2 + h\, y_2)(x_1 - h\, y_1) \equiv x_1 x_2 - y_1 y_2 + h(x_1 y_2 - x_2 y_1). \qquad (4.2.11)$$

Comparing (4.2.10) with (4.2.11) we obtain (4.2.7) and (4.2.8). $\qquad \square$

The real term of (4.2.11) represents the Cartesian expression of the *scalar product* in the hyperbolic plane. We note that, due to the distance definition (4.1.1), we have a different sign, with respect to Euclidean scalar product. The classical hyperbolic trigonometric functions are defined for $(x,\, y) \in Rs$; now we show

Theorem 4.5. *Equations (4.2.7) and (4.2.8) allow us to extend the hyperbolic trigonometric functions in the complete (x, y) plane.*

Proof. If we put $v_1 \equiv (1; 0)$ and $(\theta_2, x_2, y_2) \rightarrow (\theta, x, y)$, (4.2.7) and (4.2.8) become

$$\cosh \theta = \frac{x}{\sqrt{|x^2 - y^2|}}, \qquad \sinh \theta = \frac{y}{\sqrt{|x^2 - y^2|}}. \qquad (4.2.12)$$

We observe that expressions (4.2.12) are valid for $\{x, y \in \mathbf{R} \mid x \neq \pm y\}$ so they allow us to extend the trigonometric hyperbolic functions in the complete (x, y) plane. This extension is the same as the one proposed in [37] and [38], that we summarize in Section 4.2.3. $\qquad\qquad\square$

In the following we will denote with $\cosh_e$, $\sinh_e$ these *extended hyperbolic functions*. In Tab. 4.2 the relations between $\cosh_e$, $\sinh_e$ and traditional hyperbolic functions are reported. By this extension the hyperbolic polar transformation, (4.1.6), is given by

$$x + h\, y \Rightarrow \rho(\cosh_e \theta + h \sinh_e \theta), \qquad (4.2.13)$$

from which, for $\rho = 1$, we obtain the *extended hyperbolic Euler's formula* [37]

$$\exp_e[h\theta] = \cosh_e \theta + h \sinh_e \theta. \qquad (4.2.14)$$

From Tab. 4.2 or Equation (4.2.12) it follows that

$$\text{for } |x| > |y| \Rightarrow \cosh_e^2 \theta - \sinh_e^2 \theta = 1; \quad \text{for } |x| < |y| \Rightarrow \cosh_e^2 \theta - \sinh_e^2 \theta = -1. \qquad (4.2.15)$$

The complete representation of the extended hyperbolic trigonometric functions can be obtained by giving to x, y all the values on the circle $x = \cos \phi$, $y = \sin \phi$ for $0 \leq \phi < 2\pi$, in this way (4.2.12) becomes

$$\cosh_e \theta = \frac{\cos \phi}{\sqrt{|\cos 2\phi|}} \equiv \frac{1}{\sqrt{|1 - \tan^2 \phi|}}, \quad \sinh_e \theta = \frac{\sin \phi}{\sqrt{|\cos 2\phi|}} \equiv \frac{\tan \phi}{\sqrt{|1 - \tan^2 \phi|}}. \qquad (4.2.16)$$

Geometrical Interpretation of Extended Hyperbolic Trigonometric Functions

We have

Theorem 4.6. *Equations (4.2.16) represent a bijective mapping between points on a unit circle (specified by ϕ) and points on a unit hyperbola (specified by θ), and from a geometrical point of view (4.2.16) represent the projection of a unit circle on a unit hyperbola, from the coordinate origin.*

Proof. Let us consider the half-line $y = x \tan \phi$, $x > 0$ which crosses the unit circle, with center $O \equiv (0, 0)$, in $P_{\mathcal{C}} \equiv (\cos \phi; \sin \phi)$. The half-line crosses the unit hyperbola with center O at point

$$P'_{\mathcal{I}} \equiv \left(\frac{1}{\sqrt{1 - \tan^2 \phi}}; \frac{\tan \phi}{\sqrt{1 - \tan^2 \phi}} \right) \in Rs \quad \text{for} \quad |\tan \phi| < 1 \qquad (4.2.17)$$

Table 4.2: Relations between functions $\cosh_e$, $\sinh_e$ obtained from (4.2.12) and classical hyperbolic functions. The hyperbolic angle θ in the last four columns is calculated referring to semi-axes x, $-x$, y, $-y$, respectively.

	$|x| > |y|$		$|x| < |y|$	
	(Rs), $x > 0$	(Ls), $x < 0$	(Us), $y > 0$	(Ds), $y < 0$
$\cosh_e \theta =$	$\cosh \theta$	$-\cosh \theta$	$\sinh \theta$	$-\sinh \theta$
$\sinh_e \theta =$	$\sinh \theta$	$-\sinh \theta$	$\cosh \theta$	$-\cosh \theta$

or at point

$$P''_{\mathcal{I}} \equiv \left(\frac{1}{\sqrt{\tan^2 \phi - 1}}; \ \frac{\tan \phi}{\sqrt{\tan^2 \phi - 1}} \right) \in Us, \ Ds \quad \text{for} \quad |\tan \phi| > 1. \quad (4.2.18)$$

The half-line $y = x \tan \phi$, $x < 0$ crosses the left side of the circle and the arms Ls, Us, Ds of the hyperbola. Since the points of unit hyperbolas are given by $P_{\mathcal{I}} \equiv (\cosh_e \theta; \sinh_e \theta)$, by comparing (4.2.17) and (4.2.18) with (4.2.16), we have the assertion. $\qquad\square$

A graph of the function $\cosh_e$ from (4.2.16) is shown in Fig. 4.2.

The fact that the extended hyperbolic trigonometric functions can be represented in terms of just one expression given by (4.2.12) allows a direct application of these functions for the solution of triangles with sides in any direction, except the directions parallel to axes bisectors.

4.2.3 Fjelstad's Extension of Hyperbolic Trigonometric Functions

In the complex Gauss–Argand plane, the goniometric circle used for the definition of trigonometric functions is expressed by $x + i\, y = \exp[i\,\phi]$. In the hyperbolic plane the hyperbolic trigonometric functions can be defined on the unit equilateral hyperbola, which can be expressed in a way similar to the goniometric circle: $x + h\,y = \exp[h\theta]$. However this expression represents only the arm of unit equilateral hyperbolas $\in Rs$. If we want to extend the hyperbolic trigonometric functions on the whole plane, we must take into account all arms of the unit equilateral hyperbola $|x^2 - y^2| = 1$, given by $x + h\,y = \pm\exp[h\theta]$ and $x + h\,y = \pm h\, \exp[h\theta]$.

Here we summarize the approach followed in [37] which demonstrates how these unit curves allow us to extend the hyperbolic trigonometric functions and to obtain the addition formula for angles in any sector.

These unit curves are the set of points U, where $U = \{z\,|\,\rho(z) = 1\}$. Clearly $U(\cdot)$ is a group, subgroups of this group are for $z \in Rs$, and for $z \in Rs + Us$, $z \in Rs + Ls$, $z \in Rs + Ds$. For $z \in Rs$ the group $U(\cdot)$ is isomorphic to $\theta(+)$ where $\theta \in \mathbf{R}$ is the angular function that for $-\infty < \theta < \infty$ traverses the Rs arm of the unit hyperbolas. Now we can have a complete isomorphism between $U(\cdot)$ and

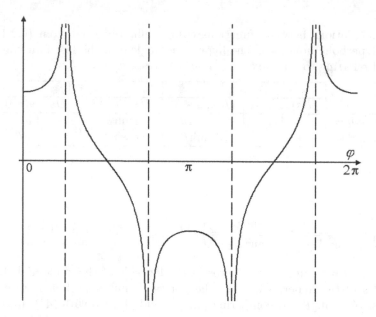

Figure 4.2: The function $\cosh_e \theta = \frac{\cos \phi}{\sqrt{|\cos 2\phi|}}$ for $0 \leq \phi < 2\pi$. The broken vertical lines represent the values for which $\cos 2\phi = 0 \Rightarrow x = \pm y$.
Because $\sin \phi = \cos(\phi - \pi/2)$ and $|\cos 2\phi| = |\cos(2\phi \pm \pi)|$, the function $\sinh_e \theta$ has the same behavior of $\cosh_e \theta$ allowing for a shift of $\pi/2$.

the angular function, extending the last one to other sectors. This can be done by Klein's four-group $k \in K = \{1, h, -1, -h\}$.

   Indeed let us consider the expressions of the four arms of the hyperbolas (Tab. 4.1 for $\rho = 1$). We can extend the angular functions as a product of $\exp[h\,\theta]$ and Klein's group, writing $U = \{k \exp[h\,\theta] \,|\, \theta \in \mathbf{R}, k \in K\}$. Calling $U_k = \{k \exp[h\,\theta] \,|\, \theta \in \mathbf{R}\}$, the hyperbola arm with the value k and, in the same way θ_k the ordered pair (θ, k), we define $\Theta \equiv \mathbf{R} \times K = \{\theta_k \,|\, \theta \in \mathbf{R}, k \in K\}$ and $\Theta_k \equiv \mathbf{R} \times \{k\} = \{\theta_k \,|\, \theta \in \mathbf{R}\}$. Θ_1 is isomorphic to $\mathbf{R}(+)$; then, accordingly, we think of $\Theta(+)$ as an extension of $\mathbf{R}(+)$. To define the complete isomorphism between $\Theta(+)$ and $U(\cdot)$, we have to define the addition rule for angles: $\theta_k + \theta'_{k'}$. This rule is obtained from the isomorphism itself,

$$\theta_k + \theta'_{k'} \Rightarrow U_k \cdot U_{k'} \equiv k \exp[h\theta] \cdot k' \exp[h\theta'] \equiv k\,k' \exp[h(\theta + \theta')] \Rightarrow (\theta + \theta')_{k\,k'}.$$
$$(4.2.19)$$

On this basis we obtain the hyperbolic angle θ and the Klein index (k) from the extended hyperbolic trigonometric functions $\sinh_e \theta$ and $\cosh_e \theta$. We have

$$\text{if } |\sinh_e \theta| < |\cosh_e \theta| \Rightarrow \quad \theta = \tanh^{-1}\left(\frac{\sinh_e \theta}{\cosh_e \theta}\right), \quad k = \frac{\cosh_e \theta}{|\cosh_e \theta|} \cdot 1;$$

$$\text{if } |\sinh_e \theta| > |\cosh_e \theta| \Rightarrow \quad \theta = \tanh^{-1}\left(\frac{\cosh_e \theta}{\sinh_e \theta}\right), \quad k = \frac{\sinh_e \theta}{|\sinh_e \theta|} \cdot h. \quad (4.2.20)$$

Application of "Klein's Index" to the Euclidean Plane

Now we see how this "extension" of hyperbolic trigonometric functions can be applied to circular angles giving well-known results.

By analogy with hyperbolic trigonometric functions we define the circular trigonometric functions just in sector Ls, i.e., for $-\pi/4 < \phi < \pi/4$, by means of Euler's formula $\cos\phi + i\sin\phi = \exp[i\,\phi]$. Let us consider the product $k\exp[i\,\phi]$ where $k \in K = \{1, i, -1, -i\}$ is a four-value group that for $-\pi/4 < \phi < \pi/4$ allows us to obtain the complete circle. The meaning of this product is well known,

$$i\exp[i\,\phi] \equiv \exp[i(\frac{\pi}{2}+\phi)]; \ -\exp[i\,\phi] \equiv \exp[i(\pi+\phi)]; \ -i\exp[i\,\phi] \equiv \exp[i(\frac{3\pi}{2}+\phi)].$$

These expressions allow one to clarify the properties of circular trigonometric functions which are determined on the whole circle, from their values for $0 < \phi < \pi/4$.

We can note the different symmetry from Euclidean and pseudo-Euclidean planes: in the former the angles in all sectors are increasing in the anticlockwise direction; for the latter the sign of the angles is symmetric with respect to axes bisectors (see also Fig. 4.1).

4.3 Geometry in the Pseudo-Euclidean Cartesian Plane

Now we restate for the pseudo-Euclidean plane some classical definitions and properties of the Euclidean plane.

Definitions. Given two points $P_j \equiv (x_j, y_j)$, $P_k \equiv (x_k, y_k)$ that are associated with the hyperbolic variables z_j and z_k, we define the square distance between them by extending (4.1.1),

$$D_{jk} = (z_j - z_k)(\tilde{z}_j - \tilde{z}_k). \quad (4.3.1)$$

As a general rule we indicate the square segment lengths by capital letters, and by the same small letters the square root of their absolute value

$$d_{jk} = \sqrt{|D_{jk}|}. \quad (4.3.2)$$

Following ([81], p. 179) a segment or line is said to be of the *first (second) kind* if it is parallel to a line through the origin located in the sectors containing the axis Ox (Oy). Then the segment $\overline{P_jP_k}$ is of the first (second) kind if $D_{jk} > 0$ $(D_{jk} < 0)$.

Straight line equations.   We shall see in Section 8.5.1 that the equations of straight lines, passing through a point (x_0, y_0), are obtained by means of the method used in differential geometry for calculating the geodesics on a "flat surface" with the metric element given by a non-definite quadratic form. The result is that they are expressed by means of hyperbolic trigonometric functions instead of the circular trigonometric functions used in the Euclidean plane. So, for $|x - x_0| > |y - y_0|$, a straight line (of the first kind) is written as

$$(x - x_0) \sinh \theta - (y - y_0) \cosh \theta = 0, \qquad\qquad (4.3.3)$$

while for $|x - x_0| < |y - y_0|$ a straight line (of the second kind), is written as (the angle θ' is referred to the y axis as stated in Tab. 4.2)

$$(x - x_0) \cosh \theta' - (y - y_0) \sinh \theta' = 0. \qquad\qquad (4.3.4)$$

The straight lines have the expressions (4.3.3) or (4.3.4) which reflect the topological characteristics of the pseudo-Euclidean plane. The use of extended hyperbolic trigonometric functions would give us just one equation for all the straight lines, but in this section we use the classical hyperbolic trigonometric functions which make more evident the peculiar characteristics of the pseudo-Euclidean plane and the differences between lines of the first and second kind.

Pseudo-orthogonality.

Theorem 4.7. *Two straight lines are pseudo-orthogonal if they are symmetric with respect to a couple of lines parallel to axes bisectors.*

Proof. As in the Euclidean plane, two straight lines in the pseudo-Euclidean plane are said to be pseudo-orthogonal when the scalar product of their unity vectors (direction cosine) is zero [34]. It is easy to show that two straight lines of the same kind, as given by (4.3.3) or (4.3.4), can never be pseudo-orthogonal. Indeed a straight line of the first kind (4.3.3) has a pseudo-orthogonal line of the second kind (4.3.4) and with the same angle ($\theta = \theta'$), and conversely. $\square$

This result, well known in special relativity ([30], p. 479) and [55], is represented in Fig. 4.3.

It is known that in complex formalism, the equation of a straight line is given by $\Re\{(x + \mathrm{i}\,y)(\exp[\mathrm{i}\,\phi]) + A + \mathrm{i}\,B = 0\}$, and its orthogonal line by $\Im\{(x + \mathrm{i}\,y)(\exp[\mathrm{i}\,\phi]) + A + \mathrm{i}\,B = 0\}$. The same result holds in the pseudo-Euclidean plane in hyperbolic formalism. Actually we can write the equation of a straight line in a hyperbolic plane as $\Re\{(x + hy)(\exp[h\theta]) + A + hB = 0\}$. If, as is usual for complex variables, we call $\mathcal{H}\{*\}$ the coefficient of the hyperbolic versor, it follows that the hyperbolic part $\mathcal{H}\{(x + hy)(\exp[h\theta]) + A + hB = 0\}$ represents its pseudo-orthogonal line. We note that the product of the angular coefficients for two pseudo-orthogonal lines is $+1$, instead of -1 as in the Euclidean plane.

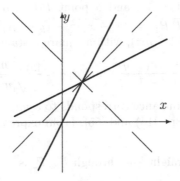

Figure 4.3: Two pseudo-orthogonal straight lines.

Axis of a segment.

Theorem 4.8. *The axis of a segment in the pseudo-Euclidean plane, as in the Euclidean plane, is pseudo-orthogonal to a segment in its middle point.*

Proof. Let us consider two points $P_1(x_1, y_1)$, $P_2(x_2, y_2)$. The points that have the same pseudo-Euclidean distance from these two points are determined by the equation

$$\overline{PP_1}^2 = \overline{PP_2}^2 \Rightarrow (x - x_1)^2 - (y - y_1)^2 = (x - x_2)^2 - (y - y_2)^2.$$

This implies that

$$(x_1 - x_2)(2x - x_1 - x_2) = (y_1 - y_2)(2y - y_1 - y_2) \qquad (4.3.5)$$

and, in canonical form,

$$y = \frac{(x_1 - x_2)}{(y_1 - y_2)} x + \frac{(y_1^2 - y_2^2) - (x_1^2 - x_2^2)}{2(y_1 - y_2)}. \qquad (4.3.6)$$

Then from (4.3.6) it follows that the axis is pseudo-orthogonal to segment $\overline{P_1 P_2}$, and from (4.3.5) that it passes through its middle point $P_M \equiv ((x_1 + x_2)/2, (y_1 + y_2)/2)$. $\square$

Distance of a point from a straight line.

Theorem 4.9. *The linear distance of a point P_1 from a straight line γ is proportional to the result of substituting the coordinates of P_1 in the equation for γ.*

Proof. Let us take a point $P(x, y)$ on a straight line of the second kind $\gamma : \{y - mx - q = 0; \; |m| > 1\}$, and a point $P_1(x_1, y_1)$ outside the straight line. The square distance $\overline{PP_1}^2 = (x - x_1)^2 - (y - y_1)^2$ has its extreme for $x \equiv x_2 = (x_1 - my_1 - mq)/(1 - m^2)$, with a square distance

$$\overline{P_2 P_1}^2 \equiv D_{12} = \frac{(y_1 - mx_1 - q)^2}{m^2 - 1} \quad \text{and} \quad d_{12} = \frac{|y_1 - mx_1 - q|}{\sqrt{|m^2 - 1|}}. \tag{4.3.7}$$

It is easy to verify that this distance corresponds to a *maximum* as is well known from special relativity ([52], p. 315) and [55]. From expression (4.3.7) follows the theorem. □

The equation of the straight line through P_1, P_2 is

$$(y - y_1) = \frac{1}{m}(x - x_1),$$

that represents a straight line *pseudo-orthogonal* to γ.

4.4 Goniometry and Trigonometry in the Pseudo-Euclidean Plane

Hyperbolic angles addition formulas. Using the hyperbolic Euler formula (4.2.14) we can derive ([37] and [38])

$$\cosh_e(\alpha \pm \beta) + h \sinh_e(\alpha \pm \beta) = (\cosh_e \alpha + h \sinh_e \alpha)(\cosh_e \beta \pm h \sinh_e \beta). \tag{4.4.1}$$

These formulas allow us to obtain for hyperbolic trigonometric functions all the expressions which are equivalent to the Euclidean goniometry ones. In particular we have

$$\cosh(\alpha \pm \beta) = \cosh \alpha \cosh \beta \pm \sinh \alpha \sinh \beta, \tag{4.4.2}$$
$$\sinh(\alpha \pm \beta) = \sinh \alpha \cosh \beta \pm \cosh \alpha \sinh \beta. \tag{4.4.3}$$

The sum product identity. Summing and subtracting the terms on the same line and setting $\theta_1 = \alpha + \beta$; $\theta_2 = \alpha - \beta \Rightarrow \alpha = (\theta_1 + \theta_2)/2$; $\beta = (\theta_1 - \theta_2)/2$, we have from (4.4.2)

$$\cosh \theta_1 + \cosh \theta_2 = 2 \cosh \frac{\theta_1 + \theta_2}{2} \cosh \frac{\theta_1 - \theta_2}{2},$$
$$\cosh \theta_1 - \cosh \theta_2 = 2 \sinh \frac{\theta_1 + \theta_2}{2} \sinh \frac{\theta_1 - \theta_2}{2}, \tag{4.4.4}$$

and from (4.4.3)

$$\sinh \theta_1 + \sinh \theta_2 = 2 \sinh \frac{\theta_1 + \theta_2}{2} \cosh \frac{\theta_1 - \theta_2}{2},$$
$$\sinh \theta_1 - \sinh \theta_2 = 2 \cosh \frac{\theta_1 + \theta_2}{2} \sinh \frac{\theta_1 - \theta_2}{2}. \tag{4.4.5}$$

4.4.1 Analytical Definitions of Hyperbolic Trigonometric Functions

Let us consider a triangle in the pseudo-Euclidean plane, with no sides parallel to axes bisectors, and call $P_n \equiv (x_n, y_n)$; $n = i, j, k \mid i \neq j \neq k$ the vertices, θ_n the hyperbolic angles. The square hyperbolic length of the side opposite to vertex P_i is defined by (4.3.1)

$$D_i \equiv D_{jk} = (z_j - z_k)(\tilde{z}_j - \tilde{z}_k) \text{ and } d_i = \sqrt{|D_i|} ; \qquad (4.4.6)$$

as pointed out before, D_i must be taken with its sign.

Following the conventions of Euclidean trigonometry we associate with the sides three vectors oriented from $P_1 \to P_2$; $P_1 \to P_3$; $P_2 \to P_3$.

From (4.2.7) and (4.2.8), taking into account the sides orientation as done in Euclidean trigonometry, we obtain

$$\cosh_e \theta_1 = \frac{(x_2 - x_1)(x_3 - x_1) - (y_2 - y_1)(y_3 - y_1)}{d_2\, d_3},$$

$$\sinh_e \theta_1 = \frac{(x_2 - x_1)(y_3 - y_1) - (y_2 - y_1)(x_3 - x_1)}{d_2\, d_3},$$

$$\cosh_e \theta_2 = -\frac{(x_3 - x_2)(x_2 - x_1) - (y_3 - y_2)(y_2 - y_1)}{d_1\, d_3},$$

$$\sinh_e \theta_2 = \frac{(x_2 - x_1)(y_3 - y_2) - (y_2 - y_1)(x_3 - x_2)}{d_1\, d_3},$$

$$\cosh_e \theta_3 = \frac{(x_3 - x_2)(x_3 - x_1) - (y_3 - y_2)(y_3 - y_1)}{d_1\, d_2},$$

$$\sinh_e \theta_3 = \frac{(x_3 - x_1)(y_3 - y_2) - (y_3 - y_1)(x_3 - x_2)}{d_1\, d_2}. \qquad (4.4.7)$$

It is straightforward to verify that all the functions $\sinh_e \theta_n$ have the same numerator. If we call this numerator

$$x_1(y_2 - y_3) + x_2(y_3 - y_1) + x_3(y_1 - y_2) = 2\,L, \qquad (4.4.8)$$

where $|L| = S$, we can write

$$2S = d_2\, d_3 \sinh_e \theta_1 = d_1\, d_3 \sinh_e \theta_2 = d_1\, d_2 \sinh_e \theta_3. \qquad (4.4.9)$$

In Euclidean geometry a quantity equivalent to S represents the triangle's area. In pseudo-Euclidean geometry, S is still *an invariant quantity linked to the triangle*. For this reason it is appropriate to call S the *pseudo-Euclidean area* [81]. We note that the expression of area (4.4.8), in terms of vertices coordinates, is exactly the same as in Euclidean geometry (Gauss' formula for a polygon area applied to a triangle).

4.4.2   Trigonometric Laws in the Pseudo-Euclidean Plane

Law of sines.   *In a triangle the ratio of the hyperbolic sine to the hyperbolic length of the opposite side is a constant*

$$\frac{\sinh_e \theta_1}{d_1} = \frac{\sinh_e \theta_2}{d_2} = \frac{\sinh_e \theta_3}{d_3}. \tag{4.4.10}$$

Proof. This theorem follows from (4.4.9) if we divide it by $d_1 \, d_2 \, d_3$.   □

We have

Theorem 4.10. *If two hyperbolic triangles have the same hyperbolic angles their sides are proportional.*

Napier's theorem.   As in Euclidean trigonometry, Napier's theorem follows at once from (4.4.10) and (4.4.5)

$$\frac{d_1 + d_2}{d_1 - d_2} = \frac{\tanh_e \frac{\theta_1 + \theta_2}{2}}{\tanh_e \frac{\theta_1 - \theta_2}{2}}. \tag{4.4.11}$$

Carnot's theorem.   From the definitions of the side lengths (4.1.2) and hyperbolic angular functions given by (4.4.7) we can verify that[2]

$$D_i = D_j + D_k - 2d_j \, d_k \cosh_e \theta_i. \tag{4.4.12}$$

Law of cosines.   As in the previous theorem we can verify

$$d_i = |d_j \cosh_e \theta_k + d_k \cosh_e \theta_j|. \tag{4.4.13}$$

Pythagoras' theorem.

Proof. Let us consider a triangle with the side $P_i \, P_k$ orthogonal to $P_i \, P_j$. We have

$$\frac{y_j - y_i}{x_j - x_i} \equiv m_{ij} = \frac{1}{m_{ik}} \equiv \frac{x_k - x_i}{y_k - y_i},$$

then

$$(x_j - x_i)(x_k - x_i) = (y_j - y_i)(y_k - y_i), \tag{4.4.14}$$

and from (4.4.7) it follows that $\cosh_e \theta_i = 0$ and from (4.4.12) the hyperbolic Pythagoras' theorem

$$D_i = D_j + D_k. \tag{4.4.15}$$

holds.   □

---

[2]Obtained by another method in [38].

It must be noted that in the right-hand sides of (4.4.12) and (4.4.15) there is a sum of the square side lengths, as in Euclidean geometry, but in pseudo-Euclidean geometry the square side lengths may be negative. In particular in (4.4.15) D_j and D_k are pseudo-orthogonal, then they always have opposite signs.

We have seen that the topology of the pseudo-Euclidean plane is more complex with respect to the Euclidean one, as well as the relations between $\sinh_e$ and $\cosh_e$ and between the side lengths and the square side lengths. For these reasons we could think that the triangle solution would require more information, nevertheless in Section 4.6 we shall see that: *All the sides and angles of a pseudo-Euclidean triangle can be determined if we know three elements (with at least one side).* We have

Theorem 4.11. *All the elements (sides and angles) of a triangle are invariant for hyperbolic rotation.*

Proof. Let us consider a triangle with vertices in points

$$P_1 \equiv (0, 0) , \quad P_2 \equiv (x_2, 0) \text{ and } P_3 \equiv (x_3, y_3); \qquad (4.4.16)$$

since $\overline{P_1 P_3} \equiv d_2$ and $\overline{P_1 P_2} \equiv d_3$ are invariant quantities, from Theorem 4.4 it follows that θ_1 is invariant too. Since these three elements determine all the other ones, all elements are invariant. $\qquad\qquad\square$

From this invariance it follows that, by a coordinate axes translation and a hyperbolic rotation, any triangle can be constructed with a vertex in $P \equiv (0, 0)$, and a side on one coordinate axis.

Then we do not lose in generality if, from now on, we consider a triangle in a position that will facilitate the demonstration of the theorems which follow. Consequently we will consider triangles with vertices in points given by (4.4.16); the sides lengths are

$$D_1 = (x_3 - x_2)^2 - y_3^2; \qquad D_2 = x_3^2 - y_3^2; \qquad D_3 = x_2^2. \qquad (4.4.17)$$

By using (4.4.6) and (4.4.7), we obtain the other elements:

$$\cosh_e \theta_1 = \frac{x_2 x_3}{d_2 d_3}; \quad \cosh_e \theta_2 = \frac{x_2(x_2 - x_3)}{d_1 d_3}; \quad \cosh_e \theta_3 = \frac{x_3(x_3 - x_2) - y_3^2}{d_1 d_2};$$

$$\sinh_e \theta_1 = \frac{x_2 y_3}{d_2 d_3}; \quad \sinh_e \theta_2 = \frac{x_2 y_3}{d_1 d_3}; \quad \sinh_e \theta_3 = \frac{x_2 y_3}{d_1 d_2} \equiv \frac{d_3 y_3}{d_1 d_2} .$$

$$(4.4.18)$$

4.4.3 The Triangle's Angles Sum

In a Euclidean triangle, given two angles (ϕ_1, ϕ_2), the third one (ϕ_3) can be found using the relation $\phi_1 + \phi_2 + \phi_3 = \pi$. This relation can be expressed in the form

$$\sin(\phi_1 + \phi_2 + \phi_3) = 0 , \qquad \cos(\phi_1 + \phi_2 + \phi_3) = -1$$

that allows us to verify

Theorem 4.12. *By means of the formalism exposed in Section 4.2.3, summarized by (4.2.19), we can state: the sum of the triangle's angles is given by*

$$(\theta_1)_k + (\theta_2)_{k'} + (\theta_3)_{k''} \equiv (\theta_1 + \theta_2 + \theta_3)_{k \cdot k' \cdot k''} = (0)_{\pm 1} . \qquad (4.4.19)$$

Proof. Exploiting (4.4.1) and using (4.4.18), we obtain

$$\sinh_e(\theta_1 + \theta_2 + \theta_3) \equiv \sinh_e \theta_1 \sinh_e \theta_2 \sinh_e \theta_3 + \sinh_e \theta_1 \cosh_e \theta_2 \cosh_e \theta_3$$
$$+ \cosh_e \theta_1 \sinh_e \theta_2 \cosh_e \theta_3 + \cosh_e \theta_1 \cosh_e \theta_2 \sinh_e \theta_3$$
$$\equiv \frac{x_2^2 y_3 [x_2\, y_3^2 + x_2\, x_3(x_2 - x_3) + x_3^2(x_3 - x_2) - x_3\, y_3^2 - x_3(x_2 - x_3)^2 - y_3^2(x_2 - x_3)]}{d_1^2 d_2^2 d_3^2} = 0,$$

$$\cosh_e(\theta_1 + \theta_2 + \theta_3) \equiv \cosh_e \theta_1 \cosh_e \theta_2 \cosh_e \theta_3 + \sinh_e \theta_1 \sinh_e \theta_2 \cosh_e \theta_3$$
$$+ \sinh_e \theta_1 \cosh_e \theta_2 \sinh_e \theta_3 + \cosh_e \theta_1 \sinh_e \theta_2 \sinh_e \theta_3$$
$$\equiv \frac{x_2^2 \{ -x_3^2(x_2 - x_3)^2 + y_3^2[-x_3(x_2 - x_3) + x_2\, x_3] + y_3^2[x_2(x_2 - x_3) - x_3(x_2 - x_3) - y_3^2] \}}{d_1^2 d_2^2 d_3^2}$$
$$\equiv \frac{x_2^2 \{ -x_3^2[(x_2 - x_3)^2 - y_3^2] + y_3^2[(x_2 - x_3)^2 - y_3^2] \}}{d_1^2 d_2^2 d_3^2} \equiv -\frac{D_1 D_2 D_3}{d_1^2 d_2^2 d_3^2} = \mp 1. \qquad (4.4.20)$$

Then

$$\sinh_e(\theta_1 + \theta_2 + \theta_3) = 0 , \qquad \cosh_e(\theta_1 + \theta_2 + \theta_3) = \pm 1, \qquad (4.4.21)$$

that are equivalent to (4.4.19). $\square$

This result allows us to state that if we know two angles, we can determine if the Klein's group index of the third angle is $\pm h : \{\theta \in Us, Ds\}$ or $\pm 1 : \{\theta \in Rs, Ls\}$. From this we obtain the relation between $\cosh_e$ and $\sinh_e$ as stated by (4.2.15). This relation and the condition (4.4.19) allow us to obtain the hyperbolic functions of the third angle. Then we have

Theorem 4.13. *Also for pseudo-Euclidean triangles, if we know two angles we can obtain the third one.*

In Section 4.6 we show some examples of solutions for general triangles.

4.5   Theorems on Equilateral Hyperbolas in the Pseudo-Euclidean Plane

The unit circle for the definition of trigonometric functions has its counterpart, in the pseudo-Euclidean plane, in the four arms of the unit equilateral hyperbolas $x^2 - y^2 = \pm 1$, as shown in [37]. Indeed the equilateral hyperbolas have many of the properties of circles in the Euclidean plane; here we point out some of them, showing some theorems in the pseudo-Euclidean plane that represent the counterpart of well-known theorems which hold for the circle in the Euclidean plane.

Definitions. If A and B are two points, lying on an equilateral hyperbola, the segment AB is called a *chord* of the hyperbola. We define two kinds of chords: if points A, B are on the same arm of the hyperbola, we have *"external chords"*, if points are in opposite arms, we have *"internal chords"*. We extend these definitions to points: given a two-arms hyperbola we call external the points inside the arms, internal the points between the arms and the axes bisectors. This definition agrees with the one for circles in which the center is an internal point.

Any internal chord which passes through the center "P_c" of the hyperbola is called a *diameter of the hyperbola*. We call p the semi-diameter and P the "square semi-diameter", with its sign ($p = \sqrt{|P|}$).

Here we extend to equilateral hyperbolas the definitions stated for segments and straight lines and call them *hyperbolas of the first (second) kind* if the tangent straight lines are of the first (second) kind.

Actually, as far as a general curve is concerned, we can not assign it, in general, a kind since general curves have tangent straight lines of both kinds and only equilateral hyperbolas have the peculiar property that the tangent straight lines to a given arm are of the same kind. This allows us to attribute a kind, depending on the P sign, to the hyperbola's arms. Then we have hyperbolas of the first (second) kind if $P < 0$ ($P > 0$), respectively.

The parametric equations of a general equilateral hyperbola are given by

$$x = x_c \pm p \, \cosh_e \theta, \quad y = y_c \pm p \, \sinh_e \theta \tag{4.5.1}$$

and depend on three parameters: the center's coordinates $P_c \equiv (x_c, y_c)$ and the half-diameter p. This hyperbola is determined by three conditions as the equations for circles. In particular these three conditions can be the passage through three non-aligned points.

Now we enunciate for equilateral hyperbolas the pseudo-Euclidean counterpart of the well-known Euclidean theorems for circles. The demonstration of these theorems is performed by elementary analytic geometry.[3]

Theorem 4.14. *The axis of two points on an equilateral hyperbola passes through the center* (x_c, y_c).

An equivalent form is: *The line joining* P_c *with the midpoint M of a chord is pseudo-orthogonal to it.*

Proof. Let us consider two points P_1, P_2 on the same arm of equilateral hyperbola (4.5.1), determined by hyperbolic angles θ_1, θ_2, and calculate the axis of $\overline{P_1 P_2}$.

---

[3]Some of these theorems are reported, without demonstration, in [81].

Substituting in (4.3.5) the coordinates given by (4.5.1), we obtain

$$(y - y_c) = (x - x_c)\frac{\cosh_e \theta_1 - \cosh_e \theta_2}{\sinh_e \theta_1 - \sinh_e \theta_2} \equiv (x - x_c)\frac{2\sinh_e \frac{\theta_1 - \theta_2}{2} \sinh_e \frac{\theta_1 + \theta_2}{2}}{2\sinh_e \frac{\theta_1 - \theta_2}{2} \cosh_e \frac{\theta_1 + \theta_2}{2}}$$

$$\equiv (x - x_c)\tanh_e \frac{\theta_1 + \theta_2}{2} . \tag{4.5.2}$$

Equation (4.5.2) demonstrates the theorem.   □

In a similar way we can find that if the points are on different arms of the hyperbola we have just to change (4.5.2) with $(y - y_c) = (x - x_c)\coth_e[(\theta_1 + \theta_2)/2]$.

This theorem is also valid in the limiting position when the points are coincident and the chord becomes tangent to the hyperbola, and we have

Theorem 4.15. *For points M on equilateral hyperbola, the tangent at M is pseudo-orthogonal to the diameter $P_c M$.*

For a demonstration of the following theorems, we do not lose in generality if we consider hyperbolas of the second kind, with their center at the coordinate origin. We also put $P_c \to O$ and Equations (4.5.1) become

$$x = \pm p \cosh_e \theta, \quad y = \pm p \sinh_e \theta. \tag{4.5.3}$$

Theorem 4.16. *The diameters of a hyperbola are the internal chord of minimum length.*

Proof. We have

$$A \equiv (p \cosh \theta_1, p \sinh \theta_1), \qquad B \equiv (-p \cosh \theta_2, -p \sinh \theta_2).$$

The square length of the chord is

$$\overline{AB}^2 = p^2[(\cosh \theta_1 + \cosh \theta_2)^2 - (\sinh \theta_1 + \sinh \theta_2)^2] \equiv 4p^2 \cosh^2[(\theta_1 - \theta_2)/2],$$

then

$$\overline{AB}^2 = 4p^2 \text{ for } \theta_1 = \theta_2; \quad \overline{AB}^2 > 4p^2 \text{ for } \theta_1 \neq \theta_2. \qquad □$$

Theorem 4.17. *If points A and B lie on the same arm of a hyperbola, for any point P between A and B, the hyperbolic angle $\widehat{APB}$ is half the hyperbolic angle $\widehat{AOB}$.*

Proof. We do not lose in generality by considering a unitary equilateral hyperbola ($p = 1$) and the arm of hyperbola $\in Rs$. Let us take points $A \equiv (\cosh \theta_A; \sinh \theta_A)$, $B \equiv (\cosh \theta_B; \sinh \theta_B)$ with $\theta_A < \theta_B$ and the hyperbola arc between them. If $P \equiv (\cosh \theta; \sinh \theta)$ is a point on this arc, i.e., $\theta_A < \theta < \theta_B$, we call α the hyperbolic angle $\widehat{APB}$. The points $A \to P \to B$ follow each other in clockwise direction, then the sign of sinh in (4.4.7) changes and it follows that

$$\tanh \alpha \qquad (4.5.4)$$
$$= \frac{(\cosh \theta_B - \cosh \theta)(\sinh \theta_A - \sinh \theta) - (\cosh \theta_A - \cosh \theta)(\sinh \theta_B - \sinh \theta)}{(\cosh \theta_A - \cosh \theta)(\cosh \theta_B - \cosh \theta) - (\sinh \theta_A - \sinh \theta)(\sinh \theta_B - \sinh \theta)}.$$

The differences between trigonometric functions in round brackets can be written, by means of (4.4.4) and (4.4.5), as products and, after simplification, we obtain

$$\tanh \alpha = \tanh \frac{\theta_B - \theta_A}{2}. \qquad (4.5.5)$$

Now we call β the angle $\widehat{AOB}$. Points $O \to A \to B$ follow each other in clockwise direction and, from (4.4.7), we have

$$\tanh \beta = -\frac{\cosh \theta_A \sinh \theta_B - \cosh \theta_B \sinh \theta_A}{\cosh \theta_A \cosh \theta_B - \sinh \theta_A \sinh \theta_B} = \tanh(\theta_B - \theta_A). \qquad (4.5.6)$$

From (4.5.5) and (4.5.6) the theorem follows. $\qquad \square$

We show *the complete equivalence with the analogous Euclidean problem for which also the complementary angles and the chords becoming tangent are considered.*

1. Let us consider a point P outside the arc AB. Points $A \to P \to B$ follow each other in anticlockwise direction, then calling γ the angle $\widehat{APB}$, we obtain

$$\tanh \gamma = \tanh \frac{\theta_A - \theta_B}{2} = -\tanh \frac{\theta_B - \theta_A}{2} \Rightarrow \alpha = -\gamma.$$

   In the language of Klein's index (Section 4.2.3), if $\tanh \alpha = -\tanh \gamma$, we have $(\alpha)_k = (\gamma)_{-k}$, then $\alpha + \gamma = (\alpha + \gamma)_{-k \cdot k} \equiv (0)_{-1}$. This relation corresponds to the Euclidean case for which $\alpha + \gamma = \pi$.

2. Let us consider $P \in Rs$, where $P \equiv (-\cosh \theta; -\sinh \theta)$. In (4.5.4) we must change the sign of θ. Then, by calling δ the angle $\widehat{APB}$, with similar calculations we obtain $\tanh \delta = \tanh \frac{\theta_A - \theta_B}{2} \Rightarrow \delta = \gamma$.

3. We complete the parallelism with Euclidean geometry by taking $P \equiv A$, then α is the angle between the tangent (τ) to the hyperbola in A and the cord (σ) $A - B$. Since

$$\text{angular coefficient of } \tau \Rightarrow \coth \theta_A \equiv \tanh \mu, \qquad (4.5.7)$$
$$\text{angular coefficient of } \sigma \Rightarrow \frac{\sinh \theta_A - \sinh \theta_B}{\cosh \theta_A - \cosh \theta_B} \equiv \tanh \nu, \qquad (4.5.8)$$

we obtain

$$\tanh \alpha \equiv \frac{\tanh \mu - \tanh \nu}{1 - \tanh \mu \tanh \nu} = \frac{1 - \cosh (\theta_A - \theta_B)}{\sinh (\theta_A - \theta_B)} \equiv -\tanh \frac{\theta_A - \theta_B}{2},$$
(4.5.9)

which completes the assertion.

As a straightforward consequence of Theorem 4.17, there follows

Theorem 4.18. *If Q is a second point between A and B, we have $\widehat{APB} = \widehat{AQB}$.*

Theorem 4.19. *If a side of a triangle inscribed in an equilateral hyperbola passes through the center of the hyperbola, the other two sides are pseudo-orthogonal.*

Proof. For two points on an equilateral hyperbola and on a line passing through the center, the coordinates are given by

$$P_1 \equiv (p \cosh_e \theta_1, \, p \sinh_e \theta_1), \quad P_3 \equiv (-p \cosh_e \theta_1, \, -p \sinh_e \theta_1).$$

Let us consider a third point on the hyperbola $P_2 \equiv (p \cosh_e \theta_2, \, p \sinh_e \theta_2)$ and the sides on straight lines $P_1 P_3$, $P_2 P_3$,

$$P_1 P_3 \Rightarrow \frac{y - p \sinh_e \theta_1}{p \,(\sinh_e \theta_2 - \sinh_e \theta_1)} = \frac{x - p \cosh_e \theta_1}{p \,(\cosh_e \theta_2 - \cosh_e \theta_1)},$$

$$P_2 P_3 \Rightarrow \frac{y + p \sinh_e \theta_1}{p \,(\sinh_e \theta_2 + \sinh_e \theta_1)} = \frac{x + p \cosh_e \theta_1}{p \,(\cosh_e \theta_2 + \cosh_e \theta_1)}.$$

The product of the angular coefficients of these straight lines is 1, then they are pseudo-orthogonal. $\square$

In Chapter 6 we generalize Euclidean theorems about circumcircles, incircles and excircles of a triangle to ellipses and general hyperbolas. Now we begin with

Theorem 4.20. *If we have three non-aligned points that can be considered the vertices of a triangle, there is an equilateral hyperbola (circumscribed hyperbola) which passes through these points, and its semi-diameter is given by*

$$p = \frac{d_1 d_2 d_3}{4 S} \equiv \frac{d_n}{2 \sinh_e \theta_n} \, , \quad n = 1, \, 2, \, 3.$$
(4.5.10)

Proof. As three points in Euclidean plane define a circumcircle, in the same way, in the pseudo-Euclidean plane, the vertices of a triangle define an equilateral hyperbola (4.5.1). Let us find its center $P_c \equiv (x_c, \, y_c)$ and the square semi-diameter P. Let us consider three points $P_1 \equiv (x_1, \, y_1)$, $P_2 \equiv (x_2, \, y_2)$, $P_3 \equiv (x_3, \, y_3)$. The parametric equation of a hyperbola passing through them is obtained by imposing this condition on (4.5.1). By calling θ_i the hyperbolic angles corresponding to points P_i, we have

$$x_i = x_c \pm p \cosh_e \theta_i, \quad y_i = y_c \pm p \sinh_e \theta_i.$$
(4.5.11)

For finding the coordinates of the hyperbola center and the condition that states if it is of the first or second kind, we do not lose in generality if we take three points in the position of (4.4.16).

Since the center is on the axis of the chord $\overline{P_1P_2}$, we have $x_c = x_2/2$. From the intersection between this axis with another one we obtain the center coordinates

$$P_c \equiv \left(\frac{x_2}{2} , \ \frac{y_3^2 - x_3^2 + x_2 x_3}{2y_3} \right).$$

Moreover, since the hyperbola passes through the coordinate origin, the square semi-diameter (P) is given by

$$P \equiv \overline{P_c P_1} = \frac{x_2^2}{4} - y_c^2 \equiv \frac{x_2^2 y_3^2 - (y_3^2 - x_3^2 + x_2 x_3)^2}{4y_3^2} \equiv \frac{(x_3^2 - y_3^2)[y_3^2 - (x_2 - x_3)^2]}{4y_3^2}.$$
(4.5.12)

Taking into account the last expressions of (4.4.9) and (4.4.18), we have that the denominator of (4.5.12) is given by $y_3^2 = (d_1 d_2 \sinh\theta_3)^2/d_3^2 \equiv S^2/D_3$; then, by means of (4.4.17), expression (4.5.12) can be written as a function of invariant quantities

$$P = -\frac{D_1 D_2 D_3}{16\, S^2};$$
(4.5.13)

for $P > 0$ we have an equilateral hyperbola of the second kind, while for $P < 0$ we have an equilateral hyperbola of the first kind. Then in relation to the hyperbola type we could say that there are two kinds of triangles depending on the sign of $D_1 \cdot D_2 \cdot D_3$. If we put $p = \sqrt{|P|}$, from (4.5.13) and (4.4.9) we obtain (4.5.10), which is the same relation that holds for the radius of a circumcircle in a Euclidean triangle. $\qquad\square$

Relation (4.5.13) is demonstrated in a more general form (6.2.10) in Chapter 6.

Theorem 4.21. *If from a non-external point $P \equiv (x_p , y_p)$ we trace a tangent and a secant line to a hyperbola, we have: the square of the distance of the tangent point is equal to the product of the distances of secant points.*

Proof. The stated theorems allow us to give a "Euclidean demonstration". Referring to Fig. 4.4, let us call T the tangent point and S_1, S_2 the intersection points between the secant and the hyperbola. We also put

$$\overline{PT} \equiv t; \quad \overline{PS_1} \equiv s_1; \quad \overline{PS_2} \equiv s_2,$$

and consider the triangles PS_2T and PTS_1, they are similar since:

- The angles $\widehat{S_2PT}$ and $\widehat{S_1PT}$ are the same.

- The angles $\widehat{PTS_1}$ and $\widehat{TS_2S_1}$ subtend the same hyperbola arc $\overset{\frown}{TS_1}$; then, from Theorem 4.18, they are equal.

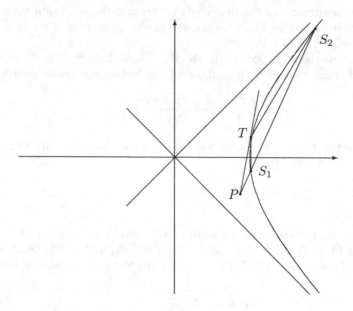

Figure 4.4: Tangent and secant lines to an equilateral hyperbola (Theorem 4.21).

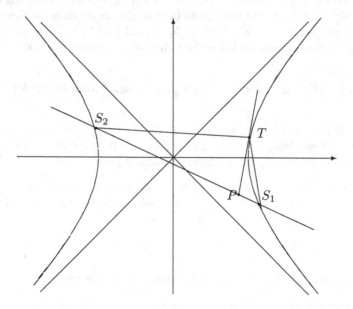

Figure 4.5: Secant line on different arms of an equilateral hyperbola.

- The third angles from Theorem 4.12 are equal.

From Theorem 4.10 the triangles PTS_1 and PS_2T shall have proportional sides; then $t^2 = s_1 \cdot s_2$ follows. $\square$

Now we demonstrate this theorem for secant lines crossing the different arms of an equilateral hyperbola.

Proof. Referring to Fig. 4.5 we have

- The angles $\widehat{S_2PT}$ and $\widehat{S_1PT}$, seen in a Euclidean way are supplementary angles. In the hyperbolic plane, if we call k the Klein index of the first angle, the index of the second one is $-k$ with the same value of the angle.

- The angles $\widehat{PTS_1}$ and $\widehat{PS_2T}$ subtend the same hyperbola arc $\widehat{TS_1}$; then, from Theorem 4.17, point 2, they are equal but for the Klein index. Actually by calling k_1 the Klein index of the first angle, the index of the second one is $-k_1$.

- The third angles are equal from Theorem 4.12.

Since the hyperbolic sine are the same, but for their sign, for angles with opposite Klein index, from Theorem 4.10 the triangles PTS_1 and PS_2T shall have proportional sides; then $t^2 = s_1 \cdot s_2$ follows. $\square$

This theorem can be demonstrated also by means of the analytic method exposed in this chapter.

Proof. Let us consider the hyperbola

$$x^2 - y^2 = 1, \tag{4.5.14}$$

the non-external point $P \equiv (x_p, y_p)$, and the straight line

$$y - y_p = m(x - x_p) \equiv (x - x_p)\tanh_e \theta_p. \tag{4.5.15}$$

Eliminating y between (4.5.14) and (4.5.15) we obtain, for the abscissas of the intersection points, the equation of degree 2,

$$(1 - m^2)x^2 + 2\,m\,x(m\,x_p + y_p) - m^2 x_p^2 - 2\,m\,x_p\,y_p - y_p^2 - 1 = 0. \tag{4.5.16}$$

Now let us see that we do not need to find the roots of this equation. Actually setting $S_{1,2} \equiv (x_{1,2}; y_{1,2})$, we have

$$s_{1,2} = \frac{|x_{1,2} - x_p|}{|\cosh_e \theta_p|} \equiv |(x_{1,2} - x_p)|\sqrt{|1 - m^2|} \tag{4.5.17}$$

and

$$
\begin{aligned}
s_1 \cdot s_2 &= |(1 - m^2)(x_1 - x_p)(x_2 - x_p)| = |(1 - m^2)[x_1 x_2 - x_p(x_1 + x_2) + x_p^2]| \\
&= \left| \frac{1 - m^2}{a}(a\,x_p^2 + b\,x_p + c) \right|
\end{aligned}
\tag{4.5.18}
$$

where, in the last passage, we use the link between the roots and the coefficients a, b, c of the equation of degree 2 (4.5.16). Substituting these coefficients and simplifying we obtain

$$s_1 \cdot s_2 = |x_p^2 - y_p^2 - 1|, \qquad (4.5.19)$$

which is independent of m. Then the product (4.5.18) is the same for coincident solutions (tangent line) or if the intersection points are in the same or different arms of (4.5.14). □

This result can be generalized and we have

Theorem 4.22. *If from a non-external point $P \equiv (x_p, y_p)$ of an equilateral hyperbola*

$$(x - x_c)^2 - (y - y_c)^2 \pm p^2 = 0, \qquad (4.5.20)$$

we trace a tangent line to the hyperbola, then the square of the distance between P and the tangent point is obtained by substituting the coordinates of P in the equation for the hyperbola

$$t^2 \equiv s_1 \cdot s_2 = |(x_c - x_p)^2 - (y_c - y_p)^2 \pm p^2|. \qquad (4.5.21)$$

Proof. The proof can be obtained, as the previous one, by considering (4.5.20) instead of (4.5.14) or by noting that the last term in the round bracket of (4.5.18) can be obtained by substituting $x \to x_p$ in (4.5.16). This substitution, directly into (4.5.15), gives $y = y_p$ and both of these substitutions into (4.5.20) give (4.5.21). □

The theorems we have seen in this section indicate that in some cases problems about equilateral hyperbolas may be solved more easily in the pseudo-Euclidean plane, by applying the above theory.

4.6   Examples of Triangle Solutions in the Minkowski Plane

The Elements in a Right-Angled Triangle

The Sum of Internal Angles

Let us consider the right-angled triangle represented in Fig. 4.6 in a Euclidean plane and call α_i the angles in the vertices P_i, and define them as "extended angles" (see Section 4.2.3). If we indicate with a subscript the "Euclidean Klein index" ($k \in K = \{1, i, -1, -i\}$), we have $\alpha_1 = (\alpha_1)_1$, $\alpha_2 \equiv \pi/2 = (0)_i$, $\alpha_3 \equiv \pi/2 - \alpha_1 = (-\alpha_1)_i$. Therefore

$$\alpha_1 + \alpha_2 + \alpha_3 \equiv (\alpha_1)_1 + (0)_i + (-\alpha_1)_i \equiv (\alpha_1 + 0 - \alpha_1)_{i \cdot i} = (0)_{-1} \equiv \pi,$$

as is well known.

Now we consider the same triangle in the pseudo-Euclidean plane. Setting in (4.4.18) $x_2 = x_3$, we have

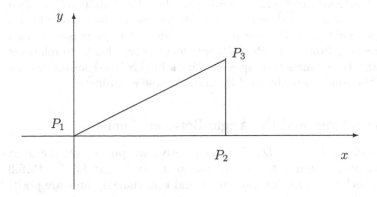

Figure 4.6: A right-angled triangle.

$$\begin{cases} D_1 = -y_3^2 \\ \cosh_e \theta_1 = x_3/d_2 \\ \sinh_e \theta_1 = y_3/d_2 \end{cases} \qquad \begin{cases} D_2 = x_3^2 - y_3^2 \\ \cosh_e \theta_2 = 0 \\ \sinh_e \theta_2 = 1 \end{cases} \qquad \begin{cases} D_3 = x_3^2 \\ \cosh_e \theta_3 = -y_3/d_2 \\ \sinh_e \theta_3 = x_3/d_2. \end{cases}$$

It results that $\theta_2 = (0)_h$ and we consider the following cases:

1. $P_3 \in Rs \Rightarrow \theta_1 = (\theta_1)_1$ and by means of Tab. 4.2 we have $\theta_3 = (-\theta_1)_h$. Therefore $\theta_1 + \theta_2 + \theta_3 \equiv (\theta_1)_1 + (0)_h + (-\theta_1)_h \equiv (\theta_1 + 0 - \theta_1)_{h \cdot h} = (0)_1$.

2. $P_3 \in Us$. Setting $\theta_3 = -\theta'$ we have $\cosh_e \theta' = -y_3/d_2$, $\sin \theta' = -x_3/d_2$. Therefore $\theta' = (\theta')_1 \Rightarrow \theta_3 = (-\theta')_{-1}$, and from Tab. 4.2 we have $\theta_1 = (\theta')_h$, and $\theta_1 + \theta_2 + \theta_3 \equiv (\theta')_h + (0)_h + (-\theta')_{-1} = (0)_{-1}$.

We have found the two possibilities of the Klein index shown in Section 4.4.3 for the sum of the triangle angles.

The Circumscribed Hyperbolas

Let us see what kind of circumscribed hyperbolas we have. From (4.5.13) and (4.4.17), we have $p^2 = x_3^2 y_3^2 (x_3^2 - y_3^2)$, then

- if $|x_3| > |y_3| \Rightarrow p^2 > 0 \Rightarrow$ second kind hyperbolas,
- if $|x_3| < |y_3| \Rightarrow p^2 < 0 \Rightarrow$ first kind hyperbolas.

Solution of Hyperbolic Triangles

In this section, in order to point out analogies and differences with Euclidean trigonometry, we report some examples in which we determine the elements of hyperbolic triangles. We shall note that the Cartesian representation can give some simplifications in the triangle solution.

The Cartesian axes will be chosen so that $P_1 \equiv (0, 0)$, $P_2 \equiv (\pm d_3, 0)$, or $P_2 \equiv (0, \pm d_3)$, where $\overline{P_1 P_2} = d_3$ is the side (or one of the sides) which we know. The two possibilities for P_2 depend on the D_3 sign, the sign plus or minus is chosen so that one goes from P_1 to P_2 to P_3 anticlockwisely. Thanks to relations (4.4.17) and (4.4.18), the triangle is completely determined by the coordinates of point P_3, therefore the solution is obtained by finding these coordinates.

Given Two Sides and the Angle Between Them

Let us be given θ_1; D_2; D_3. For D_3 positive we put P_2 on the x axis, then $P_2 \equiv (\pm d_3, 0)$, the sign of the square root being such that P_1, P_2, P_3 follow each other anticlockwisely. In this way $\sinh \theta_i$ and the triangle's area are positive, as in Euclidean trigonometry in a Cartesian representation. So we have

$$x_3 = d_2 \cosh_e \theta_1; \quad y_3 = d_2 \sinh_e \theta_1, \tag{4.6.1}$$

then Equation (4.4.18) allow us to determine the other elements.

For D_3 negative, we must take P_2 on the y axis and in (4.6.1) we must change $\sinh_e \theta_1 \leftrightarrow \cosh_e \theta_1$. Grouping together both examples, we have

$$\begin{aligned} \text{if } D_3 > 0 \quad & x_3 = d_2 \cosh_e \theta_1; \quad y_3 = d_2 \sinh_e \theta_1; \\ \text{if } D_3 < 0 \quad & x_3 = d_2 \sinh_e \theta_1; \quad y_3 = d_2 \cosh_e \theta_1. \end{aligned} \tag{4.6.2}$$

Given Two Angles and the Side Between Them

Let us be given θ_1; θ_2; D_3. In Euclidean trigonometry the solution of this problem is obtained by using the condition that the sum of the three angles is π. We use this method which allows us to use the Klein group defined in Section 4.2.3, as well as a method peculiar to analytical geometry.

Let us start with the classical method: the Klein indexes (k_1, k_2) of known angles θ_1, θ_2 are obtained by means of (4.2.20). For the third angle we have $\theta_3 = -(\theta_1 + \theta_2)$ and k_3 is such that $k_1 k_2 k_3 = \pm 1$. Then the hyperbolic trigonometric functions are given by $\sinh_e \theta_3 = \sinh |\theta_1 + \theta_2|$ if $k_3 = \pm 1$ and by $\sinh_e \theta_3 = \cosh(\theta_1 + \theta_2)$ if $k_3 = \pm h$. Now we can apply the law of sines and obtain $d_2 = d_3 \frac{\sinh_e \theta_2}{\sinh_e \theta_3}$. Equations (4.6.1) allow us to obtain the P_3 coordinates.

In the Cartesian representation we can use the following method: let us consider the straight lines $P_1 P_3$ and $P_2 P_3$. Solving the algebraic system between these straight lines we obtain the P_3 coordinates.

For $D_3 > 0$ the straight line equations are $P_1 P_3 \Rightarrow y = x \tanh_e \theta_1$ and $P_2 P_3 \Rightarrow y = -(x - x_2) \tanh_e \theta_2$, and we obtain

$$P_3 \equiv \left(x_2 \frac{\tanh_e \theta_2}{\tanh_e \theta_2 + \tanh_e \theta_1}, \ x_2 \frac{\tanh_e \theta_1 \tanh_e \theta_2}{\tanh_e \theta_2 + \tanh_e \theta_1} \right). \tag{4.6.3}$$

For $D_3 < 0$ the straight line equations are $y = x \coth_e \theta_1$ and $y - y_2 = -x \coth_e \theta_2$, with the solution

$$P_3 \equiv \left(y_2 \frac{\tanh_e \theta_1 \tanh_e \theta_2}{\tanh_e \theta_2 + \tanh_e \theta_1}, \; y_2 \frac{\tanh_e \theta_2}{\tanh_e \theta_2 + \tanh_e \theta_1} \right). \tag{4.6.4}$$

Given Two Sides and one Opposite Angle

Let us be given θ_1; D_1; D_3, with $D_3 > 0$. Applying Carnot's theorem (4.4.12) to the side d_1 we have $D_2 - 2d_2 d_3 \cosh_e \theta_1 + D_3 - D_1 = 0$, from which we can obtain d_2.
Actually for $\cosh_e \theta_1 > \sinh_e \theta_1$,

$$D_2 = d_2^2 \Rightarrow d_2 = d_3 \cosh_e \theta_1 \pm \sqrt{d_3^2 \sinh_e^2 \theta_1 + D_1},$$

for $\cosh_e \theta_1 < \sinh_e \theta_1$,

$$D_2 = -d_2^2 \Rightarrow d_2 = -d_3 \cosh_e \theta_1 \pm \sqrt{d_3^2 \sinh_e^2 \theta_1 - D_1}. \tag{4.6.5}$$

So, as for the Euclidean counterpart, we can have, depending on the value of the square root argument, two, one or no solutions. The coordinates of the vertex P_3 are given by (4.6.1).

Now we use an analytic method typical of the Cartesian plane. The coordinates of P_3 can be obtained by intersecting the straight line $y = x \tanh_e \theta_1$ with the hyperbola centered in P_2 and having square semi-diameter $P = D_1$, i.e., by solving the system

$$y = x \tanh_e \theta_1; \quad (x - d_3)^2 - y^2 = D_1. \tag{4.6.6}$$

The results are the same as those of (4.6.5), but now it is easy to understand the geometrical meaning of the solutions which can be compared with the equivalent Euclidean problem with a circle instead of an equilateral hyperbola.

Actually if $D_1 > 0$ and $d_1 > d_3$, the point P_1 is included in a hyperbola arm and we have always two solutions. Otherwise, if $\sinh \theta_1 < d_1/d_3$, there are no solutions, if $\sinh \theta_1 = d_1/d_3$, there is just one solution and, if $\sinh \theta_1 > d_1/d_3$, two solutions.

If $D_3 < 0$, the P_2 vertex must be put on the y axis and we have the system

$$x = y \tanh_e \theta_1; \quad (y - d_3)^2 - x^2 = -D_1.$$

Comparing this result with the solutions of system (4.6.6) we have to change $x \leftrightarrow y$.

Given Two Angles and One Opposite Side

Let us be given θ_1; θ_3; D_3 with $D_3 > 0$; we have

$$d_1 = d_3 \frac{\sinh_e \theta_1}{\sinh_e \theta_3}; \quad |d_2| = |d_1 \cosh_e \theta_3 + d_3 \cosh_e \theta_1|. \tag{4.6.7}$$

From (4.6.1) we find P_3 coordinates.

Given Three Sides

From Carnot'a theorem (4.4.12), we have

$$\cosh_e \theta_1 = \frac{D_2 + D_3 - D_1}{2\, d_2\, d_3}.$$ (4.6.8)

For $D_3 > 0$, we take $P_1 \equiv (0,0), P_2 \equiv (\pm\sqrt{D_3}, 0)$ and $\sinh_e \theta_1$ is given by,

for $D_2 > 0 \Rightarrow \sinh_e \theta_1 = \sqrt{\cosh_e^2 \theta_1 - 1}$, for $D_2 < 0 \Rightarrow \sinh_e \theta_1 = \sqrt{\cosh_e^2 \theta_1 + 1}.$

From (4.6.1) we find P_3 coordinates.

Chapter 5

Uniform and Accelerated Motions in the Minkowski Space-Time (Twin Paradox)

In this chapter we show how the formalization of trigonometry in the pseudo-Euclidean plane allows us to treat exhaustively all kinds of motions and to give a complete formalization to what is called today the "twin paradox". After a century this problem continues to be the subject of many papers, not only relative to experimental tests [1] but also regarding physical and epistemological considerations [51]. We begin by recalling how this "name" originates.

The final part of Section 4 of Einstein's famous 1905 special relativity paper [31] contains sentences concerning moving clocks on which volumes have been written: "... If we assume that the result proved for a polygonal line is also valid for a continuously curved line, we obtain the theorem: If one of two synchronous clocks at A is moved in a closed curve with constant velocity until it returns to A, the journey lasting t seconds, then the clock that moved runs $\frac{1}{2} t \left(\frac{v}{c}\right)^2$ seconds[1] slower than the one that remained at rest".

About six years later, on 10 April 1911, at the Philosophy Congress at Bologna, Paul Langevin replaced the clocks A and B with human observers and the "twin paradox" officially was born. Langevin, using the example of a space traveler who travels a distance L (measured by someone at rest on the Earth) in a straight line to a star in one year and than abruptly turns around and returns on the same line, wrote: "... Revenu à la Terre ayant vielli deux ans, il sortira de son arche et trouvera notre globe vielli deux cents ans si sa vitesse est restée dans l'interval inférieure d'un vingt-millième seulement à la vitesse de la lumière."[2] We must remark that Langevin, besides not rejecting ether's existence, stresses the point which will be the subject of subsequent discussions, that is the asymmetry between the two reference frames. The space traveler undergoes an acceleration halfway through his journey, while the twin at rest in the Earth's reference frame always remains in an inertial frame. For Langevin, every acceleration has an ab-

---

[1]Obviously neglecting magnitudes of fourth and higher order.

[2]Langevin's address to the Congress of Bologna was published in *Scientia* **10**, 31-34 (1911). As reported by Miller [31], the popularisation of relativity theory for philosophers had an immediate impact which we can gauge from the comment of one of the philosophers present. Henry Bergson (1922)wrote: "... *it was Langevin's address to the Congress of Bologna on 10 April 1911 that first drew our attention to Einstein's ideas. We are aware of what all those interested in the theory of relativity owe to the works and teachings of Langevin*".

solute meaning. Even though the effect foreseen by Einstein's theory has received several experimental confirmations, the contribution of accelerated stretches of the path still stands as a subject of discussion and controversies.

In order to avoid misunderstandings, we stress that the discussions we allude to are rigorously confined to the ambit of special relativity, that is to the space-time of special relativity. It is in this ambit that Rindler says: "... If an ideal clock moves *non-uniformly* through an inertial frame, we shall *assume* that acceleration as such has no effect on the rate of the clock, i.e., that its instantaneous rate depends only on its instantaneous speed v according to the above rule. Unfortunately, there is no way of *proving* this. Various effects of acceleration on a clock would be consistent with S. R. Our assumption is one of simplicity – or it can be regarded as a definition of an "ideal" clock. We shall call it the clock hypothesis." [62].

We think that a conclusion on the role of accelerated motions and, most of all, an evaluation of the amount of slowing down of an accelerated clock, can only be reached through a rigorous and exhaustive exploitation of the mathematics of special relativity. If the theory has no logical inconsistencies, the theory itself must thereby provide a completely accurate account of the asymmetrical aging process.

Even if Minkowski gave a geometrical interpretation of the special relativity space-time shortly after (1907–1908) Einstein's fundamental paper, a mathematical tool exploitable in the context of the Minkowski space-time began to be utilized only a few decades ago [81]. This mathematical tool is based on the use of hyperbolic numbers, as we have seen in Chapter 4. The self-consistency of this formalization allows us to solve problems for every motion in the Minkowski space-time through elementary mathematics as if we were working on the Euclidean plane. We note that in some examples, even if we could obtain the result by using directly the hyperbolic counterpart of Euclidean theorems, we have preferred not to use this approach. Actually, the results are obtained by means of elementary mathematics and the obtained correspondences represent alternative demonstrations with respect to the ones given in Chapter 4.

5.1   Inertial Motions

In a representative (t, x) plane let us start with the following example:

- the first twin is steady at the point $x = 0$, his path is represented by the t axis,

- the second twin, on a rocket, starts with speed v from $O \equiv (0, 0)$ and after a time τ_1, at the point T, he reverses its direction and comes back arriving at the point $R \equiv (\tau_2, 0)$.

From a physical point of view the speed cannot change in a null time, but this time can be considered negligible with respect to τ_1, τ_2 ([55], p. 41). An experimental result with only uniform motion is reported in [1]. In this experiment the lifetime of the muon in the CERN muon storage ring was measured. In Fig. 5.1 we represent this problem by means of the triangle OTR.

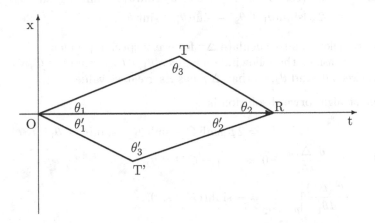

Figure 5.1: The twin paradox for uniform motions.

Solution. From a geometrical point of view we can compare the elapsed travel times for the twins by comparing the "lengths" (proper times) of the sum $\overline{OT} + \overline{TR}$ with the side $\overline{OR}$.

A qualitative interpretation is reported in many books and is easily explained by means of the reverse triangle inequality in space-time geometry with respect to Euclidean geometry ([12] p. 130). Also a graphical visualization can be easily performed considering that a segment must be superimposed on another by means of an equilateral hyperbola, instead of the Euclidean circle ([81] p. 190). Now we see that the Euclidean formalization of space-time trigonometry allows us to obtain a simple quantitative formulation of the problem.

Let us call $\theta_1 \equiv \tanh v$ the hyperbolic angle $\widehat{ROT}$, θ_2 the hyperbolic angle $\widehat{ORT}$ and θ_3 the hyperbolic angle $\widehat{OTR}$. Given their physical meaning, the angles θ_1 and θ_2 are such that the straight-lines $\overline{OT}$ and $\overline{TR}$ are time-like [55] (in a Euclidean representation the angle of straight-lines with the t axis must be less than $\pi/4$). Applying the law of cosines (4.4.13) to $\overline{OR}$; we have

$$\overline{OR} = \overline{OT} \cosh \theta_1 + \overline{TR} \cosh \theta_2. \qquad (5.1.1)$$

It follows that the difference between the twins' proper times $\Delta \tau$ is

$$\Delta \tau \equiv \overline{OR} - \overline{OT} - \overline{TR} = \overline{OT} \left(\cosh \theta_1 - 1 \right) + \overline{TR} \left(\cosh \theta_2 - 1 \right). \qquad (5.1.2)$$

If we call p the semi-diameter of the equilateral hyperbola circumscribed to triangle $\widehat{OTR}$, from (4.4.10) we have $\overline{OT} = 2\,p\,\sinh\theta_2$; $\overline{TR} = 2\,p\,\sinh\theta_1$, and

$$
\begin{aligned}
\Delta\tau &= 2\,p\,(\cosh\theta_1\,\sinh\theta_2 + \cosh\theta_2\,\sinh\theta_1 - \sinh\theta_1 - \sinh\theta_2) \\
&\equiv 2\,p\,[\sinh(\theta_1 + \theta_2) - \sinh\theta_1 - \sinh\theta_2].
\end{aligned}
\tag{5.1.3}
$$

This equation allows us to calculate $\Delta\tau$ for every specific problem.

Now we consider the following one: given $\theta_1 + \theta_2 = const \equiv C$, what is the relation between θ_1 and θ_2, so that $\Delta\tau$ has its greatest value?

Proof. The straightforward solution is

$$
\Delta\tau = 2\,p\,[\sinh C - \sinh\theta_1 - \sinh(C - \theta_1)],
$$

$$
\frac{d\,(\Delta\tau)}{d\,\theta_1} = 0 \Rightarrow \quad \theta_1 = C/2 \equiv \theta_2,
\tag{5.1.4}
$$

$$
\left.\frac{d^2\,(\Delta\tau)}{d\,\theta_1^2}\right|_{\theta_1 = C/2} = -\sinh(C/2) < 0.
$$

$\square$

We have obtained the "intuitive Euclidean" solution that the greatest difference between elapsed times, i.e., the shortest proper-time for the moving twin, is obtained for $\theta_1 = \theta_2$. For this value (5.1.3) corresponds to the well-known solution [31]

$$
\tau_{\overline{OR}} = \tau_{(\overline{OT} + \overline{TR})}\cosh\theta_1 \equiv \frac{\tau_{(\overline{OT} + \overline{TR})}}{\sqrt{1 - v^2}}.
\tag{5.1.5}
$$

Now we give a geometrical interpretation of this problem. From (4.4.19) we know that if $\theta_1 + \theta_2 = C$, θ_3 is constant too, then the posed problem is equivalent to: what can be the position of vertex T if starting and final points and angle θ_3 are given?

The problem is equivalent to having, in a triangle, a side and the opposite angle. In an equivalent problem in Euclidean geometry we know immediately that the vertex T does move on a circle's arc. Then, from the established correspondence of circles in Euclidean geometry with equilateral hyperbolas in pseudo-Euclidean geometry, we have that in the present space-time problem the vertex T moves on an arc of an equilateral hyperbola.

Now let us generalize the twin paradox to the case in which both twins change their state of motion: their motions start in O, both twins move on (different) straight-lines and cross again in R. The graphical representation is given by a quadrilateral figure and we call the other two vertices T and T'. Since a hyperbolic rotation does not change hyperbolic angles between the sides and hyperbolic side lengths (Chapter 4), we can rotate the figure so that the vertex R lies on the t axis (see Fig. 5.1). The problem can be considered as a duplicate of the previous one in the sense that we can compare proper times of both twins with side $\overline{OR}$. If we indicate by $(')$ the quantities referred to the triangle under the t axis, we apply

(5.1.3) twice and obtain $\Delta\tau - \Delta\tau'$ for every specific example. In particular, if we have $\theta_1 + \theta_2 = \theta_1' + \theta_2' = C$, from the result of (5.1.4) it follows that the youngest twin is the one for which θ_1 and θ_2 are closer to $C/2$.

5.2 Inertial and Uniformly Accelerated Motions

Now we consider some "more realistic" examples in which uniformly accelerated motions are taken into account. The geometrical representation of a motion with constant acceleration is given by an arm of an equilateral hyperbola with the semi-diameter p linked to the acceleration a by the relation $p^{-1} = a$ ([52], p. 166) and ([55], p. 58). In Chapter 10 the hyperbolic motion is obtained as a straightforward consequence of the invariance of the wave equation with respect to Lorentz transformations.

Obviously, the geometrical representation of a motion with non-uniform acceleration is given by a curve which is the envelope of the equilateral hyperbolas corresponding to the instantaneous accelerations. Or, vice versa, we can construct in every point of a curve an "osculating hyperbola" which has the same properties of the osculating circle in Euclidean geometry. Actually the semi-diameter of these hyperbolas is linked to the second derivative with respect to the line element ([12] Section 3.3) as the radius of osculating circles in Euclidean geometry. This general problem is formalized in Section 5.3; now we return to uniformly accelerated motions represented by equilateral hyperbolas. For these hyperbolas we indicate by $C \equiv (t_C, x_C)$ its center and with θ a parameter that, from a geometrical point of view, represents a hyperbolic angle measured with respect to an axis passing through C and parallel to the x axis (Section 4.2.2). Then its equation, in parametric form, is

$$\mathcal{I} \equiv \begin{cases} t = t_C \pm p \sinh\theta \\ x = x_C \pm p \cosh\theta \end{cases} \quad \text{for } -\infty < \theta < +\infty, \tag{5.2.1}$$

where the $+$ sign refers to the upper arm of the equilateral hyperbola and the $-$ sign to the lower one. We also have

$$dx = \pm p \sinh\theta \, d\theta, \qquad dt = \pm p \cosh\theta \, d\theta \tag{5.2.2}$$

and the proper time on the hyperbola

$$\tau_{\mathcal{I}} = \int_{\theta_1}^{\theta_2} \sqrt{dt^2 - dx^2} \equiv \int_{\theta_1}^{\theta_2} p \, d\theta \equiv p\,(\theta_2 - \theta_1). \tag{5.2.3}$$

This relation states the link between proper time, acceleration, and hyperbolic angle and also shows that hyperbolic angles are given by the ratio between the "lengths" of hyperbola arcs and semi-diameter, as circular angles in Euclidean trigonometry are given by the ratio between circle arcs and radius. Moreover, as in Euclidean geometry, the magnitude of hyperbolic angles is equal to twice the

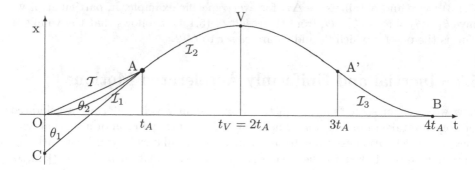

Figure 5.2: The uniform and accelerated motions of the first example.

area of the hyperbolic sector (Chapter 6) and, taking into account that the "area" is the same quantity in Euclidean and pseudo-Euclidean geometries, it can be calculated in a simple Euclidean way ([81] p. 183).

At the point P, determined by $\theta = \theta_1$, the velocity is given by $v \equiv dx/dt = \tanh \theta_1$ and the straight-line tangent to the hyperbola for $\theta = \theta_1$ is given by (Chapter 4):

$$x - (x_C \pm p \cosh \theta_1) = \tanh \theta_1 \left[t - (t_C \pm p \sinh \theta_1) \right]$$
$$\Rightarrow \quad x \cosh \theta_1 - t \sinh \theta_1 = x_C \cosh \theta_1 - t_C \sinh \theta_1 \mp p. \qquad (5.2.4)$$

From this equation we see that θ_1 also represents the hyperbolic angle of the tangent to hyperbola with t axis. This last property means that semi-diameter $\overline{CP}$ is pseudo-orthogonal to the tangent in P (see also Figs. 4.3 and 5.3). This property corresponds, in the Euclidean counterpart, to the well-known property of a circle where the radius is orthogonal to the tangent-lines.

First Example

We start with the following example in which the first twin, after some accelerated motions, returns to the starting point with vanishing velocity. The problem is represented in Fig. 5.2.

- The first twin (I) starts with a constant accelerated motion with acceleration p^{-1} (indicated by $\mathcal{I}_1$) from O to A and then a constant decelerated (p^{-1}) motion up to V and then accelerated with reversed velocity up to A' ($\mathcal{I}_2$), then another decelerated (p^{-1}) motion ($\mathcal{I}_3$) up to $B \equiv (4t_A,\, 0)$.

- The second twin (II) moves with a uniform motion ($\mathcal{T}_1$) that, without loss of generality, can be represented as stationary at the point $x = 0$.

Solution. Hyperbola $\mathcal{I}_1$ has its center at $C \equiv (0, -p)$. Then we have

$$\mathcal{I}_1 \equiv \begin{cases} t = p \sinh \theta \\ x = p(\cosh \theta - 1) \end{cases} \quad \text{for } 0 < \theta < \theta_1. \qquad (5.2.5)$$

We also have $A \equiv (p \sinh \theta_1, \ p \cosh \theta_1 - p)$.

The symmetry of the problem indicates that for both twins the total elapsed times are four times the elapsed times of the first motion. The proper time of twin I is obtained from (5.2.3),

$$\tau_I \equiv 4\,\tau_{\mathcal{I}_1} = 4 \int_0^{\theta_1} \sqrt{dt^2 - dx^2} \equiv 4 \int_0^{\theta_1} p\,d\theta \equiv 4\,p\,\theta_1; \qquad (5.2.6)$$

the proper time of twin II is four times the abscissa of point A,

$$\tau_{II} \equiv 4\,t_A = 4\,p \sinh \theta_1. \qquad (5.2.7)$$

The difference between the elapsed times is $\Delta\tau = 4\,p\,(\sinh \theta_1 - \theta_1)$, and their ratio is

$$\frac{\tau_I}{\tau_{II}} = \frac{\theta_1}{\sinh \theta_1}. \qquad (5.2.8)$$
$$\square$$

For $\theta_1 \equiv \tanh^{-1} v \ll 1$, we have $\Delta\tau \simeq 0$, and for $\theta \gg 1 \Rightarrow \sinh \theta \propto \exp[\theta]$: *The proper time for the accelerated motions is linear in θ and the stationary (inertial) one is exponential in θ.*

- Now we compare the motion on hyperbola $\mathcal{I}_1$ with the motion T on the side $\overline{OA}$.

Solution. Let us call θ_2 the hyperbolic angle between straight-line OA and the t axis; the equation of the straight-line OA is

$$T \equiv \{x = t \tanh \theta_2\}, \qquad (5.2.9)$$

and we calculate θ_2, imposing that this straight-line crosses the hyperbola (5.2.5) for $\theta = \theta_1$. By substituting (5.2.9) in (5.2.5), we have

$$\begin{cases} t = p \sinh \theta_1 \\ t \tanh \theta_2 = p(\cosh \theta_1 - 1) \end{cases} \Rightarrow \frac{\sinh \theta_2}{\cosh \theta_2} = \frac{\cosh \theta_1 - 1}{\sinh \theta_1} \Rightarrow \theta_1 = 2\,\theta_2, \quad (5.2.10)$$

i.e., the central angle (θ_1) is twice the hyperbola angle (θ_2) on the same chord (Theorem 4.17 p. 46). Then we have

$$\overline{OA} = \frac{\overline{Ot_A}}{\cosh \theta_2} \equiv \frac{p \sinh \theta_1}{\cosh \theta_2} \equiv 2\,p \sinh \theta_2 \qquad (5.2.11)$$

and, taking into account the proper time on the hyperbola given by (5.2.3), we obtain

$$\frac{\tau_I}{\tau_T} = \frac{\theta_2}{\sinh \theta_2}. \qquad (5.2.12)$$
$$\square$$

Relation 5.2.12 has a simple "Euclidean" interpretation. Actually it can be interpreted by means of the correspondence (Section 4.1) with Euclidean geometry, where it represents the ratio between the length of a circle arc and its chord. We have to note how, in hyperbolic geometry, this ratio is less than 1 whereas in Euclidean geometry it is greater than 1, as the reverse triangle inequality requires. We shall look into this property more comprehensively in Section 5.2. It can be verified that the ratio between (5.2.8) and (5.2.12), gives the ratio (5.1.5).

Second example

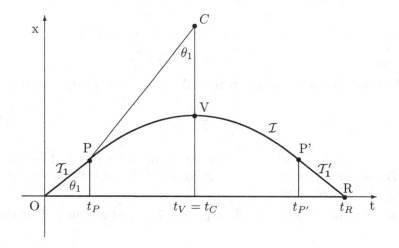

Figure 5.3: The uniform and accelerated motions of the second example.

In this example we connect the two sides of the triangle of Fig. 5.1 by means of an equilateral hyperbola, i.e., we consider the decelerated and accelerated motions too (Fig. 5.3).

- The twin I moves from $O \equiv (0, 0)$ to $P \equiv (p \sinh \theta_1, \ p \cosh \theta_1 - p)$ with a uniform motion, indicated as $\mathcal{T}_1$, then he goes on with a constant decelerated motion up to V, and then he accelerates with reversed velocity up to P', where he has the same velocity as the initial one, and moves again with uniform velocity up to $R \equiv (t_R, 0)$ ($\mathcal{T}_1'$).

- The twin II moves with a uniform motion ($\mathcal{T}_2$) which, without loss of generality, can be represented as stationary in the point $x = 0$.

Solution. A mathematical formalization can be the following: let us consider the decelerated and accelerated motions that can be represented by the equilateral hyperbola (5.2.1) for $-\theta_1 < \theta < \theta_1$ and the tangent to the hyperbola for $\theta = \theta_1$ as given by (5.2.4). This straight-line represents the motion $\mathcal{T}_1$ if it passes through O.

This happens if the center $C \equiv (x_C, t_C)$ of the hyperbola lies on the straight line $x_C \cosh \theta_1 - t_C \sinh \theta_1 - p = 0$, where t_C is given by (5.2.1): $t_C = t_P + p \sinh \theta_1$. If we write down the straight line (5.2.4) in parametric form

$$\mathcal{T}_1 \quad \equiv \begin{cases} t = \tau \cosh \theta_1 \\ x = \tau \sinh \theta_1 \end{cases}, \tag{5.2.13}$$

where τ is the proper time on the straight line, we have at the end of the uniform motion $P \equiv (t_P, x_P)$, with $t_P = \tau \cosh \theta_1$. Then the proper time for twin I is $\tau_{\overline{OP}} = \tau$, and from P to the vertex of the hyperbola $\tau_{\mathcal{I}} = p\theta_1$. The proper times of the other lines are a duplicate of these times.

For twin II we have: $\tau_{II} \equiv 2t_C = 2 (\tau \cosh \theta_1 + p \sinh \theta_1)$. Then we have

$$\Delta \tau = 2 \left[\tau (\cosh \theta_1 - 1) + p (\sinh \theta_1 - \theta_1) \right]. \tag{5.2.14}$$

$\square$

The proper time on this rounded off triangle is greater than the one on the triangle, as we shall better see in the next example. The physical interpretation is that the velocity on the hyperbola arc is less than the one on straight-lines OT, TR of Fig. 5.1.

Third Example

In the following example we consider the motions

1. stationary motion in $x = 0$,

2. on the upper triangle of Fig. 5.1 with sides $\overline{OT} = \overline{TR}$,

3. on hyperbola $\mathcal{I}$ tangent in O and in R to sides $\overline{OT}$ and $\overline{TR}$, respectively,

4. on hyperbola $\mathcal{I}_c$ circumscribed to triangle $\overset{\triangle}{OTR}$.

The problem is represented in Fig. 5.4. In this example we can also note a formalization of the triangle inequality, reversed with respect to the Euclidean plane ([12] p. 130). Actually we shall see that the shorter the lines (trajectories) look in a Euclidean representation (Fig. 5.4), the longer they are in space-time geometry (5.2.16).

Solution. Side $\overline{OT}$ lies on the straight-line represented by the equation

$$x \cosh \theta_1 - t \sinh \theta_1 = 0. \tag{5.2.15}$$

Hyperbola $\mathcal{I}$ is obtained requiring that it be tangent to straight-line (5.2.15) in O. We obtain from (5.2.1) and (5.2.4) $t_C = p \sinh \theta_1$, $x_C = p \cosh \theta_1$ and, from the definitions of hyperbolic trigonometry, $\overline{OT} = t_C / \cosh \theta_1 \equiv p \tanh \theta_1$.

For hyperbola $\mathcal{I}_c$, with vertex in T, its semi-diameter is given by (4.4.10): $p_c = \overline{OT} / (2 \sinh \theta_1) \equiv p / (2 \cosh \theta_1)$. If we call C_c its center and $2\theta_c$ the angle

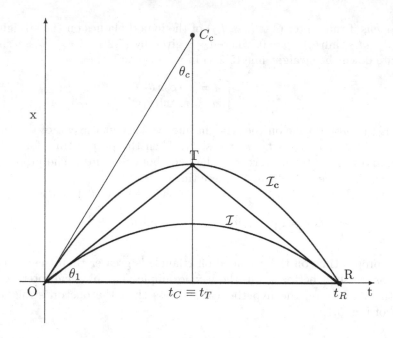

Figure 5.4: The uniform and accelerated motions of the third example.

$\widehat{OC_cR}$, we note that θ_c is a central angle of chord $\overline{OT}$ while θ_1 is a hyperbola angle on the chord $\overline{TR} = \overline{OT}$. Then, as has been shown in the first example, we have $\theta_c = 2\theta_1$.

Summarizing the lengths (proper times) for the motions are:

1. $\overline{OR} \equiv 2t_T = 2p \sinh\theta_1$;

2. from (5.2.15) it follows that

$$T \equiv (p \sinh\theta_1, \, p \sinh\theta_1 \tanh\theta_1), \text{ so that } \overline{OT} = \overline{TR} = p \tanh\theta_1;$$

3. from (5.2.3) the length of arc of hyperbola $\mathcal{I}$ between O and R is given by

$$\tau_{\mathcal{I}} = 2p\,\theta_1;$$

4. from (5.2.3) the length of arc of $\mathcal{I}_c$ from O and R is given by

$$\tau_{\mathcal{I}_c} = 2p_c\,\theta_c \equiv 2p\,\theta_1/\cosh\theta_1.$$

Then the following relations hold:

$$
\begin{aligned}
\overline{OR} &\equiv 2p \sinh\theta_1 > \text{arc}(\mathcal{I}) \equiv 2p\,\theta_1 > \overline{OT} + \overline{TR} \\
&\equiv 2p \tanh\theta_1 > \text{arc}(\mathcal{I}_c) \equiv 2p\,\theta_1/\cosh\theta_1.
\end{aligned}
\tag{5.2.16}
$$

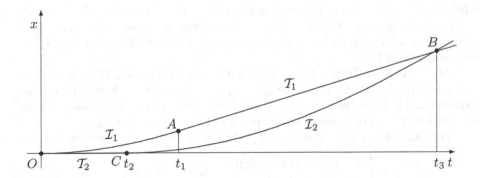

Figure 5.5: The uniform and accelerated motions of the fourth example.

We also observe that $\overline{OR}$ is a chord of $\mathcal{I}$, $\overline{OT}$ is a chord of $\mathcal{I}_c$ and their ratios are given by (5.2.12):

$$\frac{\tau_{\overline{OR}}}{\tau_{\mathcal{I}}} = \frac{\tau_{\overline{OT}}}{\tau_{\mathcal{I}_c}} \equiv \frac{\sinh\theta_1}{\theta_1}. \qquad (5.2.17)$$

□

As a corollary of this example we consider the following one: given the side $\overline{OR} = \tau$ (proper time of the stationary twin) what is the proper time of twin I moving on an equilateral hyperbola, as a function of the acceleration p^{-1}?

The answer is: as acceleration p^{-1} does increase, proper time $\tau_{\mathcal{I}}$ can be as little as we want ([52] p. 167).

Proof. From the hyperbolic motion of (5.2.1) we have $t = \tau/2 - p\sinh\theta$ and for $t = 0$ we have $2p\sinh\theta_1 = \tau$, and for relativistic motions ($\theta_1 \gg 1$) we obtain

$$\tau \simeq p\exp[\theta_1] \Rightarrow \theta_1 \simeq \ln\frac{\tau}{p}. \qquad (5.2.18)$$

Then from relation (5.2.3), $\tau_{\mathcal{I}} \equiv 2p\,\theta_1 = 2p\ln[\tau/p] \xrightarrow{p\to 0} 0.$ □

Fourth Example

We conclude this section with a more general example in which both twins have a uniform and an accelerated motion.

- The first twin (I) starts with a constant accelerated motion and then goes on with a uniform motion.

- The second twin (II) starts with a uniform motion and then goes on with a constant accelerated motion.

The problem is represented in Fig. 5.5.

Solution. We can represent this problem in the (t, x) plane in the following way:

(I) starts from point $O \equiv (0, 0)$ with an acceleration given by p^{-1} ($\mathcal{I}_1$) up to point $A \equiv (p \sinh \theta_1; \, p \cosh \theta_1 - p)$, then goes on with a uniform motion ($\mathcal{T}_1$) up to time t_3 (point B).

(II) starts from the point $O \equiv (0, 0)$ with a uniform motion ($\mathcal{T}_2$), (stationary in $x = 0$) up to point C, in a time $t_2 = \alpha p \sinh \theta_1$ which we have written proportional to t_A. Then he goes on with an accelerated motion ($\mathcal{I}_2$), with the same acceleration p^{-1} up to crossing the trajectory of twin (I) at time t_3.

The analytical representation of $\mathcal{I}_1$ is given by (5.2.5). $\mathcal{I}_2$ is represented by

$$\mathcal{I}_2 \quad \equiv \quad \begin{cases} t = p\,(\alpha \sinh \theta_1 + \sinh \theta) \\ x = p\,(\cosh \theta - 1) \end{cases} \quad \text{for } 0 < \theta < \theta_2, \qquad (5.2.19)$$

where θ_2 represents the value of the hyperbolic angle in the crossing point B between $\mathcal{I}_2$ and $\mathcal{T}_1$.

$\mathcal{T}_1$ is given by the straight-line tangent to $\mathcal{I}_1$ in θ_1:

$$\mathcal{T}_1 \equiv \{x \cosh \theta_1 - t \sinh \theta_1 = p\,(1 - \cosh \theta_1)\}. \qquad (5.2.20)$$

From (5.2.19) and (5.2.20) we calculate the crossing point between $\mathcal{I}_2$ and $\mathcal{T}_1$. We have

$$\cosh(\theta_2 - \theta_1) = \alpha \sinh^2 \theta_1 + 1. \qquad (5.2.21)$$

This equation has an explicit solution for $\alpha = 2$. Actually we have $\cosh(\theta_2 - \theta_1) = 2 \sinh^2 \theta_1 + 1 \equiv \cosh 2\,\theta_1 \Rightarrow \theta_2 = 3\,\theta_1$.

Let us calculate the proper times.

- The proper times relative to the accelerated motions are obtained from (5.2.3) $\tau_{\mathcal{I}_1} = p\,\theta_1$, $\tau_{\mathcal{I}_2} = p\,\theta_2$.

- The proper time relative to $\mathcal{T}_2$ is given by $t_2 = \alpha p \sinh \theta_1$.

- On straight-line $\mathcal{T}_1$, between points A and $B \equiv p\,(\alpha \sinh \theta_1 + \sinh \theta_2, \, \cosh \theta_2 - 1)$, the proper time is obtained by means of hyperbolic trigonometry,

$$\tau_{\mathcal{T}_1} \equiv \overline{AB} = (x_B - x_A)/\sinh \theta_1 \equiv p\,(\cosh \theta_2 - \cosh \theta_1)/\sinh \theta_1. \qquad (5.2.22)$$

Then the complete proper times of the twins are

$$\tau_I = p\,[\theta_1 + (\cosh \theta_2 - \cosh \theta_1)/\sinh \theta_1], \qquad \tau_{II} = p\,(\alpha \sinh \theta_1 + \theta_2). \qquad (5.2.23)$$

$\square$

Let us consider relativistic velocities ($v = \tanh \theta_i \simeq 1 \Rightarrow \theta_1, \theta_2 \gg 1$); in this case we can approximate the hyperbolic functions in (5.2.21) and (5.2.23) with the positive exponential term and, for $\alpha \neq 0$, we obtain from (5.2.21): $\exp[\theta_2 - \theta_1] \simeq \alpha \exp[2\,\theta_1]/2$, and from (5.2.23)

$$\tau_I \simeq p\,(\theta_1 + \exp[\theta_2 - \theta_1]) \simeq p\,(\theta_1 + \alpha \exp[2\,\theta_1]/2), \qquad \tau_{II} \simeq p\,(\alpha \exp[\theta_1]/2 + \theta_2). \qquad (5.2.24)$$

The greatest contributions to the proper times are given by the exponential terms that derive from uniform motions. If we neglect the linear terms with respect to the exponential ones, we obtain a ratio of the proper times *independent of the* $\alpha \neq 0$ *value,*

$$\tau_I \simeq \tau_{II} \exp[\theta_1]. \tag{5.2.25}$$

The twin that moves for a shorter time with uniform motion has the shortest proper time. Since the total time is the same, a shorter time with uniform motion means a longer time with accelerated motion, so this result is the same as (5.2.12).

With regard to the result of this example we could ask: how is it possible that a uniform motion close to a light-line is the longer one? We can answer this question by a glance at (5.2.22). Actually in this equation the denominator $\sinh \theta_1 \gg 1$ takes into account that the motion is close to a light-line, but in the numerator $\cosh \theta_2 \gg \cosh \theta_1$ indicates that the crossing point B is so far away that its contribution is the determining term of the result we have obtained.

5.3 Non-uniformly Accelerated Motions

In the representative Minkowski plane with coordinates $(t,\, x)$ we can represent points P by means of the hyperbolic-polar coordinates (Chapter 4), i.e., by means of a radius vector $\boldsymbol{r} \equiv (t,\, x)$ and a hyperbolic angle θ,

$$|\boldsymbol{r}| = \sqrt{t^2 - x^2}\,, \qquad \tanh \theta = \frac{x}{t}, \tag{5.3.1}$$

and consider a time-like curve, in parametric form, from the origin of the coordinates to point A, given by

$$\boldsymbol{r} = \boldsymbol{r}(\tau)\,,\ t = t(\tau),\ \ x = x(\tau) \tag{5.3.2}$$

and

$$\boldsymbol{v} = \boldsymbol{v}(\tau)\,;\ v_t \equiv \dot{t} = \frac{dt}{d\tau}\,,\ \ v_x \equiv \dot{x} = \frac{dx}{d\tau}. \tag{5.3.3}$$

If τ is the proper time on the curve, $\boldsymbol{v} \equiv (\dot{t},\, \dot{x})$ is the relativistic velocity unit vector. For every parametrization, (5.3.3) allows us to obtain the proper time on the curve if we know its equation. Actually the length of the curve (proper time on the curve) is given by

$$\tau_A = \int_0^A \sqrt{(\dot{t})^2 - (\dot{x})^2}\, d\tau. \tag{5.3.4}$$

Now we consider another problem: to calculate the proper time in an inertial frame if we know the acceleration on the curve as a function of the proper time on the curve. Thanks to the established correspondence between Euclidean and space-time geometry the problem can be solved in a straightforward way in two steps. As a first step we state Frenet's formulas for the space-time plane.

5.3.1 Frenet's Formulas in the Minkowski Space-Time

Let us consider a parametrization by means of proper time: we have

Theorem 5.1. *In the space-time plane the following* **Frenet-like formulas** *hold:*

$$\frac{d^2 r}{d\tau^2} \equiv \frac{d v}{d\tau} = K(\tau)\, n, \qquad (5.3.5)$$

$$\frac{d n}{d\tau} = K(\tau)\, v, \qquad (5.3.6)$$

where n *represents the unit vector pseudo-orthogonal to the curve,* $K(\tau)$ *the modulus of acceleration and* $K(\tau)^{-1}$ *the semi-diameter of osculating hyperbolas.*

We have changed the symbol p, introduced in Chapter 4, with $K(\tau)$, that recalls the non-constant curvature of lines in the Euclidean plane.

Proof. If the parameter τ is the proper time, $\ddot{t}$ and $\ddot{x}$ are the components of a vector a that represents the relativistic acceleration ([55], p. 56). It can be shown that v and a are orthogonal ([12], p. 132), [55]. Then a is a space-like vector and we can write the proportionality equation (5.3.5). Equation (5.3.5) is equivalent to the first Frenet formula in the Euclidean plane ([28], p. 20). Now we show that also the second Frenet formula holds.

By indicating with a dot ($\cdot$) the scalar product in space-time [81] and taking into account that n, pseudo-orthogonal to v, is space-like (its modulus is negative) we have

$$n \cdot n = -1 \qquad \Rightarrow \qquad n \cdot \frac{d n}{d\tau} = 0; \qquad (5.3.7)$$

as for Equation (5.3.5) we can write

$$\frac{d n}{d\tau} = \alpha\, v \qquad (5.3.8)$$

and taking into account that $v \cdot n = 0$, from (5.3.5) and (5.3.8) we obtain α,

$$0 = \frac{d}{d\tau}(v \cdot n) \equiv \left(\frac{dv}{d\tau} \cdot n\right) + \left(\frac{dn}{d\tau} \cdot v\right) \equiv -K + \alpha \Rightarrow \alpha = K. \qquad (5.3.9)$$

So we have the second Frenet-like formula (5.3.6). □

5.3.2 Proper Time in Non-Uniformly Accelerated Motions

Now we apply Frenet space-time formulas for comparing the proper times in the following motions:

1. motion I, a non-uniformly accelerated motion of which we know the acceleration as a function of the proper time on the curve ($K(\tau)$),

2. motion II, stationary in $x = 0$,

3. motion III, uniform motion from O to A.

Solution. Calling $\theta(\tau)$ the hyperbolic angle between the tangent-line to the curve and t axis, we can write[3]

$$v_t \equiv \frac{dt}{d\tau} = |\boldsymbol{v}| \cosh \theta(\tau) \equiv \cosh \theta(\tau); \quad v_x \equiv \frac{dx}{d\tau} = |\boldsymbol{v}| \sinh \theta(\tau) \equiv \sinh \theta(\tau).$$
$$(5.3.10)$$

Then $\boldsymbol{v}$ can be written, in the language of "hyperbolic vectors", as

$$\boldsymbol{v} \equiv v_t + h\,v_x = \exp[h\,\theta(\tau)] \qquad (5.3.11)$$

and the pseudo-orthogonal unit vector $\boldsymbol{n}$ (Chapter 4) as

$$\boldsymbol{n} = h \, \exp[h\,\theta(\tau)]. \qquad (5.3.12)$$

Substituting (5.3.11) and (5.3.12) into (5.3.5), we obtain

$$\frac{d \exp[h\,\theta(\tau)]}{d\tau} \equiv \{h \exp[\theta(\tau)]\} \frac{d\theta(\tau)}{d\tau} = K(\tau)\{h \exp[\theta(\tau)]\}$$

$$\Rightarrow \quad \theta(\tau) = \theta_0 + \int K(\tau)\,d\tau. \qquad (5.3.13)$$

The same result could be obtained by means of (5.3.6).

By means of the value of $\theta(\tau)$ obtained by (5.3.13), we obtain the components of $\boldsymbol{v}$ and $\boldsymbol{n}$, from (5.3.11) and (5.3.12), respectively.

If we also write $\boldsymbol{r}$ in the language of hyperbolic vectors, we have

$$\boldsymbol{r} \equiv t + h\,x \equiv \int [v_t + h\,v_x]\,d\tau = \int \exp[h\,\theta(\tau)]\,d\tau. \qquad (5.3.14)$$

By means of the value of $\theta(\tau)$ obtained by (5.3.13), from (5.3.14) we obtain the expressions equivalent to the ones which in the Euclidean plane give the parametric equation of a curve from its curvature ([28], p. 24)

$$t(\tau_A) = t_0 + \int_0^{\tau_A} \cosh \theta(\tau')\,d\tau' \equiv t_0 + \int_0^{\tau} \cosh \left[\int_0^{\tau'} K(\tau'')\,d\tau'' \right] d\tau', \quad (5.3.15)$$

$$x(\tau_A) = x_0 + \int_0^{\tau_A} \sinh \theta(\tau')\,d\tau' \equiv x_0 + \int_0^{\tau} \sinh \left[\int_0^{\tau'} K(\tau'')\,d\tau'' \right] d\tau'. \quad (5.3.16)$$

These expressions have been recently obtained in a different way ([51], (11 a, b)).

Now we consider the straight-line OA which forms with the t axis the hyperbolic angle

$$\theta = \tanh^{-1} \frac{x(\tau_A)}{t(\tau_A)} \qquad (5.3.17)$$

and we have

---

[3]The following procedure is formally equal to the ones which, in the Euclidean plane, are developed by means of complex numbers.

1. (I) proper time, τ_A,

2. (II) proper time, $t(\tau_A)$ from (5.3.15),

3. (III) proper time $\tau(\overline{OA}) = t(\tau_A)/\cosh\theta$.

Motions (III)+(I) can be considered a round trip.

The motions (I)+(II) are not a closed curve, however we can close it by a reversed motion of twin (I) up to crossing again the stationary twin. Since the curvature (acceleration) is given by a second derivative, it is independent of the parameter direction ([28], p. 17), then the result is to double the proper times and obtain the ratio

$$\frac{\tau_{(I)}}{\tau_{(II)}} = \frac{\tau_A}{t(\tau_A)}, \qquad (5.3.18)$$

where $t(\tau_A)$ is given by (5.3.15).

Equation (5.3.18) allows us to construct an inertial clock (e.g., the time on the Earth) from data taken on a rocket, as is shown in [51].

Conclusions

As we know, the twin paradox spreads far and wide, having been considered the most striking exemplification of the space-time "strangeness" of Einstein's theory of special relativity. What we have striven to show is that hyperbolic trigonometry supplies us with an easy tool by which one can deal with any kinematic problem in the context of special relativity. Otherwise it allows us to obtain the *quantitative solution* of any problem and dispels all doubts regarding the role of acceleration in the flow of time.

Actually, if we consider "true" the space-time symmetry stated by Lorentz transformations, in whatever way obtained, the hyperbolic numbers provide the right mathematical structure inside which the two-dimensional problems must be dealt with. Finally, we remark that the application of hyperbolic trigonometry to relativistic space-time turns out to be a "Euclidean way" of dealing with the pseudo-Euclidean plane.

Chapter 6

General Two-Dimensional Hypercomplex Numbers

In this chapter we study the Euclidean and pseudo-Euclidean geometries associated with the general two-dimensional hypercomplex variable, i.e., the algebraic ring (see Section 2.2)

$$\{z = x + uy; \; u^2 = \alpha + u\beta; \; x, y, \alpha, \beta \in \mathbf{R}; \; u \notin \mathbf{R}\}, \qquad (6.0.1)$$

and we show that in geometries generated by these numbers, ellipses and general hyperbolas play the role which circles and equilateral hyperbolas play in Euclidean and in pseudo-Euclidean planes, respectively.

Moreover we show that these geometries can be studied simultaneously by using a two-dimensional variable without specification if it is elliptic or hyperbolic. In this way all the discussed geometries, are "unified" and the theorems are demonstrated by just one analytical development. By means of this approach we demonstrate, in an unusual algebraic way, Hero's formula and Pythagoras' theorem for the general Euclidean and pseudo-Euclidean plane geometries.

Definitions

Referring to the definitions of Section 2.2, we call

- *Euclidean*, the geometry associated with complex numbers;
- *pseudo-Euclidean*, the geometry associated with canonical hyperbolic numbers.

In the general case we call

- *elliptic*, the geometries associated with general systems with $\Delta < 0$;
- *hyperbolic*, the geometries associated with general systems with $\Delta > 0$.

In the following developments we do not consider the geometry associated with parabolic numbers which has been widely studied by Yaglom [81], who has shown its link with the Galileo group of classical kinematics.

6.1 Geometrical Representation

Here we briefly extend to general number some definitions given in Chapter 4.

Let us introduce a plane by analogy with the Gauss–Argand plane of the complex variable. In these *representative planes*, which are called the names of the number systems, i.e., elliptic or hyperbolic, we associate points $P \equiv (x, y)$ with the general two-dimensional hypercomplex numbers $z = x + u\,y$. Their topology is the same as the Euclidean plane for elliptic numbers, and as the pseudo-Euclidean (space-time) plane for hyperbolic numbers [38]. The hyperbolic plane is divided into four sectors [47] by straight lines (*null lines* [81]) $x + \frac{1}{2}(\beta \pm \sqrt{\Delta})y = 0$. With the same definitions of Section 4.3, a segment or line is called of the *first (second) kind* if it is parallel to a straight line passing through the origin located in the sectors containing the axis Ox (Oy).

Let us also consider the hypercomplex conjugate of z, which we indicate by $\bar{z}$, given by (Section 2.2)

$$\bar{z} = x + (\beta - u)y. \tag{6.1.1}$$

Therefore we have $z + \bar{z} \equiv 2\,x + \beta\,y \in \mathbf{R}$. The "square module" of z is given by

$$D \equiv z\bar{z} = x^2 - \alpha\,y^2 + \beta\,x\,y. \tag{6.1.2}$$

This real quantity is a definite quadratic form for $\Delta \le 0$ and it is not definite (it is zero for x, y on the null-lines) for $\Delta > 0$. It defines, in the representative planes, the square distance of a point from the origin of the coordinates. It must be taken in the quadratic form of (6.1.2), which can also be negative (see Chapter 4).

If we consider two points $P_i \equiv (x_i, y_i)$, $P_j \equiv (x_j, y_j)$, their square distance D_{ij} is given by an extension of (6.1.2)

$$
\begin{aligned}
D_{ij} = D_{ji} &= (z_i - z_j)(\bar{z}_i - \bar{z}_j) \equiv z_i\bar{z}_i + z_j\bar{z}_j - z_i\bar{z}_j - z_j\bar{z}_i \\
&\equiv (x_i - x_j)^2 - \alpha(y_i - y_j)^2 + \beta(x_i - x_j)(y_i - y_j).
\end{aligned} \tag{6.1.3}
$$

We set

$$D_{i\,O} \equiv z_i\bar{z}_i, \tag{6.1.4}$$

which represents the square distance from the origin of coordinate axes. For hyperbolic numbers this square distance is positive for segments of the first kind, negative for segments of the second kind. In Section 6.2 we exclude the points in a position for which $D_{ij} = 0$ (*null distance*).

The definition of distances (metric element) is equivalent to introducing the bilinear form of the *scalar product*. The scalar product and the properties of hypercomplex numbers allow us to state suitable axioms ([81], p. 245) and to give the structure of a vector space to the representative planes of elliptic and hyperbolic numbers. When the linear distance (a segment length or radial coordinate) has to be used, we follow the conventions of Section 4.3 and put

$$d_{ij} = \sqrt{|D_{ij}|}. \tag{6.1.5}$$

From (6.1.3) it follows that the equation of the locus of $P \equiv (x, y)$ which has the same distance from a fixed point $P_c \equiv (x_c, y_c)$ is

$$(x - x_c)^2 - \alpha(y - y_c)^2 + \beta(x - x_c)(y - y_c) = K, \tag{6.1.6}$$

which represents a conic section that we call *the fundamental conic section*. Therefore in the representative planes of general algebras, these conic sections can be considered as extensions of Euclidean circles (*the locus of points which are at a fixed distance (r) from a given point P_c.*). Moreover, as it has been shown in Chapter 4 for equilateral hyperbolas in the pseudo-Euclidean plane and later in this chapter for the general geometries, the conic section (6.1.6) retains many peculiar properties of Euclidean circles. Then, following Yaglom ([81], p. 181) we call *"circles"* the conic sections represented by (6.1.6). We call K the square *"semi-diameter"*, and $P_c \equiv (x_c, y_c)$ the center of the conic section, and in general we extend the Euclidean terminology by using the italic style included in quotation marks.

For $\Delta > 0$ (the conic sections are hyperbolas) K can be positive or negative and, as it is shown in Section 4.5, we can assign a kind to these hyperbolas, and to call *hyperbolas of the first (second) kind if their tangent straight lines are of the first (second) kind* ([55], p. 52). For hyperbolas of the first kind $K < 0$, for hyperbolas of the second kind $K > 0$.

We define *"semi-diameter"* of the hyperbolas to be the quantity

$$k = \sqrt{|K|}. \tag{6.1.7}$$

The hyperbolas with common center and square *"semi-diameter"* k^2 and $-k^2$ are said to be *conjugate*.

For fixed values of α, β we have a family of conics. These conics have the same coefficients of the quadratic terms; then, as it is known from analytical geometry, they have the same eccentricity, directions of axes and, as far as hyperbolas are concerned, the null-lines as asymptotes. From another point of view we can associate with a given conic an (α, β) algebra so that this conic, in the representative plane, has the properties of a circle.

As for complex and hyperbolic planes, the angular coefficient of a straight line, determined by two points P_i, P_j, is given by (Chapter 4)

$$m_{ij} = \frac{y_i - y_j}{x_i - x_j}. \tag{6.1.8}$$

Indicating by m the angular coefficient of straight lines, from the previously given definition, we have

- if $|2\,\alpha\,m - \beta| < \sqrt{\Delta}$ the straight line is of the first kind,

- if $|2\,\alpha\,m - \beta| > \sqrt{\Delta}$ the straight line is of the second kind.

Now we are in position to generalize well-known theorems of Euclidean geometry.

6.2   Geometry and Trigonometry in Two-Dimensional Algebras

6.2.1   The "Circle" for Three Points

As in Euclidean and in pseudo-Euclidean planes, circles and equilateral hyperbolas are determined by three points, so the general *"circle"*, given by (6.1.6) is determined by three points. We have

Theorem 6.1. *Given three non-aligned points $P_i \equiv (x_i, y_i)$, $i = 1, 2, 3$, in the representative plane of a generalized two-dimensional hypercomplex number, there is a "circle" (6.1.6), passing for these points. This "circle" has square semi-diameter given by*

$$K = -\frac{D_{12}D_{13}D_{32}}{4\,\Delta \cdot S^2}, \tag{6.2.1}$$

and the "center" in a point P_c, associated with number z_c. We have

$$z_c = \frac{(-\beta + 2\,u)[D_{1\,0}1(z_3 - z_2) + D_{2\,0}(z_1 - z_3) + D_{3\,0}(z_2 - z_1)]}{2\,\Delta \cdot S}. \tag{6.2.2}$$

Proof. These three points can be considered as the vertexes of a triangle, then this problem is the generalization of the well-known Euclidean problem of finding a circumcircle of a triangle. In the representation on a Cartesian plane, the problem is solved if we know the center $P_c \equiv (x_c, y_c)$ and the circle radius. Now the problem is the same and we find x_c, y_c, K, by solving the system of three equations derived from (6.1.6)[1]

$$D_{n\,c} = (z_n - z_c)(\bar{z}_n - \bar{z}_c) \equiv z_n \bar{z}_n + z_c \bar{z}_c - z_n \bar{z}_c - z_c \bar{z}_n = K \text{ for } n = 1, 2, 3. \tag{6.2.3}$$

Subtracting the first equation ($n = 1$) from the other two ($n = 2, 3$), we obtain two equations for the unknown z_c, $\bar{z}_c$. In matrix form

$$\begin{pmatrix} \bar{z}_2 - \bar{z}_1 & z_2 - z_1 \\ \bar{z}_3 - \bar{z}_1 & z_3 - z_1 \end{pmatrix} \cdot \begin{pmatrix} z_c \\ \bar{z}_c \end{pmatrix} = \begin{pmatrix} \bar{z}_2 z_2 - \bar{z}_1 z_1 \\ \bar{z}_3 z_3 - \bar{z}_1 z_1 \end{pmatrix}, \tag{6.2.4}$$

with the solution

$$z_c = \frac{z_1 z_2 \bar{z}_2 - z_1 z_2 \bar{z}_1 + z_2 z_3 \bar{z}_3 - z_3 z_2 \bar{z}_2 + z_1 z_3 \bar{z}_1 - z_1 z_3 \bar{z}_3}{z_1 \bar{z}_2 + z_3 \bar{z}_1 + z_2 \bar{z}_3 - z_2 \bar{z}_1 - z_1 \bar{z}_3 - z_3 \bar{z}_2}. \tag{6.2.5}$$

The solution for $\bar{z}_c$ is the conjugate of (6.2.5). From one of (6.2.3), (6.2.5) and the equation for $\bar{z}_c$, we find

$$
\begin{aligned}
K &= -\frac{(z_2 - z_1)(\bar{z}_2 - \bar{z}_1)(z_3 - z_1)(\bar{z}_3 - \bar{z}_1)(z_2 - z_3)(\bar{z}_2 - \bar{z}_3)}{(z_1 \bar{z}_2 + z_3 \bar{z}_1 + z_2 \bar{z}_3 - z_2 \bar{z}_1 - z_1 \bar{z}_3 - z_3 \bar{z}_2)^2} \\
&\equiv -\frac{D_{12}D_{13}D_{32}}{(z_1 \bar{z}_2 + z_3 \bar{z}_1 + z_2 \bar{z}_3 - z_2 \bar{z}_1 - z_1 \bar{z}_3 - z_3 \bar{z}_2)^2}. \tag{6.2.6}
\end{aligned}
$$

---

[1]The following calculations have been performed with the scientific software *Mathematica* [79].

Let us call Q the term in round brackets of the denominator, then

$$\overline{Q} = -Q, \tag{6.2.7}$$

and, calculating Q as a function of the points' coordinates, we have

$$\begin{aligned} Q &\equiv z_1\bar{z}_2 + z_3\bar{z}_1 + z_2\bar{z}_3 - z_2\bar{z}_1 - z_1\bar{z}_3 - z_3\bar{z}_2 \\ &= (-\beta + 2u)[x_1(y_3 - y_2) + x_2(y_1 - y_3) + x_3(y_2 - y_1)]. \end{aligned} \tag{6.2.8}$$

The content of the square brackets can be recognized, but for the sign (Chapter 4), as the double of the area (S) of the triangle and, by means of (2.2.10), we have

$$Q^2 = 4\Delta \cdot S^2, \tag{6.2.9}$$

and from (6.2.5) and (6.2.6) we obtain (6.2.2) and (6.2.1), respectively. $\qquad\square$

We note that the area of the triangle is independent of the particular geometry. It is shown in Chapter 4 that it has the same value for Euclidean and for pseudo-Euclidean geometries. In the same way it can be shown that it is independent of α, β too.

Equation (6.2.1) represents the generalized expression for the square *"semi-diameter"* of the circumcircle of a Euclidean triangle and for the circumscribed equilateral hyperbola in pseudo-Euclidean geometry. For the canonical systems (6.2.1) becomes

$$K = \begin{cases} \dfrac{D_{12}D_{13}D_{32}}{16\,S^2} \;\Rightarrow\; k = \dfrac{d_{12}d_{13}d_{32}}{4\,S} & (u^2 = -1,\; \Delta = -4) \\[4mm] -\dfrac{D_{12}D_{13}D_{32}}{16\,S^2} & (u^2 = 1,\; \Delta = 4). \end{cases} \tag{6.2.10}$$

For hyperbolic numbers, K can be positive or negative, i.e., the circumscribed hyperbola is of the second or first kind, depending on the sides' kinds.

6.2.2 Hero's Formula and Pythagoras' Theorem

In this section we demonstrate in an unusual, algebraic way Hero's formula and, as a consequence, the theorem of Pythagoras. These demonstrations hold for elliptic geometries, as well as for hyperbolic ones.

We know that in Euclidean geometry the triangle area can be expressed as a function of the triangle sides lengths by means of Hero's formula. Now, since the characteristic quantities are the square distances, we have

Theorem 6.2. *For the general two-dimensional geometries the area of a triangle as a function of the square side lengths can be obtained by the following generalized Hero's formula:*

$$\begin{aligned} S^2 &= \frac{D_{12}^2 + D_{13}^2 + D_{32}^2 - 2(D_{12}D_{13} + D_{12}D_{32} + D_{13}D_{32})}{4\Delta} \\[3mm] &\equiv \frac{(D_{13} + D_{23} - D_{12})^2 - 4D_{13}D_{23}}{4\Delta}. \end{aligned} \tag{6.2.11}$$

Proof. We start from the identity

$$(z_1 - z_2)(\bar{z}_1 - \bar{z}_2) \equiv [(z_1 - z_3) - (z_2 - z_3)][(\bar{z}_1 - \bar{z}_3) - (\bar{z}_2 - \bar{z}_3)]. \qquad (6.2.12)$$

Expanding the right-hand side and taking into account (6.1.3) and (6.2.8), we obtain

$$D_{12} = D_{32} + D_{31} - 2(\bar{z}_2 - \bar{z}_3)(z_1 - z_3) - Q$$
$$\Rightarrow 2(\bar{z}_2 - \bar{z}_3)(z_1 - z_3) = -D_{12} + D_{32} + D_{31} - Q. \qquad (6.2.13)$$

Multiplying (6.2.13) by its hypercomplex conjugate and taking into account (6.2.7), we have

$$4\,D_{32}\,D_{31} = (-D_{12} + D_{32} + D_{31})^2 - Q^2. \qquad (6.2.14)$$

Substituting (6.2.9) into (6.2.14) we obtain Hero's formula (6.2.11).   □

We note that this demonstration is another example of how the proposed approach allows us to demonstrate theorems by means of identities as shown in Chapter 4. For complex numbers (Euclidean geometry) this expression can be easily reduced to the product of four linear terms which represents the well-known *Hero's formula*.

As a consequence of (6.2.11) we obtain

Theorem 6.3. *For a right-angled triangle the following generalized theorem of Pythagoras holds between the square side lengths,*

$$D_{32} = D_{13} + D_{12}. \qquad (6.2.15)$$

Proof. Let us solve (6.2.11) in terms of the quadratic distance D_{32},

$$D_{32} = D_{13} + D_{12} \mp 2\sqrt{D_{12}D_{13} + \Delta \cdot S^2}; \qquad (6.2.16)$$

if $D_{12}D_{13} + \Delta \cdot S^2 = 0$, (6.2.16) becomes (6.2.15) which for Euclidean and pseudo-Euclidean geometries is the theorem of Pythagoras for a right-angle triangle.

For a general algebra, if we have $D_{12}D_{13} + \Delta \cdot S^2 = 0$, we say that sides $\overline{P_1 P_2}$ and $\overline{P_1 P_3}$ are "orthogonal", and *the theorem of Pythagoras holds for all two-dimensional hypercomplex systems.*

We note that the definition of *"orthogonal"* line, here reported, is the same as that of differential geometry, i.e., two vectors are *"orthogonal"* if their scalar product (which depends on the metric) is null.   □

As already pointed out in [38] this theorem holds, in the form of (6.2.15), for all systems, i.e., with a sum in the right-hand side but for hyperbolic geometries, since the sides are of different kinds, the square distances have opposite signs.

6.2.3 Properties of "Orthogonal" Lines in General Algebras

We have

Theorem 6.4. *For the general two-dimensional geometry the angular coefficients* (m_N) *and "m" of "orthogonal" lines are related by*

$$m_N = \frac{\beta\, m + 2}{2\alpha\, m - \beta}. \tag{6.2.17}$$

Proof. The relation we are looking for can be obtained using Pythagoras' theorem (6.2.15). Let us consider the straight lines determined by points P_1, P_2 and P_1, P_3; if these straight lines are orthogonal, substituting in (6.2.15) the coordinates of points, we have

$$(x_3 - x_2)^2 - \alpha(y_3 - y_2)^2 + \beta(x_3 - x_2)(y_3 - y_2) = (x_3 - x_1)^2 - \alpha(y_3 - y_1)^2$$
$$+ \beta(x_3 - x_1)(y_3 - y_1) + (x_1 - x_2)^2 - \alpha(y_1 - y_2)^2 + \beta(x_1 - x_2)(y_1 - y_2). \tag{6.2.18}$$

After simplifying this equation and substituting the angular coefficients given by (6.1.8), we obtain the relation (6.2.17). □

Equation (6.2.17) can be written in a form which shows immediately the reciprocity relation between m_N and m,

$$2\,\alpha\, m_N\, m - \beta(m_N + m) = 2\,; \text{ or } (2\,\alpha\, m_N - \beta)(2\,\alpha\, m - \beta) = \Delta. \tag{6.2.19}$$

From the second relation it follows that *if a straight line is of one kind* (Section 6.1), *its "orthogonal" straight line is of the other kind.*

For canonical systems $\beta = 0$ and $\alpha = \mp 1$, we obtain the well-known relations of Euclidean and pseudo-Euclidean geometries

$$m_N = \frac{1}{\alpha\, m}. \tag{6.2.20}$$

6.3 Some Properties of Fundamental Conic Sections in General Two-Dimensional Algebras

6.3.1 "Incircles" and "Excircles" of a Triangle

Let there be given three points which can be considered the vertexes of a triangle, and the three straight lines passing through them. We put the problem of finding a *"circle"* inside the triangle *"(incircle)"* and three *"circles"*, outside the triangle and tangent to the straight line prolongations of the three sides *"(excircles)"*. This problem represents an extension to geometries associated with a general two-dimensional algebra of a well-known Euclidean problem. The solution of this problem shows the properties which are preserved when going from Euclidean geometry to general geometries, which properties must be considered peculiar to the Euclidean geometry. On the other hand we cannot construct a hyperbola inside a triangle, but we see that, solving together problems concerning *"incircle"* and *"excircles"*, we find four *"circles"* with the properties of the four Euclidean circles (Figs. 6.2–6.4).

By utilizing a property of Euclidean incircles and excircles, we change the problem to the follwing one: *to find "circles" with their centers equidistant from three straight lines.*

This problem has a solution; moreover the *"circles"* which we find, also have other properties of the corresponding Euclidean circles. These other properties can be demonstrated by simple calculations, as it is shown in Chapter 4 for equilateral hyperbolas in the pseudo-Euclidean plane. We note that for hyperbolic geometry we must consider, in general, both conjugate arms of the hyperbolas, since the square distance from a straight line and the center of a hyperbola can be positive or negative depending on the side (straight lines) kind, as we also see in the numerical examples of Section 6.4.

For the solution of the problem we must find the center $P_c \equiv (x_c, y_c)$ and *"semi-diameter"* $k \equiv \sqrt{|K|}$ of *"incircle"* and *"excircles"*. This problem is solved by means of two linear equations, as in Euclidean geometry. Let us begin with pseudo-Euclidean geometry.

Distance Between a Point and a Straight Line in a Pseudo-Euclidean Plane

In Euclidean geometry the distance $d_{c\,\gamma_E}$, between a point $P_c \equiv (x_c, y_c)$ and a straight line $\gamma_E : \{\sin\phi\, x + \cos\phi\, y + q = 0\}$, is equal to the value which we obtain after substituting the coordinates of P_c in the equation of γ_E. Moreover we usually take the positive quantity

$$d_{c\,\gamma_E} = |\sin\phi\, x_c + \cos\phi\, y_c + q| \qquad (6.3.1)$$

as the distance between the point P_c and the straight line γ_E. Actually the quantity $\sin\phi\, x_c + \cos\phi\, y_c + q$ can be positive or negative depending on the position of the point with respect to the straight line.

By means of Theorem 4.9 we know that the same relation holds for pseudo-Euclidean geometry. In particular if we consider a straight line in canonical form $\gamma_H : \{\sinh_e \theta\, x + \cosh_e \theta\, y + q = 0\}$, (4.3.7) becomes the linear function of x_c, y_c given by

$$d_{c\,\gamma_H} = |\sinh_e \theta\, x_c + \cosh_e \theta\, y_c + q|, \qquad (6.3.2)$$

and, as in Euclidean geometry, we take the absolute value which removes the sign depending on the position of the point with respect to the straight line.

For the solution of the problem we must preserve the sign with a simple geometrical consideration. Actually, for Euclidean geometry, we note that the centers of excircles, with respect to the incircle, are on opposite sides of one straight line, then since we are finding the centers of the four circles by means of the same linear equations, we have to equate the distances but for the sign. Therefore, let us consider the straight lines determined by three points P_1, P_2, P_3 and let us denote $\gamma_i : \{\sinh_e \theta_i\, x + \cosh_e \theta_i\, y + q_i = 0\}$ the straight line between points P_j, P_k, $i \neq j$, k. By introducing two quantities ϵ_1, ϵ_2 equal to ± 1, which take into account the elimination of the absolute values in (6.3.2), we find the centers of the inscribed and circumscribed hyperbolas by means of equations

$$d_{c\,\gamma_1} = \epsilon_1\, d_{c\,\gamma_2}, \qquad d_{c\,\gamma_1} = \epsilon_2\, d_{c\,\gamma_3}. \tag{6.3.3}$$

The semi-diameters shall be obtained from the center coordinates and one of (6.3.2).

The Solution for General Algebras

All the exposed considerations can be applied to general algebras, and we have

Theorem 6.5. *Given a straight line $\gamma_i : \{y - m_i\, x - c_i = 0\}$ and a point $P_c \equiv (x_c, y_c)$ outside the straight line, the distance between them is a linear function of the point coordinates, given by*

$$d_{c\gamma_i} = \frac{1}{2}\sqrt{\left|\frac{\Delta}{f(m_i)}\right|}\,|(y_c - m_i\, x_c - c_i)|, \tag{6.3.4}$$

where

$$f(m_i) = 1 - \alpha(m_i)^2 + \beta m_i. \tag{6.3.5}$$

Proof. With the definition (6.3.5) we obtain, as for canonical systems, that the square distance from P_c and γ_i is proportional to the result of substituting the coordinates of P_c in the equation of γ_i, in particular

$$D_{c\gamma_i} = -\frac{\Delta \cdot (y_c - m_i\, x_c - c_i)^2}{4\, f(m_i)} \tag{6.3.6}$$

and, also in this case, the distance can be put in the linear form (6.3.4) □

As for Euclidean and pseudo-Euclidean planes, the sign of $y_c - m_i\, x_c - c_i$ is determined by the position of the point P_c with respect to straight line γ_i.

In our specific problem the straight line passing through points P_k and P_j is given by

$$\gamma_i : \{y - y_j = \frac{y_k - y_j}{x_k - x_j}(x - x_j)\ |\ i \neq j \neq k\}, \tag{6.3.7}$$

and the angular coefficient of the straight line *"orthogonal"* to γ_i, is given by (6.2.17)

$$m_N = \frac{2\,(x_j - x_k) + \beta\,(y_j - y_k)}{-\beta\,(x_j - x_k) + 2\,\alpha\,(y_j - y_k)}.$$

Therefore for the starting problem, we have

Theorem 6.6. *The centers and "semi-diameters" of "incircle" and "excircles" of a triangle are given, as functions of the side lengths, by*

$$x_c = \frac{\epsilon_1\, x_1\, d_{23} - \epsilon_2\, x_2\, d_{13} + \epsilon_3\, x_3\, d_{12}}{\epsilon_1\, d_{23} - \epsilon_2\, d_{13} + \epsilon_3\, d_{12}},$$

$$y_c = \frac{\epsilon_1\, y_1\, d_{23} - \epsilon_2\, y_2\, d_{13} + \epsilon_3\, y_3\, d_{12}}{\epsilon_1\, d_{23} - \epsilon_2\, d_{13} + \epsilon_3\, d_{12}}, \qquad\qquad (6.3.8)$$

$$K = -\frac{\Delta \cdot S^2}{(\epsilon_1\, d_{23} - \epsilon_2\, d_{13} + \epsilon_3\, d_{12})^2} \;\Rightarrow\; k = \frac{\sqrt{|\Delta|} \cdot S}{|\epsilon_1\, d_{23} - \epsilon_2\, d_{13} + \epsilon_3\, d_{12}|},$$

where e_1, $\epsilon_2 = \pm 1$ and $\epsilon_3 = \epsilon_1\, \epsilon_2$.

Proof. The results (6.3.8) are obtained by solving the system (6.3.3), with $d_{c\,\gamma_i}$ given by (6.3.4) and m_i, γ_i given by (6.3.7) and setting $\epsilon_3 = \epsilon_1\, \epsilon_2$. $\qquad\square$

We have obtained the centers and the semi-diameters of *"incircle"* and *"excircles"* by means of the same equations. By analogy with Euclidean geometry, we define as inscribed the *"circle"* with the smallest semi-diameter ($\epsilon_1 = -\epsilon_2 = \epsilon_3 = -1$).

We note that the expressions for k as a function of the side lengths are the same as those of Euclidean geometry.

In Section 6.4 we show two numerical examples relative to elliptic and hyperbolic planes, and their graphical representation.

6.3.2   The Tangent Lines to the Fundamental Conic Section

By means of the analytic methods used up to now and by the support of an adequate software, the theorems of Section 4.5 can be demonstrated for the fundamental conic section associated with the general two-dimensional algebra. In particular Theorem 4.22 becomes

Theorem 6.7. *If from a point $P \equiv (x_p, y_p)$ we trace a straight line which crosses the fundamental conic section (6.1.6) in points S_1 and S_2, the product of the square distances from P to S_1 and S_2 is constant and is obtained by substituting the coordinates of P in the equation for the fundamental conic section*

$$D_{\overline{P\,S_1}} \cdot D_{\overline{P\,S_2}} = [(x_c - x_p)^2 - \alpha(y_c - y_p)^2 + \beta(x_c - x_p)(y_c - y_p) - K]^2. \quad (6.3.9)$$

Because the theorem holds for all the secants, in particular it is true for tangent lines, if they exist. In this case we have $S_1 \equiv S_2 \equiv T$ and

$$D_{\overline{PT}}^2 \equiv (D_{\overline{PC}} - K)^2 \Rightarrow D_{\overline{PC}} = D_{\overline{PT}} + K, \qquad\qquad (6.3.10)$$

which represents Pythagoras' theorem for the right-angled triangle PTP_C. Then we have the extension of Theorem 4.15

Theorem 6.8. *The tangent line to a fundamental conic section is "orthogonal" to the diameter.*

6.4 Numerical Examples

Figs. 6.2–6.4 represent two applications of the results obtained in section (6.2) about the *"circumcircle"*, *"incircle"* and *"excircles"* of a triangle[2].

In particular let us consider three points $P_1 \equiv (3, 4)$, $P_2 \equiv (6, 8)$, $P_3 \equiv (7, 3)$; the first application is pertinent to elliptic geometry and in Fig. 6.2 we report the five ellipses, obtained with the values of the parameters $\alpha = -3$, $\beta = -1$, $\Delta = -11 < 0$.

In the second application we refer to hyperbolic geometry and in Figs. 6.3 and 6.4, the five hyperbolas, obtained with the values of the parameters $\alpha = 3$, $\beta = -1$, $\Delta = 13 > 0$, are reported.

The *"circumcircles"* have the coordinates of the centers obtained by (6.2.2) and the square *"semi-diameters"* given by (6.2.6). The centers and *"semi-diameters"* of the *"incircle"* and *"excircles"* are obtained by (6.3.8). Following the convention of Section 6.1 we have $m_{12} = 4/3$, $m_{13} = -1/4$, $m_{23} = -5$ and $|2\alpha m_{12} - \beta| \equiv 9 > \sqrt{\Delta} \equiv \sqrt{13}$, $|2\alpha m_{13} - \beta| \equiv 1/2 < \sqrt{\Delta}$, $|2\alpha m_{23} - \beta| \equiv 29 > \sqrt{\Delta}$.

Then the sides $\overline{P_1 P_2}$ and $\overline{P_2 P_3}$ (the straight lines γ_3, γ_1) are of the second kind, the side $\overline{P_1 P_3}$ (the straight line γ_2) is of the first kind. The hyperbola tangent to straight lines γ_3, γ_1 is of the second kind $(x-x_c)^2 - (y-y_c)^2 = k^2$, the hyperbola tangent to straight line γ_2 is of the first kind $(x - x_c)^2 - (y - y_c)^2 = -k^2$.

As far as the circumscribed hyperbola is concerned, we report both the conjugate arms, tough the problem is solved by a hyperbola of a specific kind. In this particular example D_{12}, $D_{13} < 0$, $D_{23} > 0$, then from (6.2.1) we have $K < 0$, i.e., as one can note from Fig. 6.3, the circumscribed hyperbola is of the first kind $(y - y_c)^2 - (x - x_c)^2 = k^2$.

As far as the inscribed and ex-inscribed hyperbolas are concerned, their parameters are obtained by means of (6.3.8) and, by analogy with Euclidean geometry, we define as inscribed the hyperbola with the smallest *"semi-diameter"*. By observing Figs. 6.3 and 6.4, we shall see that these hyperbolas also have other properties of the corresponding Euclidean circles.

---

[2]These figures were obtained with the software *Mathematica* [79].

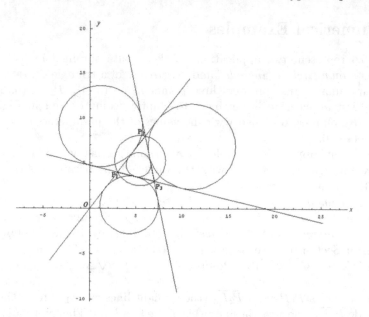

Figure 6.1: Given, in the Euclidean plane, three points $P_1 \equiv (3, 4)$, $P_2 \equiv (6, 8)$, $P_3 \equiv (7, 3)$, these points can be considered the vertices of a triangle. In this figure we report the incircle, circumcircle and ex-circles of the triangle, associated with the canonical complex algebra $\{z = x + i\,y\}$.

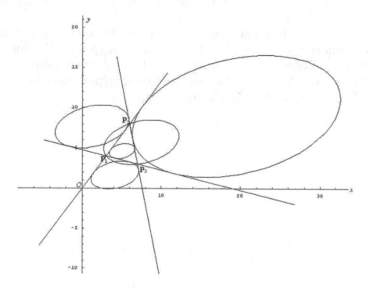

Figure 6.2: Given three points $P_1 \equiv (3, 4)$, $P_2 \equiv (6, 8)$, $P_3 \equiv (7, 3)$, these points can be considered the vertices of a triangle. In this figure we report the *inscribed, circumscribed,* and *ex-inscribed* ellipses associated with the elliptic algebra $\{z = x + u\,y;\ u^2 = \alpha + \beta\,u;\ x, y \in \mathbf{R},\ \alpha = -3,\ \beta = -1,\ \Delta = -11\}$. This example and the following ones are generalisations of the Euclidean circles shown in Fig. 6.1.

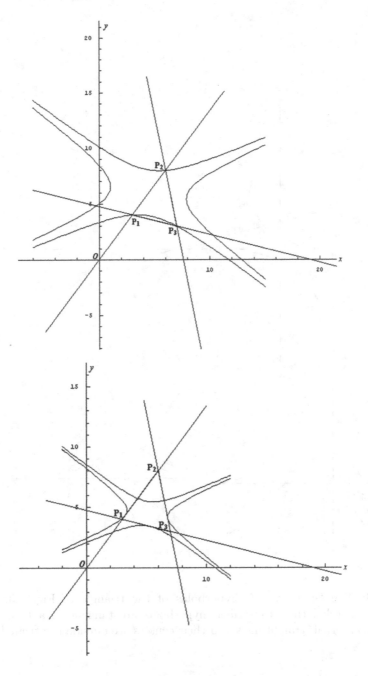

Figure 6.3: Given three points $P_1 \equiv (3, 4)$, $P_2 \equiv (6, 8)$, $P_3 \equiv (7, 3)$, these points can be considered the vertices of a triangle. In these figures and in the following ones we report the hyperbolas associated with the hyperbolic algebra $\{z = x + u\,y;\ u^2 = \alpha + \beta\,u;\ x, y \in \mathbf{R},\ \alpha = 3,\ \beta = -1,\ \Delta = 13\}$.
- Up: The "*circumscribed*" hyperbola.
- Down: The "*inscribed*" hyperbola.

As for Euclidean incircles, the inscribed hyperbola is tangent to the triangle sides (and not to the external straight line) and its center is inside the triangle.

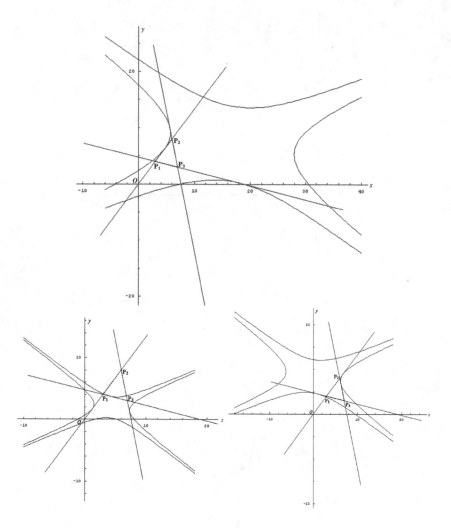

Figure 6.4: The "*ex-inscribed*" hyperbolas of the triangle of Fig. 6.3. As for Euclidean excircles, the ex-inscribed hyperbolas are tangent to a triangle side and to two external straight lines and their centers are outside the triangle.

Chapter 7

Functions of a Hyperbolic Variable

7.1 Some Remarks on Functions of a Complex Variable

For real variables, the definition of polynomials (linear combinations of powers) stems from the definitions of elementary algebraic operations. Since for complex variables the same algebraic rules hold, also for them the polynomial can be defined and, grouping together the terms with and without the coefficient i, we can always express them as $P(z) = u(x,y) + iv(x,y)$, where u, v are real functions of the real variables x, y.

As far as the transcendental functions of a complex variable are concerned, we cannot "a priori" assume that they can be written in a similar form, i.e., that the versor i can be taken out from the function for obtaining two real functions linked by i. The Euler formula for the exponential function has been considered as a secret of nature up to the formalization of the theory of analytic functions given by absolutely convergent power series to which the same algebraic rules of the polynomial can be applied. Then for these functions we can always write $w \equiv f(z) \equiv f(x + iy) = u(x,y) + iv(x,y)$. Since u and v are obtained in an algebraic way from the binomial $x + iy$, they cannot be arbitrary functions but must satisfy definite conditions. These conditions can be obtained in many ways. Following Riemann, we say that a function $w \equiv f(z) = u(x,y) + iv(x,y)$ is a function of the complex variable z if its derivative is independent of the direction with respect to which the incremental ratio is taken. From these conditions two partial differential equations, named after Cauchy and Riemann, link u and v.

A formal way for obtaining these equations is the following one. Let us consider a generic expression $w = u(x,y) + iv(x,y)$ and perform the bijective substitution $z = x + iy$, $\bar{z} = x - iy$; we say that w is a function of z if it does not depend on $\bar{z}$, i.e., its partial derivative $w_{,\bar{z}} = 0$[1]. By means of the derivative chain rule we carry out the calculation and obtain

$$w_{,\bar{z}} = u_{,x}\, x_{,\bar{z}} + u_{,y}\, y_{,\bar{z}} + i\left(v_{,x}\, x_{,\bar{z}} + v_{,y}\, y_{,\bar{z}}\right) \equiv [u_{,x} - v_{,y} + i\left(u_{,y} + v_{,x}\right)]/2 = 0,$$

from which the Cauchy–Riemann conditions follow;

$$u_{,x} = v_{,y}, \qquad u_{,y} = -v_{,x}. \tag{7.1.1}$$

The functions of a complex variable have the property that, even if they are functions of x and y, they are just functions of z.

---

[1]The comma stands for differentiation with respect to the variables which follow.

The same conclusion is obtained if we consider functions of just $\bar{z}$. In this case u, v must satisfy the partial differential equations

$$u_{,x} = -v_{,y}, \qquad\qquad u_{,y} = v_{,x}. \tag{7.1.2}$$

These equations have a physical relevance. Actually if we give to u, v the physical meaning of the components of a two-dimensional vector field $E \equiv (u, v)$, (7.1.2) represent the conditions rot E=0, div E =0. E is a harmonic field since, thanks to (7.1.1) and (7.1.2), u and v satisfy the Laplace equation $U_{,xx} + U_{,yy} = 0$. Moreover the Laplace equation is also invariant under the transformations $x, y \Rightarrow u, v$, where u, v are given by the real and imaginary part of a function of a complex variable [69]. These functional transformations are called a *conformal group*.

This link between functions of a complex variable and relevant physical fields is outstanding, as we can see with the following examples. We know that the study of a central field is simplified if we use polar coordinates which have the same "symmetry" of the problem and reduce the dependence of the function we have to calculate (as an example the potential) just on one variable (the distance from the source). The same result occurs with functions of a complex variable and a harmonic field. Actually two functions $(u(x, y), v(x, y))$ of the variables x, y become one function of one variable $w(z)$. A practical result is that some problems related to Laplace or Poisson partial differential equations can be reduced to integrals or solved by means of functional transformations.

Now let us state a link between the multiplicative group of complex numbers and the conformal group of functional transformations. In the language of the classical Lie groups, the two-dimensional Euclidean group (recalled in Section 3.1.2) is said to be a finite group because it depends on three parameters, whilst the conformal group is said to be infinite because it depends on arbitrary functions (see Section 7.2.3).

The Euclidean group is very important in itself. Moreover, if considered as addition and multiplication of complex constants and variables, it represents the simplest subgroup of the conformal group. Both groups derive from the symmetries related to the "operator" "i" of a complex variable. A connection between these two groups can be found if one looks for a field which satisfies the Laplace equation and is invariant for the rotation group. Having set these requirements the problem is equivalent to calculating the potential of a central field, which is a function only of distance from the source. This problem is usually solved ([69], p. 341) by means of a polar coordinate transformation $(x, y \Rightarrow \rho, \phi)$, which transforms the Laplace equation into $u_{,\rho\rho} + \rho^{-1}u_{,\rho} + \rho^{-2}u_{,\phi\phi} = 0$.

The use of a complex exponential transformation, which has the same symmetries as the polar one, leaves the Laplace equation invariant; then with the transformation

$$x = \exp X \cos Y, \; y = \exp X \sin Y \text{ or } X = \ln \sqrt{x^2 + y^2}, \; Y = \tan^{-1}(y/x),$$

we have to solve the equation

$$U_{,XX} + U_{,YY} = 0. \tag{7.1.3}$$

The U invariance for rotation means independence of the rotation angle, represented here by Y. Therefore $U_{,YY} \equiv 0$, and U depends only on the variable X. The partial differential equation, (7.1.3), becomes an ordinary differential equation $d^2U/dX^2 = 0$, with the elementary solution

$$U = aX + b \equiv a \ln \sqrt{x^2 + y^2} + b, \tag{7.1.4}$$

which represents the potential of a point charge. Then in the (X, Y) plane the straight lines $X = const.$ give the equipotential, and, as is better known, the circles $x^2 + y^2 = const.$ are the equipotentials in the x, y plane. We can note that, from the "symmetry" of the finite rotation group, we have obtained *the Green function for the partial differential Laplace equation.*

From this example there follows a practical and simple solution for particular problems, as an example: *if we know an equipotential of a plane harmonic field and a conformal transformation u, v $\Leftrightarrow$ x, y, which maps the given equipotential into a circle, the other equipotentials are given by* $\ln \sqrt{u^2 + v^2}$.

7.2 Functions of Hypercomplex Variables

7.2.1 Generalized Cauchy–Riemann Conditions

In Appendix C we shall introduce the functions of a hypercomplex variable by means of matrix algebra; here, following an approach that is similar to the one used for introducing the functions of a complex variable, we obtain the *Generalized Cauchy–Riemann (GCR) conditions*, for functions of N-dimensional commutative systems.[2] Scheffers' theorem quoted in Section 2.1 allows us to define the functions of commutative hypercomplex variables by extending Riemann's ideas concerning the functions of complex variables. Let us denote by $x = x^0 + e_i x^i$ (the unity versor e_0 has been omitted) the hypercomplex variable and $w = w^0 + e_i w^i$ a hypercomplex function. w^α are real and differentiable functions of the real variables $x^0, \ldots, x^{N-1}$.

We say that w is a function of the variable x in a domain D if for all $P_0 \equiv (x_0^\alpha) \in D$, the limit of the incremental ratio Δw does not depend on the manner in which the point $P_0 + \Delta P$ approaches the point P_0, with the additional condition that Δx is not a divisor of zero. If these conditions are satisfied w is called a *holomorphic function* [60].

---

[2]The functions of hypercomplex variables allow us to associate the group of functional transformations (*conformal mapping*) with the hypercomplex numbers. In Section 7.2.3 we explain their relevance.

This requirement implies that the components must satisfy some Cauchy–Riemann-like equations given by a system of first-order partial differential equations. These equations, as Cauchy–Riemann conditions, can be obtained in many ways and with the same mathematical rigor as for the theory of functions of a complex and bicomplex variable [60] and [73]. Here we give two equivalent formulations.

Theorem 7.1. *w is a holomorphic function of the variable x if the following Generalized Cauchy–Riemann conditions (GCR) are satisfied:*

$$\text{(GCR)} \qquad \frac{\partial w}{\partial x^m} \equiv w_{,m} = e_m w_{,0}, \qquad\qquad (7.2.1)$$

Proof. Let us denote by dw and dx the differentials and by $p \equiv e_\alpha p^\alpha$ the hypercomplex derivative of w with respect to the hypercomplex variable x, so that

$$dw = p\, dx. \qquad\qquad (7.2.2)$$

We can write

$$p = \frac{dw}{dx} \equiv \frac{w_{,\gamma}\, dx^\gamma}{e_\gamma\, dx^\gamma}. \qquad\qquad (7.2.3)$$

Since the incremental ratio must be independent of the dx direction, we can take $dx^\gamma \equiv (dx^0, 0, \ldots, 0)$ obtaining

$$dw/dx \equiv w_{,x} = w_{,0}. \qquad\qquad (7.2.4)$$

I.e., *The derivative of w with respect to the hypercomplex variable x is given by the partial derivative with respect to the variable x^0.*

If we take $dx \equiv (0, \ldots, dx^m, \ldots, 0)$ in the direction of the coordinate axis m, we obtain the *Generalized Cauchy–Riemann conditions* (7.2.1) which are an extension of the one which holds for functions of a complex variable $w_{,y} = i\, w_{,x}$. □

The partial differential equations (7.2.1) are $N - 1$ differential conditions between hypercomplex functions; then they correspond to $N(N - 1)$ relations between real functions. Since all the partial derivatives of the N functions w^α with respect to the N variables x^α are N^2, only N *partial derivatives are independent*, i.e., the number of derivatives of a component with respect to the N variables. Then we have

Theorem 7.2. *All the partial derivatives can be expressed as functions of the partial derivatives of just one component.*

A straightforward consequence of this property is: *if we know, in a simply connected domain, a component of a holomorphic function, we can calculate, but for a constant term, the other components by means of line integrals.* This property is similar to the well-known property of functions of a complex variable, and in Section A.3.1 we see the application to quaternions.

Now let us show a relation between the partial derivatives of the components w^α and the characteristic matrix. This relation represents an alternative formulation for (GCR). In [73] it is demonstrated that this condition is necessary and sufficient for the definition of the functions of a hypercomplex variable.

Theorem 7.3. *The Jacobian matrix of the components of a hypercomplex function is the same as the characteristic matrix.*

Proof. By substituting the components in (7.2.2) we have

$$dw \equiv e_\beta \, dw^\beta \equiv e_\beta \frac{\partial w^\beta}{\partial x^\gamma} \, dx^\gamma; \qquad p \, dx \equiv e_\alpha p^\alpha e_\gamma dx^\gamma \equiv e_\beta C^\beta_{\alpha\gamma} p^\alpha \, dx^\gamma. \quad (7.2.5)$$

Equating the last terms of these equations, by means of (2.1.4), we have

$$\frac{\partial w^\beta}{\partial x^\gamma} = C^\beta_{\alpha\gamma} p^\alpha \equiv P^\beta_\gamma. \qquad (7.2.6)$$

By noting that the elements on the left-hand side are the ones of the Jacobian matrix, we obtain the demonstration of the theorem and a second formulation of (GCR) conditions. □

From this theorem follows

Theorem 7.4. *The characteristic determinant of the derivative of a hypercomplex function is the same as the Jacobian determinant.*

7.2.2 The Principal Transformation

As for complex numbers, let us consider the bijective transformation between the components of a hypercomplex number and its principal conjugations:
$x^0, x^1, \ldots, x^{(N-1)} \Leftrightarrow x, {}^1\bar{x}, \ldots {}^{(N-1)}\bar{x}$ where, as in Sections 2.1.6 and C.3.5, ${}^k\bar{x}$ indicates the kth conjugation. We call this transformation the *principal transformation*. As well as for complex variables we equate real with hypercomplex variables and this is algebraically correct since the determinant of transformation is different from zero. So we write the hypercomplex function w as a function of ${}^k\bar{x}$,

$$w = w(x, {}^1\bar{x}, \ldots, {}^{(N-1)}\bar{x}).$$

Calculating by the derivative chain rule, the partial derivatives of w with respect to the components x^μ, we have

$$w_{,\mu} = w_{,x} x_{,\mu} + \sum_{k=1}^{N-1} w_{,({}^k\bar{x})} {}^k\bar{x}_{,\mu}, \qquad (7.2.7)$$

where we have pointed out the dependence on hypercomplex variable x. Since $x_{,\mu} = e_\mu$, by comparing (7.2.7) with (7.2.1), we have $w_{,({}^k\bar{x})} \equiv 0$, and it follows that w *depends only on* x.

This result allows us to obtain (7.2.1) in the same way as for the functions of a complex variable, i.e., by requiring that $w_{k\bar{x}} = 0$ for all $^k\bar{x}$. The derivative can be formally obtained as a derivative of a function of x. Therefore we can state

Theorem 7.5. *The derivative of a function of a hypercomplex variable is a function just of x, and not of its conjugations; then it is a holomorphic function as well.*

These considerations lead us to believe that it is important, for both mathematical analysis and for applied sciences, to develop a function theory for hypercomplex variables analogous to the one for a complex variable. In this chapter we begin these studies as regards the hyperbolic variable and in Chapters 8–10, we give some geometrical and physical applications.

7.2.3   Functions of a Hypercomplex Variable as Infinite-Dimensional Lie Groups

S. Lie showed [50] that continuous groups can be related to systems of differential equations. This property can be inverted in the following way: *If the "composition" of solutions of a differential system is yet a solution, this system defines a continuous group* [5]. Two possibilities can occur;

1. solutions depend on arbitrary constants,

2. solutions depend on arbitrary functions.

In the first case we have a finite continuous group and its order is given by the number of arbitrary constants. In the second one (the system is a partial differential system) the group is an *infinite-dimensional Lie group*.

Let us consider the well-known example of complex numbers. We know that the variable z is used for representing plane vectors in Euclidean plane, i.e., in a plane with its own geometry, summarized by the invariance with respect to roto-translations group. The functions of a complex variable satisfy the partial differential Cauchy–Riemann system and so do their compositions (the functions of a function). Therefore they can be related to an infinite-dimensional Lie group (conformal group), which represents the two-dimensional harmonic field. Among these functions there is the identity function $w = z$ and we can say: *vectors and vector fields, belong to the same group; then they have the same symmetries.* Otherwise the functions of a complex variable can be considered as a function of just one variable (z) with the symmetry between components stated by the structure constants. This property characterizes all the commutative hypercomplex systems.

Actually GCR conditions (7.2.1) represent a partial differential system and the composition of solutions (functions of a function) is yet a solution of this system; then the functions of commutative hypercomplex variables define an infinite-dimensional Lie group. We have

1. every hypercomplex number can be associated with the finite group of a geometry,

2. the functions of a hypercomplex variable can be associated with the infinite-dimensional group of functional mappings.

The first group can represent vectors, the second one can be associated with physical fields. Therefore it follows

Theorem 7.6. *If vectors and vector fields can be represented by hypercomplex variables and hypercomplex functions, they have automatically the same invariant quantities. The structure constants can be considered as symmetry preserving operators not only from an algebraic point of view, but also for functions.*

Vector functions of vectors

The just exposed considerations allow us to introduce a concept which relates vectors to vector fields: *vector functions of a vector*. Now let us explain what this statement means.

Vectors are usually represented in a three-dimensional Euclidean space by the expression $\vec{r} = \vec{i}\,x + \vec{j}\,y + \vec{k}\,z$ and are determined by modulus and direction (a scalar quantity and angles). As it is known, the modulus represents the invariant quantity with respect to the Euclidean group of roto-translations.

In addition to vectors also vector fields are of physical relevance. They are given by three functions of the point (x, y, z) and are represented as $\vec{E}(x, y, z) = \vec{i}\,E_1(x, y, z) + \vec{j}\,E_2(x, y, z) + \vec{k}\,E_3(x, y, z)$. The three components are independent one from the others except that, for relevant physical fields, they must satisfy some (partial) differential equations. As a consequence of their representation in Euclidean space, we require that components are transformed as components of vectors and have the same invariant (the modulus given by a quadratic expression). We do not give any meaning to an expression like $\vec{E}(\vec{r})$. On the contrary, for representations by means of complex numbers we give a meaning to $w = f(z) \equiv f(\vec{1}\,x + \vec{i}\,y)$ and the same meaning can be given to functions of commutative hypercomplex variables. Moreover, *the functions of a hypercomplex variable can be considered as functions of just one variable*, as shown in Section 7.2.2, *with the symmetry between the components stated by the structure constants*.

7.3 The Functions of a Hyperbolic Variable

7.3.1 Cauchy–Riemann Conditions for General Two-Dimensional Systems

Now we apply the methods exposed in Section 7.2.1 to general two-dimensional systems considered in Section 2.2; we have $w = w^0 + u\,w^1$ and $x = x^0 + u\,x^1$. The other symbols have the meaning of Section 2.2. The conditions (7.2.6) become

$$
\begin{pmatrix} \dfrac{\partial w^0}{\partial x^0} & \dfrac{\partial w^0}{\partial x^1} \\[2mm] \dfrac{\partial w^1}{\partial x^0} & \dfrac{\partial w^1}{\partial x^1} \end{pmatrix} = \begin{pmatrix} z^0 & \alpha\, z^1 \\[1mm] z^1 & z^0 + \beta\, z^1 \end{pmatrix} \tag{7.3.1}
$$

and we obtain the GCR

$$
\frac{\partial w^0}{\partial x^1} = \alpha\, \frac{\partial w^1}{\partial x^0}; \quad \frac{\partial w^0}{\partial x^0} = \frac{\partial w^1}{\partial x^1} - \beta\, \frac{\partial w^1}{\partial x^0}. \tag{7.3.2}
$$

The same conditions are obtained from (7.2.1)

$$
\frac{\partial w^0}{\partial x^1} + u\, \frac{\partial w^1}{\partial x^1} = u\left(\frac{\partial w^0}{\partial x^0} + u\, \frac{\partial w^1}{\partial x^0} \right) \equiv \alpha\, \frac{\partial w^1}{\partial x^0} + u\left(\frac{\partial w^0}{\partial x^0} + \beta\, \frac{\partial w^1}{\partial x^0} \right).
$$

By equating the coefficients of the versors we obtain (7.3.2).

7.3.2 The Derivative of Functions of a Canonical Hyperbolic Variable

Let us consider in a domain D of the (x, y) plane two real differentiable functions $u \equiv u(x, y)$, $v \equiv v(x, y)$ of the real variables x, y. We associate with these functions a function of a canonical hyperbolic variable $w \equiv f(z) \equiv f(x + h\,y) = u + h\,v$. We say that the function $w = u + h\,v$ is a function of z if the GCR conditions are satisfied. Equations (7.3.2) become

$$
u_{,x} = v_{,y}; \qquad u_{,y} = v_{,x}. \tag{7.3.3}
$$

These functions are called hyperbolic analytic or more simply h-analytic [47], and the derivative is given by

$$
\frac{d\,w}{d\,z} \equiv f'(z) = u_x + h\, v_x \equiv u_x + h\, u_y. \tag{7.3.4}
$$

Differentiating (7.3.3) with respect to x (or y) and equating the mixed partial derivatives $v_{,xy} = v_{,yx}$ we see that $u(x, y)$ (and $v(x, y)$) satisfy the wave equation. This result is analogous to that obtained for functions of a complex variable with respect to the Laplace equation. In particular the system (7.3.3) represents the canonical form of hyperbolic partial differential equations. Moreover we have

Theorem 7.7. *The transformations with functions of a hyperbolic variable leave the wave equation unchanged.*

Proof. It can be demonstrated with the procedure followed for the Laplace equation with respect to the conformal group.   □

As well as for functions of a complex variable it can be easily shown [47] that rational functions are h-analytic with the additional condition that the denominator is not a zero-divisor. This last condition represents a peculiar characteristic of hyperbolic functions and of a representative hyperbolic plane. Actually, if we approach z_0 on an arbitrary line, the tangent to this line must not be parallel to axes bisectors. In particular, if we call τ a real parameter and consider the line

$$C \equiv x(\tau) + h\,y(\tau), \tag{7.3.5}$$

this line, in addition to being " a Jordan line", must satisfy this condition which also means that in the hyperbolic plane we can consider just some closed cycles.

7.3.3 The Properties of H-Analytic Functions

Equations (7.3.3), as functions of decomposed variables ξ, η (Section 2.1.7), become

$$u_{,\xi} - v_{,\xi} \equiv (u - v)_{,\xi} \equiv \psi_\xi = 0; \quad u_{,\eta} + v_{,\eta} \equiv (u + v)_{,\eta} \equiv \phi_{,\eta} = 0, \tag{7.3.6}$$

where we have put , $\psi = u - v$ and $\phi = u + v$. Then ψ is a function of just η and ϕ of ξ and can be put as arbitrary functions of $\xi = x + y$, $\eta = x - y$, respectively. As a consequence the general expression of the functions of hyperbolic variables can be expressed as the linear combination [47]

$$u = \frac{1}{2}[\phi(x + y) + \psi(x - y)], \quad v = \frac{1}{2}[\phi(x + y) - \psi(x - y)]. \tag{7.3.7}$$

If we give to x the physical meaning (see Section 4.1.2 at p. 30) of a normalized time variable, (7.3.7) show that ϕ, ψ represent the onward and backward solutions of the wave equation and u, v are a linear combination of them (Section C.7). Vice versa we can say that the real decomposition of a hyperbolic system is equivalent to D'Alembert transformation of the wave equation [72].

7.3.4 The Analytic Functions of Decomposable Systems

From (2.2.22) and (2.2.23) it follows that the positive and negative powers of decomposable systems are obtained as the powers of the decomposed systems [43]. Then by calling e_1, e_2 the versors of an idempotent basis, ξ, η the hyperbolic decomposed variables (Section 2.1.7), we have: *The elementary functions, expressed by power series of $\zeta = e_1\,\xi + e_2\,\eta$ with coefficients $a_k \in \mathbf{R}$, are given by*

$$f(\zeta) = \sum_k a_k\,\zeta^k \equiv \sum_k a_k\,(e_1\,\xi + e_2\,\eta)^k = e_1\,f(\xi) + e_2\,f(\eta). \tag{7.3.8}$$

Then the elementary functions of these systems are an analytic continuation of the functions of a real variable. From (2.2.20) it follows: *Analytic functions, defined by series with real coefficients, of conjugate variables are conjugate analytic functions*

$$\overline{f(z)} = f(\bar{z}). \tag{7.3.9}$$

In particular, following the same procedure shown in ([60], Section 17) for the bicomplex functions and in Appendix C for the commutative hypercomplex systems, it can be demonstrated that analytic continuations of circular, exponential and logarithm functions have the same properties as the corresponding real and complex functions. Then the decomposability of hyperbolic systems allows us to set up their functions

- in the same way of functions of a complex variable [47],

- or by means of (7.3.8) which represents an analytic continuation of real functions of a real variable.

7.4   The Elementary Functions of a Canonical Hyperbolic Variable

Exponential Function

This function has been already introduced in Section 4.1.1 where it was the starting point for introducing hyperbolic trigonometry.

$$\exp[z] \equiv \exp[x + h\,y] = \exp[x]\,(\cosh y + h\,\sinh y). \qquad (7.4.1)$$

This function can be extended as is shown in Section 4.2.2. We can easily verify that "real" and "hyperbolic" components of (7.4.1) satisfy (7.3.3).

Logarithm Function

It is defined as the inverse of the exponential function [47]

$$\text{for } |x| > |y| \Rightarrow \ln z \equiv \ln[x + h\,y] = \frac{1}{2}\ln[x^2 - y^2] + h\tanh^{-1}\frac{y}{x}, \qquad (7.4.2)$$

$$\text{for } |x| < |y| \Rightarrow \ln z \equiv \ln[x + h\,y] = \frac{1}{2}\ln[y^2 - x^2] + h\tanh^{-1}\frac{x}{y}.$$

Hyperbolic Trigonometric Function

Setting in (7.4.1) $x = 0$, we have

$$\exp[h\,y] = \cosh y + h\,\sinh y\,; \qquad \exp[-h\,y] = \cosh y - h\,\sinh y$$

and

$$\cosh y = \frac{\exp[h\,y] + \exp[-h\,y]}{2}\,; \qquad \sinh y = \frac{\exp[h\,y] - \exp[-h\,y]}{2\,h}. \qquad (7.4.3)$$

As well as for functions of a complex variable, the substitution $y \Rightarrow z$ allows us to define $\cosh z$, $\sinh z$. Comparing the series development of (7.4.3) with the ones of real functions $\cosh y$, $\sinh y$ we obtain

$$\cosh[h\,y] = \cosh[y],\ \sinh[h\,y] = h\,\sinh[y]. \qquad (7.4.4)$$

Then from (4.4.2), (4.4.3) and (7.4.4) we get

$$\cosh z \equiv \cosh[x + h\,y] = \cosh x \cosh y + h\,\sinh x \sinh y,$$
$$\sinh z \equiv \sinh[x + h\,y] = \sinh x \cosh y + h\,\cosh x \sinh y. \tag{7.4.5}$$

Circular Trigonometric Functions

In complex analysis the circular trigonometric functions are defined by a continuation into the complex field by means of Euler's formula, exactly as we did in the previous section when we defined the hyperbolic trigonometric functions. From this point of view circular trigonometric functions seem not to be in agreement with the properties of functions of a hyperbolic variable. Nevertheless we know that the circular trigonometric functions are usually considered for describing the solution of the wave equation, and thus their introduction appears important. They can be introduced by means of (7.3.8) or, comparing absolutely convergent series, we extend (7.4.4) to circular trigonometric functions and apply the usual properties of the functions of the sum of angles. Following ([25], p. 72) we define

$$\cos z \equiv \cos[x + h\,y] = \cos x \cos y - h\,\sin x \sin y,$$
$$\sin z \equiv \sin[x + h\,y] = \sin x \cos y + h\,\cos x \sin y. \tag{7.4.6}$$

We note that in the complex field the components of hyperbolic and circular trigonometric functions are given by products of both circular and hyperbolic functions and they can be obtained one from the others. Now from (7.4.5) and (7.4.6) we see that circular and hyperbolic trigonometric functions hold the same properties that they have in the real field.

Derivatives

It is easy to verify that the functions just defined satisfy conditions (7.3.3) and their derivatives can be obtained by the same differentiation formulas which hold for real and complex variables. As an example, we have

$$\frac{d\,\exp[z]}{d\,z} = \exp[z], \quad \frac{d\,\ln z}{d\,z} = \frac{1}{z}.$$

7.5 H-Conformal Mappings

As well as for functions of a complex variable we call [47] *h-conformal mappings* the functional bijective mappings x, $y \Leftrightarrow u$, v with u, v satisfying conditions (7.3.3). The first property must be that the Jacobian determinant (J) of the mapping

$$J(x,\ y) \equiv \left\| \frac{\partial(u,\ v)}{\partial(x,\ y)} \right\| \equiv \left(\frac{\partial u}{\partial x} \right)^2 - \left(\frac{\partial v}{\partial x} \right)^2 \equiv |f'(z)|^2, \tag{7.5.1}$$

must be $\neq 0$.

Therefore, since analytic functions are defined if $f'(z) \neq 0$ and is not a zero divisor, we have

Theorem 7.8. *For a domain in which $f(z)$ is differentiable the mapping is bijective.*

Writing (7.3.3) in decomposed form (7.3.7), the Jacobian determinant becomes

$$J(x,\, y) \equiv \left\| \frac{\partial(\phi,\, \psi)}{\partial(\xi,\, \eta)} \right\| \equiv \left(\frac{d\phi}{d\xi} \right) \left(\frac{d\psi}{d\eta} \right) \equiv \phi'(x+y)\, \psi'(x-y), \qquad (7.5.2)$$

which is zero when the derivatives with respect to ξ or η are zero, i.e., on the straight lines parallel to axes bisectors $x \pm y = const$. These straight lines are called *characteristic lines* of the differential system (7.3.3). Unlike for conformal mappings of a complex variable for which J is never < 0, for hyperbolic mappings it changes sign when crossing the characteristic lines which represent [47] *ramification lines*. From (7.3.7), it follows that the characteristic lines of system (7.3.3) are mapped into the new characteristic lines

$$u + v = \phi(x+y); \;\; u - v = \psi(x-y), \qquad (7.5.3)$$

i.e., for $x \pm y = const$, it is $u \pm v = const$. From this we have [47]

Theorem 7.9. *In an h-conformal mapping, lines parallel to characteristic lines are transformed into lines parallel to characteristic lines: the topologies of starting and transformed planes are the same.*

In other words both planes are divided into the four sectors which, in Section 4.1, we have called Rs, Us, Ls, Ds. Since a Jacobian determinant cannot change sign in bijective mappings, the borderlines of starting and transformed domains must be on the same side with respect to characteristic lines.

As an example, this implies that we cannot map the internal points of a circle into a half-plane. Vice versa we shall see that we can map the points of an equilateral hyperbola on the x axis of the transformed plane (Theorem 7.20).

Now we see that if we represent hyperbolic functions in a plane in which the geometry is the one generated by hyperbolic numbers (Section 3.2), in every point z_0 in which $f(z_0)$ is holomorphic we have

Theorem 7.10. *The hyperbolic conformal mappings have exactly the same properties as the conformal mappings of a complex variable, in particular:*

1. *The stretching is constant in hyperbolic geometry and is given by $\|f'(z)\|$ which is equal to the Jacobian determinant of the mapping.*

2. *The hyperbolic angles between any two curves passing through z_0, are preserved.*

Proof. 1. From the definition of the derivative we have

$$\lim_{\substack{\Delta z \to 0 \\ |\Delta z| \neq 0}} \frac{\Delta w}{\Delta z} \equiv \frac{dw}{dz} = f'(z) \tag{7.5.4}$$

and

$$\frac{d\overline{w}}{d\overline{z}} = \overline{f'(z)}. \tag{7.5.5}$$

If we multiply the sides of (7.5.4) by the corresponding ones of (7.5.5) we have

$$\|dw\| = \|f'(z)\| \cdot \|dz\|, \tag{7.5.6}$$

then in hyperbolic geometry (distance defined as $\|z\|$) the stretching does not depend on direction, moreover $\|f'(z)\|$ is the characteristic determinant of the derivative of $f(z)$, then it is equal to the Jacobian determinant of $f(z)$.

2. Let us consider, in the representative hyperbolic plane, a regular curve $\lambda(\tau)$ given by (7.3.5), which passes through the point $z_0 \equiv z(\tau_0)$. This curve is mapped into a curve Λ of the w plane passing through the point $w_0 \equiv f(z_0) = f[z(\tau_0)]$. For the derivative chain rule we have

$$w_0' = f'(z_0) \, z'(\tau_0); \tag{7.5.7}$$

since differentiability requires that $f'(z_0) \neq 0$ and for regular curves $z'(\tau_0) \neq 0$, we can put all terms in (7.5.7) in exponential form and by calling ϑ, ϕ, γ the arguments of w', $f'(z)$, $z'(\tau)$, respectively, we have

$$\exp[\vartheta] = \exp[\phi] \exp[\gamma] \quad \Rightarrow \quad \vartheta - \gamma = \phi. \tag{7.5.8}$$

Then *the mapping $w = f(z)$ rotates all curves passing through the point z_0 of the same hyperbolic angle equal to the argument of the derivative at the point z_0.* $\quad\square$

7.5.1 H-Conformal Mappings by Means of Elementary Functions

Exponential and Hyperbolic Trigonometric Functions

Exponential function $w = \exp[x + hy]$. Let us consider the mapping between the planes $(x, y) \Leftrightarrow (u, v)$ by means of the exponential function (7.4.1)

$$u + hv = \exp[x + hy] \equiv \exp[x](\cosh y + h \sinh y).$$

- Whereas the variables x, y assume all the values in their plane, the functions u, v assume just the values $\in Rs$.

- The zero divisors of the (x, y) plane (axes bisectors $x = \pm y$) are mapped into the half-lines $u \pm v = 1$ for $\{u \pm v = 1\} \in Rs$.

- Since $J = \exp[2\,x] > 0$, sectors $(Rs,\ Ls)$ of the $(x,\ y)$ plane are mapped into the same sectors of the $(u,\ v)$ plane starting from the crossing point of lines $u + v = 1$; $u - v = 1$, i.e., in point $u = 1$, $v = 0$.

- The border-lines of sector (Ls) of the $(u,\ v)$ plane are given by the square inside the lines $u \pm v = 1$ and the axes bisectors.

- Lines $\mathcal{H}(z) : \{y = y_0\}$ are mapped into the half-lines $u = v \cdot \coth y_0$; $u > 0$.

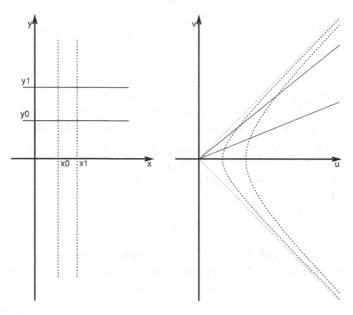

Figure 7.1: *Mapping of a rectangle in the (x,y) plane by means of function $u+h\,v = \exp[x+h\,y]$. The lines $\mathcal{H}z : \{y = y_0\}$ are mapped into the half-lines $u = v \cdot \coth y_0$ $u > 0$. The lines $\Re z : \{x = x_0,\ y = \tau; \tau \in \mathbf{R}\}$ are mapped into the arm $(\in Rs)$ of equilateral hyperbolas $u+h\,v = \exp[x_0+h\,\tau]$. So the rectangle in the $(x,\ y)$ plane, $c_1 < x < c_2$; $b_1 < y < b_2$, is mapped into the domain of the $(u,\ v)$ plane between the equilateral hyperbolas $\exp[c_1+h\,\tau] < u+h\,v < \exp[c_2+h\,\tau]$ and the half-lines $u = v \cdot \coth b_1$; $u = v \cdot \coth b_2$; $u > 0$. We note, once again, that this mapping is equivalent to a sector of an annulus obtained in the complex plane, mapping a rectangle by the complex exponential.*

- Lines $\Re(z) : \{x = x_0,\ y = \tau; \tau \in \mathbf{R}\}$ are mapped into the arm $(\in Rs)$ of equilateral hyperbolas $u + h\,v = \exp[x_0 + h\,\tau]$.

- The rectangle in the $(x,\ y)$ plane $c_1 < x < c_2$; $b_1 < y < b_2$ is mapped into the domain of the $(u,\ v)$ plane between the equilateral hyperbolas $\exp[c_1+h\,\tau] < u + h\,v < \exp[c_2+h\,\tau]$ and the half-lines $u = v \cdot \coth b_1$; $u = v \cdot \coth b_2$ (Fig. 7.1).

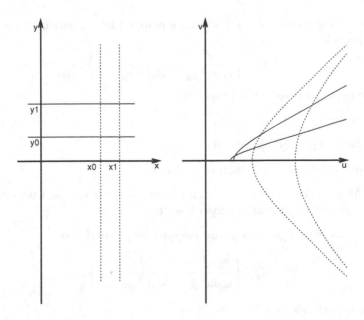

Figure 7.2: *Mapping of a rectangle in the (x, y) plane by means of the function* $u + h v = \cosh[x + hy]$. Lines $\in Rs$ $x = const \equiv x_0$, $y < x_0$ are mapped into a part of the arm $\in Rs$ of hyperbolas $u^2/\cosh^2 x_0 - v^2/\sinh^2 x_0 = 1$ with vertex on the u axis in point $u_V = \cosh x_0$, focuses in points $u_F \equiv (\pm \cosh 2x_0, 0)$ and asymptotes $u = \pm v \cdot \coth x_0$. As x_0 increases the vertex goes far from the axes origin, asymptotes go near the axes bisectors. Lines $y = const \equiv y_0$, $x > y_0$ are mapped into hyperbolas $u^2/\cosh^2 y_0 - v^2/\sinh^2 y_0 = 1$ with vertex on the u axis in point $u_V = \cosh y_0$, focuses in points $u_F \equiv (\pm \cosh 2y_0, 0)$ and asymptotes $u = \pm v \cdot \coth y_0$. As y_0 increases the vertex goes far from the axes origin, asymptotes go near the axes bisectors. For the bijective mappings it must be $x > y$.

Hyperbolic cosine $w = \cosh[x + h y]$. We have

$$u = \cosh x \cosh y, \quad v = \sinh x \sinh y. \qquad (7.5.9)$$

- In decomposed variables we have $\phi = \cosh \xi, ;$ $\psi = \cosh \eta \Rightarrow \phi, \psi \geq 1$ and $J = \sinh \xi \sinh \eta$ changes its sign on the axes bisectors $x + y = 0$ and $x - y = 0$ which are mapped into the lines $u + v = 1$ and $u - v = 1$, respectively.

- All points of the (ξ, η) plane are mapped into points of the first quadrant for $\phi, \psi \geq 1$.

- In the (x, y) plane, points in all sectors are mapped into points $\in Rs$ with $|u| > 1$. This is evident from (7.5.9) since u, v have the same value by changing $x \leftrightarrow y$ and $x, y \leftrightarrow -x, -y$.

- Lines $\in Rs$ $x = const \equiv x_0$, $y < x_0$ are mapped into a part of the arm $\in Rs$ of hyperbolas

$$\mathcal{I}_x : \left\{ \frac{u^2}{\cosh^2 x_0} - \frac{v^2}{\sinh^2 x_0} = 1 \right\}.$$

These hyperbolas have the properties

- *vertex* $V_{\mathcal{I}_x} \equiv (\cosh x_0, 0)$,

- *focus* $F_{\mathcal{I}_x} \equiv (\cosh 2 x_0, 0)$,

- *asymptotes* $u = \pm v \cdot \coth x_0 \in (Ls, Rs)$.

- As x_0 increases the vertex and focus go far from the axes origin, the asymptotes go near the axes bisectors.

- Lines $y = const \equiv y_0$, $x > y_0$ are mapped into hyperbolas

$$\mathcal{I}_y : \left\{ \frac{u^2}{\cosh^2 y_0} - \frac{v^2}{\sinh^2 y_0} = 1 \right\}.$$

These hyperbolas have

- *vertex* $V_{\mathcal{I}_y} \equiv (\cosh y_0, 0)$,

- *focus* $F_{\mathcal{I}_y} \equiv (\cosh 2 y_0, 0)$,

- *asymptotes* $u = \pm v \cdot \coth y_0 \in (Ls, Rs)$.

- As y_0 increases the vertex goes far from the axes origin, asymptotes go near the axes bisectors (Fig. 7.2).

- Now we verify that all hyperbolas are $\in Rs$ of the (u, v) plane inside the lines $u \pm v = 1$.

Proof. If we look for the intersections of these lines with hyperbolas, we have $u \pm v \equiv \cosh x \cosh y \pm \sinh x \sinh y \equiv \cosh[x \pm y] = 1$; since this condition is satisfied just for zero divisors, it never happens.   □

- The hyperbolas $\mathcal{I}_x : \{x = x_0\}$ and $\mathcal{I}_y : \{y = y_0\}$ are the transformed curves of two orthogonal lines; therefore for the second property of h-conformal mapping (Section 7.5), they must cross orthogonally. Now we verify this property.

Proof. In the Euclidean plane the cosine of the angle between two crossing curves is proportional to the scalar product between the gradients to the curves. This property also holds for curves in the hyperbolic plane by the formal changes of cos ⇒ cosh and Euclidean scalar product ⇒ hyperbolic scalar product (see p. 33). Since the components of the versors of the gradients in the crossing point $P_{\mathcal{I}} \equiv (\cosh x_0 \cosh y_0, \ \sinh x_0 \sinh y_0)$ are proportional to

$$\text{for } \mathcal{I}_x : \left\{ \frac{\cosh x_0}{\cosh y_0}, \ \frac{\sinh x_0}{\sinh y_0} \right\}, \quad \text{for } \mathcal{I}_y : \left\{ \frac{\cosh y_0}{\cosh x_0}, \ \frac{\sinh y_0}{\sinh x_0} \right\},$$

the hyperbolic scalar product is zero.   □

Hyperbolic sine $w = \sinh[x + h\,y]$. Unlike the previously considered functions, $\sinh[x + h\,y]$ maps the complete (x, y) plane into the complete (u, v) plane. Actually functions $\phi = \sinh \xi$, $\psi = \sinh \eta$ are an increasing bijective mapping $\xi, \eta \leftrightarrow \phi, \psi$. Coming back to x, y variables, we have

$$u = \sinh x \cosh y, \quad v = \cosh x \sinh y. \tag{7.5.10}$$

$J = \cosh(x + y) \cosh(x - y) > 0$, i.e., the whole (x, y) plane can be mapped into the whole (u, v) plane.

- Lines $x \pm y = 0$ are mapped into lines $u \pm v = 0$.

- Lines $x = const \equiv x_0$ for $x_0 > 0$ are mapped into the arm $\in Rs$ and for $x_0 < 0$ in the arm $\in Ls$ of hyperbolas

$$\mathcal{I}_x : \left\{ \frac{u^2}{\sinh^2 x_0} - \frac{v^2}{\cosh^2 x_0} = 1 \right\}.$$

These hyperbolas have

  - *vertex* $V_{\mathcal{I}_x} \equiv (\pm \sinh x_0, \ 0)$,

  - *asymptotes* $u = \pm v \cdot \tanh x_0 \in (Us, \ Ds)$.

  - As x_0 increases the vertex and focus go far from the axes origin, asymptotes go near the axes bisectors.

- Lines $y = const \equiv y_0$ are mapped, for $y_0 > 0$ into the arm $\in Us$, and for $y_0 < 0$ into the arm $\in Ds$ of hyperbolas

$$\mathcal{I}_y : \left\{ \frac{v^2}{\sinh^2 y_0} - \frac{u^2}{\cosh^2 y_0} = 1 \right\}.$$

These hyperbolas have

  - *vertex* $V_{\mathcal{I}_y} \equiv (0, \ \pm \sinh y_0)$,

  - *asymptotes* $u = \pm v \cdot \coth y_0 \in (Ls, \ Rs)$.

  - As y_0 increases the vertex goes far from the axes origin, the asymptotes go near the axes bisectors.

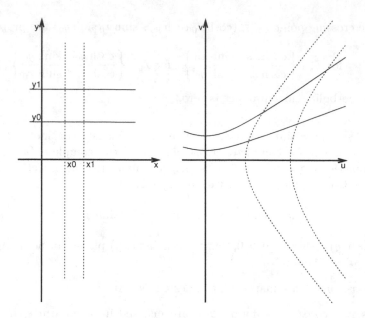

Figure 7.3: *Mapping of a rectangle in the (x, y) plane by means of the function $u + h\,v = \sinh[x + h\,y]$.* Lines $x = const \equiv x_0$ are mapped for $x_0 > 0$ into the arm $\in Rs$ and for $x_0 < 0$ in the arm $\in Ls$ of hyperbolas $(u^2/\sinh^2 x_0) - (v^2/\cosh^2 x_0) = 1$. The vertexes of these hyperbolas are on the u axis at the point $u_V = \pm \sinh x_0$. The asymptotes $u = \pm v \cdot \tanh x_0$ are in sectors Us, Ds. As x_0 increases the vertex goes far from the axes origin, the asymptotes go near the axes bisectors. The lines $y = const \equiv y_0$ are mapped for $y_0 > 0$ into the arm $\in Us$ and for $y_0 < 0$ in the arm $\in Ds$ of hyperbolas $(v^2/\sinh^2 y_0) - (u^2/\cosh^2 y_0) = 1$. These hyperbolas have the vertexes on v axis, in points $v_V = \pm \sinh y_0$ and asymptotes $u = \pm v \cdot \coth y_0$.

• The rectangle of the (x, y) plane $\pm c_1 < x < \pm c_2$; $\pm b_1 < y < \pm b_2$ is mapped into the domain of the (u, v) plane, between the hyperbolas (see Fig. 7.3)

$Ls - Rs$

$\Rightarrow \pm \sinh c_1 \cosh \tau + h \, \cosh c_1 \sinh \tau < u + h\,v$

$\qquad\qquad\qquad\qquad < \pm \sinh c_2 \cosh \tau + h \, \cosh c_2 \sinh \tau;$

$Us - Ds$

$\Rightarrow \pm \sinh b_1 \cosh \tau + h \, \cosh b_1 \sinh \tau < v + h\,u$

$\qquad\qquad\qquad\qquad < \pm \sinh b_2 \cosh \tau + h \, \cosh b_2 \sinh \tau.$

A short summary

Let us summarize the elements of the three just exposed mappings which can be considered peculiar to the hyperbolic plane.

We have seen that for both $\exp[z]$ and $\sinh[z]$, J is always > 0. This property means that the whole z plane can be bijectively mapped into the w plane with the following difference: $\sinh[z]$ is mapped into the whole w plane, $\exp[z] \in Rs$.

In particular, let us consider $\exp[z]$ in the (ψ, ϕ) plane; we have:

- the $\eta = 0$ axis is mapped into the line $\psi = 1$,

- the $\xi = 0$ axis is mapped into the line $\phi = 1$;

then, numbering anticlockwisely the quadrants in the Cartesian representation of the decomposed variables, we have:

- I Q. of the (ξ, η) plane is mapped into the sector $\psi > 1$, $\phi > 1$,

- II Q. of the (ξ, η) plane is mapped into the band $0 < \psi < 1$, $\phi > 1$,

- III Q. of the (ξ, η) plane is mapped into the square $0 < \psi < 1, 0 < \phi < 1$,

- IV Q. of the (ξ, η) is mapped into the band $0 < \phi < 1$, $\psi > 1$.

Now let us consider $\cosh[z]$. We have $J = 0$ on axes bisectors of the (x, y) plane. Therefore no domain which crosses the axes bisector can be bijectively mapped. Actually sectors (Us) and (Rs) of the (x, y) plane are mapped into the (Rs) sector of the (u, v) plane. Further since $\cosh$ is an even function, the sectors (Ls) and (Ds) too, are mapped into (Rs) of the (u, v) plane. Moreover characteristic lines of the (x, y) plane are mapped into lines $u \pm v = 1$, the sector (Rs) of the (u, v) plane is $|u| > |v|$, $u > 1$.

Logarithm Function

We have $J = 1/(z\tilde{z}) \equiv 1/(x^2 - y^2)$, then the axes bisectors are the characteristic lines. The mapping can be bijective just for domains inside one of the four sectors.

Circular Trigonometric Functions

Sine $w = \sin[x + h\,y]$. We have

$$u = \sin x \cos y \; ; \qquad v = \cos x \sin y \qquad (7.5.11)$$

and $u + v = \sin(x + y)$, $u - v = \sin(x - y)$, then $|u + v| < 1$, $|u - v| < 1$. $J = \cos(x + y)\cos(x - y)$ so $J = 0$ for $x \pm y = \pi/2$. These lines limit the domain for bijective mapping to points $x + y < \pi/2$ and $x - y < \pi/2$. The domain of the (x, y) plane, given by $|x + y| < \pi/2$, $|x - y| < \pi/2$, is the square with vertexes at the points $P_{1,3} \equiv (0, \pm\pi/2)$ e $P_{2,4} \equiv (\pm\pi/2, 0)$. This square is mapped into the square of the (u, v) plane with vertexes at $P'_{1,3} \equiv (0, \pm 1)$ and $P'_{2,4} \equiv (\pm 1, 0)$ (see. Fig. 7.4).

The zero divisors $x = \pm y$ are mapped into $u = \pm v$.

Segments $x = const = \gamma_1 < \pi/2$, $|y| < \pi/2 - \gamma_1$, are mapped into ellipse arcs[3] $u = \sin\gamma_1 \cos y$, $v = \sqrt{1 - \sin^2\gamma_1} \sin y$. In the same way, segments $y = const = \gamma_2 < \pi/2$, $|x| < \pi/2 - \gamma_2$, are mapped into ellipse arcs.

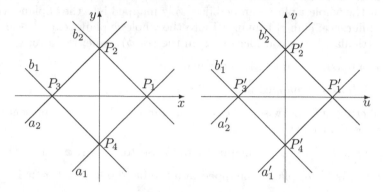

Figure 7.4: *The square domain h-conformally mapped by the function $u + hv = \sin[x + hy]$. Since $u + v = \sin(x + y)$, $u - v = \sin(x - y)$, then $|u + v| < 1$, $|u - v| < 1$. $J = \cos(x + y)\cos(x - y)$ so $J = 0$ for $x \pm y = \pi/2$. These lines limit the domain for bijective mappings to points $x + y < \pi/2$ and $x - y < \pi/2$. The domain of the (x, y) plane, given by $|x + y| < \pi/2$, $|x - y| < \pi/2$, is the square with vertexes at the points $P_{1,3} \equiv (0, \pm\pi/2)$ and $P_{2,4} \equiv (\pm\pi/2, 0)$. This square is mapped into the square of the (u, v) plane with vertexes at the points $P'_{1,3} \equiv (0, \pm 1)$ and $P'_{2,4} \equiv (\pm 1, 0)$. The zero divisors $x = \pm y$ are mapped into $u = \pm v$.*

Cosine $w = \cos[x + hy]$.   We have

$$u = \cos x \cos y ; \qquad\qquad v = -\sin x \sin y. \qquad (7.5.12)$$

The zero divisors $x \pm y = 0$ are mapped into the lines $u \pm v = 1$. We have $J = \sin(x + y)\sin(x - y)$, then $J = 0$ for $x = \pm y$; it changes its sign on the axes bisectors which represent the limit for bijective mappings of domains. We can also see, from (7.5.12), that points symmetric with respect to axes bisectors $P_1 \equiv (x_1, y_1)$ and $P_2 \equiv (y_1, x_1)$ are mapped into the same point of the (u, v) plane. Then we can have bijective mappings for points in the triangle $|x| > |y|$, $0 < x < \pi/2$, $-\pi/2 < y < \pi/2$. The triangle with vertexes at the origin $P_1 \equiv (0, 0)$ and at $P_{2,3} \equiv (\pi/2, \pm\pi/2)$ on axes bisectors are bijectively mapped into points of the triangle with vertexes on the axes at $P'_1 \equiv (1, 0)$ and $P'_{2,3} \equiv (0, \mp 1)$ of the (u, v) plane (see Fig. 7.5).

---

[3] Circles for $\gamma_1 = \pi/4$, segments of v axes between $-1 < v < 1$ for $\gamma_1 = 0$.

We also have $u + v = \cos(x + y)$, $u - v = \cos(x - y)$, then $|u + v| < 1$, $|u - v| < 1$.

Segments $x = const = \gamma_1 < \pi$, $y < \gamma_1$ are mapped into ellipse arcs[4] $u = \cos \gamma_1 \cos y$, $v = -\sqrt{1 - \cos^2 \gamma_1} \sin y$.

Segments $|y| = const = \gamma_2 < \pi$, $x > \gamma_2$ are again mapped into ellipse arcs.

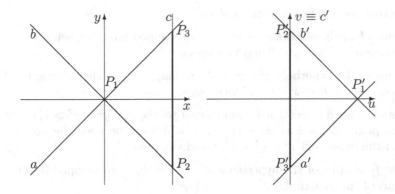

Figure 7.5: *The triangular domain h-conformally mapped by the function* $u + h\,v = \cos[x + h\,y]$. *We have bijective mappings for points in the triangle* $|x| > |y|$, $0 < x < \pi/2$, $-\pi/2 < y < \pi/2$. *The triangle with vertexes at the origin* $P_1 \equiv (0, 0)$ *and at* $P_{2,3} \equiv (\pi/2, \pm\pi/2)$ *of* (x, y) *plane is bijectively mapped into the triangle with vertexes at* $P_1' \equiv (1, 0)$ *and* $P_{2,3}' \equiv (0, \mp1)$ *of the* (u, v) *plane.*

The Function w=1/z

We have $J = 1/(x^2 - y^2)^2 > 0$, then the whole hyperbolic z plane can be bijectively mapped and, from the symmetry of the transformation, it goes into the whole w plane.

In decomposed variables we have

$$\frac{1}{\zeta} = \frac{1}{e_0\,\xi + e_1\,\eta} \equiv e_0\frac{1}{\xi} + e_1\frac{1}{\eta}, \tag{7.5.13}$$

then the mapping is equivalent to two inversions with respect to the unit points on the axes of decomposed variables. The point $P(\xi, \eta)$ is mapped into the point $P_T(x_T^0 \equiv \frac{1}{\xi}, x_T^1 \equiv \frac{1}{\eta})$ and, coming back to x, y variables, we have

$$x_T = \frac{x}{x^2 - y^2}, \qquad y_T = -\frac{y}{x^2 - y^2}. \tag{7.5.14}$$

---

[4]In particular, we have circles for $\gamma_1 = \pi/4$, segments of v axes between $-1 < v < 1$ for $\gamma_1 = \pi/2$.

If we put the variables in exponential form $x + hy = \rho_h \exp_e[h\,\theta]$, we find in the hyperbolic plane the well-known results for complex variables,

$$(\rho_h)_T = \frac{1}{\rho_h}, \qquad \theta_T = -\theta. \tag{7.5.15}$$

Now we look for the geometrical meaning of (7.5.14) and (7.5.15).

1. Sectors are mapped into themselves.

2. Arms of the hyperbola $x^2 - y^2 = 1$ are mapped into themselves, but for the inversion $y > 0 \Rightarrow y < 0$ and vice versa.

3. Arms of the hyperbola $y^2 - x^2 = 1$ are mapped into themselves, but for the inversion $x > 0 \Rightarrow x < 0$ and vice versa.

4. Points defined in Section 4.5 as *external points*, $x^2 - y^2 \geq p^2 > 1$, are mapped into points defined as *internal points*, which are between the axes bisectors and the hyperbola $x^2 - y^2 = 1/p^2$ and vice versa.

5. The four arms of the hyperbolas $x^2 - y^2 = \pm p^2$ are mapped into the same arms of the hyperbolas $x^2 - y^2 = \pm 1/p^2$.

Points symmetric with respect to equilateral hyperbolas.   Two points P, P^* in Euclidean geometry are symmetric with respect to a straight line if the segment between them is orthogonal to the straight line and their distance from the straight line is the same. A similar definition is extended to symmetry with respect to a circle: two points P, P^*, in Euclidean geometry, are symmetric with respect to a circle with center at $O \equiv (0,\,0)$ and radius R if the three points O, P, P^*, lie on a single ray which start from point O and if $\overline{OP} \times \overline{OP}^* = R^2$.

The complex function $w = 1/z$ is related with this symmetry ([69], p. 284). Actually, if z represents the complex coordinate of P, the coordinate of P^* is $R^2/\bar{z}$.

We obtain the same result in a hyperbolic plane by considering the symmetry with respect to equilateral hyperbolas.

Let us consider the equilateral hyperbolas with center in $O \equiv (0,\,0)$:

$$\Gamma : [x^2 - y^2 = p^2]. \tag{7.5.16}$$

We say that *points* $P \equiv (z)$, $P^* \equiv (z^*)$ *are symmetric with respect to* Γ, *if the three points* O, P, P^*, *lie on a single ray and* $\overline{OP} \times \overline{OP}^* = p^2$. The same as in complex analysis, from (7.5.15) we obtain $z^* = p^2/\bar{z}$. For symmetric points the following theorem holds.

Theorem 7.11. *By considering points* $P \equiv (z)$ *and* $P^* \equiv 1/(\bar{z})$ *symmetric with respect to the hyperbolas* (7.5.16), *all equilateral hyperbolas passing through* P *and* P^* *are pseudo-orthogonal to* Γ.

Proof. Thanks to Theorem 4.20 on p. 48, the proof of this theorem is the same as for functions of a complex variable ([69], p. 285). $\square$

7.5.2 Hyperbolic Linear-Fractional Mapping

In the complex field the following functions are also called *circular (or Möbius), bilinear or homographic mappings*. We extend to hyperbolic variables the well-known studies about complex variables ([69], p. 282 and [48], p. 128).

As for complex variables, we call *hyperbolic linear-fractional or bilinear*, the mappings

$$w = \frac{\alpha z + \beta}{\gamma z + \delta} \equiv \frac{\alpha}{\gamma} - \frac{\alpha\delta - \beta\gamma}{\gamma^2(z + \delta/\gamma)} \Rightarrow w - \frac{\alpha}{\gamma} = \frac{\beta\gamma - \alpha\delta}{\gamma^2(z + \delta/\gamma)} \equiv \frac{B}{z + \delta/\gamma} \quad (7.5.17)$$

where w, z are hyperbolic variables and α, β, γ, δ, $B = (\beta\gamma - \alpha\delta)/\gamma^2$ hyperbolic constants which must satisfy the conditions $\gamma \neq 0$; $\|\alpha\delta - \beta\gamma\| \neq 0$. These conditions imply that w is not a constant and $J \neq 0$. We have

$$J = \frac{(\alpha\delta - \beta\gamma)(\tilde{\alpha}\tilde{\delta} - \tilde{\beta}\tilde{\gamma})}{[(\gamma z - \delta)(\tilde{\gamma}\tilde{z} - \tilde{\delta})]^2} \equiv \frac{\|\alpha\delta - \beta\gamma\|}{\|\gamma z - \delta\|^2}.$$

The sign of J, which is the same for all z, is determined by $(\alpha\delta - \beta\gamma)(\tilde{\alpha}\tilde{\delta} - \tilde{\beta}\tilde{\gamma}) \neq 0$. If we include the "points at infinity" ($w = \infty$ for $z = -\delta/\gamma$ and $\lim_{z \to \infty} w = \alpha/\gamma$) and let the points at infinity of axes bisectors ([81], p. 278) correspond to zero divisors, the mapping from the closed z plane and the closed w planes is bijective. Then we have

Theorem 7.12. *The linear-fractional function maps the extended hyperbolic plane h-conformally into the extended hyperbolic plane.*

As for complex variables we have *the group property*

Theorem 7.13. *The bilinear-mappings constitute a (non-commutative) group, i.e.,*

1. *a linear-fractional transformation of a linear-fractional transformation is a linear-fractional transformation;*

2. *the inverse of a linear-fractional transformation is a linear-fractional transformation.*

Proof. The demonstrations are the same as for complex variables. □

From (7.5.17) follows: a linear-fractional transformation is equivalent to the transformations

$$\text{I)}\ z' = z + \frac{\delta}{\gamma}, \quad \text{II)}\ w' = \frac{B}{z'}, \quad \text{III)}\ w = \frac{\alpha}{\gamma} + w' \quad (7.5.18)$$

performed in the specified order.

- Transformations I) and III) correspond to axes translations in z and w planes, respectively.

- Transformations II) correspond to an inversion and a homothety. Actually $w'\tilde{w}' = B\tilde{B}/(z'\tilde{z}')$, and unit hyperbola $z'\tilde{z}' = 1$ is transformed into hyperbola $w'\tilde{w}' = B\tilde{B}$.

Now let us see how z and w planes are divided in sectors. The transformed sectors can be determined by considering the three mappings of (7.5.18). We have to note that this division is not a limit to bijectivity of transformation but means: if a line in the z plane passes through many sectors, the transformed line will pass through the transformed sectors of the w plane.

Actually the *centers of inversion* are at the points $z_0 \equiv \{\Re(-\delta/\gamma),\ \mathcal{H}(-\delta/\gamma)\}$ and $w_0 \equiv \{\Re(\alpha/\gamma),\ \mathcal{H}(\alpha/\gamma)\}$ of z and w planes, respectively; then unitary hyperbolas with center in z_0 are transformed into hyperbolas with center at w_0 and square semi-diameter $p_w^2 = B\tilde{B}$. The sign of p_w^2 is the same as the one of J.

Sectors in the two planes are defined by lines from points z_0 and w_0 and parallel to axes bisectors. These lines correspond to zero divisors of $z + \delta/\gamma$, $w - \alpha/\gamma$.

If $J \propto B\tilde{B} > 0$, the sectors of the z plane are mapped into the same sectors of the w plane.

If $J < 0$, the sectors (Rs; Ls) of the z plane are mapped into sectors (Us, Ds) of the w plane and vice versa.

In these sectors the internal parts of hyperbolas $(z - z_0)(\tilde{z} - \tilde{z}_0) = \pm 1$ are mapped into the external parts of hyperbolas $(w - w_0)(\tilde{w} - \tilde{w}_0) = \pm p_w^2$ and vice versa. This means: equilateral hyperbolas $(z - z_0)(\tilde{z} - \tilde{z}_0) = \pm p_i^2 < \pm 1$ are mapped into the equilateral hyperbolas $(w - w_0)(\tilde{w} - \tilde{w}_0) = \pm p_w^2/p_i^2 > \pm p_w^2$.

Now let us extend to hyperbolic plane some well-known theorems of complex analysis ([69], p. 282).

Theorem 7.14. *The image of a straight line or an equilateral hyperbola under a linear-fractional mapping is a straight line or an equilateral hyperbola.*

In particular, straight lines and equilateral hyperbolas which pass through $z = -\delta/\gamma$ are mapped into straight lines; all other straight lines and equilateral hyperbolas are mapped into equilateral hyperbolas. Moreover if $J > 0$, the hyperbolas are mapped into hyperbolas of the same kind, if $J < 0$, into hyperbolas of the other kind.

Proof. The theorem is evident for the linear mappings I) and III) of (7.5.18).

For mapping II) we begin to show the property that for complex variables is called *circular property*: *the function $w = 1/z$ maps straight lines and equilateral hyperbolas into straight lines or equilateral hyperbolas.*

The equations of straight lines and equilateral hyperbolas are given by

$$a(x^2 - y^2) + b\,x + c\,y + d = 0. \qquad (7.5.19)$$

If $a = 0$, we have straight lines. As a function of hyperbolic variables we have

$$x^2 - y^2 \equiv (x + hy)(x - hy) \equiv z\,\tilde{z}, \quad x = (z + \tilde{z})/2,\ y = h(z - \tilde{z})/2$$

and (7.5.19) becomes

$$a\,z\,\tilde{z} + D\,z + \tilde{D}\tilde{z} + d = 0 \text{ with } D = (b + h\,c)/2. \qquad (7.5.20)$$

Substituting $z \Rightarrow 1/w$ the form of (7.5.20) does not change, then we have again straight lines and equilateral hyperbolas. □

Theorem 7.15. *A pair of points z_1, z_2 symmetric with respect to an equilateral hyperbola (Γ) is mapped by a linear-fractional transformation into a pair of points z_1^*, z_2^* symmetric with respect to the image of the hyperbola (Γ^*).*

Proof. Actually equilateral hyperbolas passing through z_1, z_2 are pseudo-orthogonal to Γ and are mapped into equilateral hyperbolas passing through points z_1^*, z_2^*. Thanks to the preservation of hyperbolic angles for h-conformal mappings, these hyperbolas are pseudo-orthogonal to Γ^*, then from Theorem 7.11, p. 108, they are symmetric with respect to Γ^*.

We note that if Γ is a straight line, the symmetry means that points P, P^* lie on a pseudo-Euclidean normal and have the same pseudo-Euclidean distance from the straight line. □

The linear-fractional mappings depend on three hyperbolic constants, i.e., on six real constants, thus they are determined by six conditions. The simplest ones are to require that three arbitrary points of the z plane correspond to three arbitrary points of the w plane. This transformation is determined by means of the following theorem.

Theorem 7.16. *There is only one linear-fractional transformation which maps three given points z_1, z_2, z_3 (so that their difference is not a divisor of zero), into three given points w_1, w_2, w_3. This transformation is given by the formula*

$$\frac{w - w_1}{w - w_2} \cdot \frac{w_3 - w_2}{w_3 - w_1} = \frac{z - z_1}{z - z_2} \cdot \frac{z_3 - z_2}{z_3 - z_1}. \qquad (7.5.21)$$

Proof. The proof is the same as for the complex variable [47] and ([69], p. 287). □

From this theorem follows

Theorem 7.17. *The function $w = f(z)$ given by (7.5.21) maps the equilateral hyperbola which passes through three points z_k $(k = 1, 2, 3)$, into the equilateral hyperbola which passes through three points w_k $(k = 1, 2, 3)$.*

With respect to the same problem of complex analysis we have to note: as it is demonstrated in Chapter 4, the equilateral hyperbola that passes through three points has square semi-diameter given by (4.5.13), from which

$$P_z \propto -(z_1 - z_2)(\tilde{z}_1 - \tilde{z}_2)(z_1 - z_3)(\tilde{z}_1 - \tilde{z}_3)(z_3 - z_2)(\tilde{z}_3 - \tilde{z}_2)$$

and the same expression holds for P_w. If the signs of P_z and P_w are the same, the kind of the corresponding hyperbolas is the same and $J > 0$.

Theorem 7.18. *The linear-fractional transformations preserve the cross ratio between four corresponding points of an equilateral hyperbola.*

Proof. The proof follows from (7.5.21). □

Examples of linear-fractional transformations

Theorem 7.19. *The linear-fractional function which maps unit hyperbola $|z| = 1$ with center at $O \equiv (0, 0)$ into unit hyperbolas is given by*

$$w = \exp[h\,\alpha]\frac{z - z_1}{1 - \tilde{z}_1 z}, \tag{7.5.22}$$

where z_1 (with $|z_1| < 1$) is the point of the z plane mapped into O of the w plane.

Proof. Since z_1 with $|z_1| < 1$ is mapped into $w \equiv (0, 0)$, the point $1/\tilde{z}_1$ is mapped into $w = \infty$; then the transformation is given by $w = C(z - z_1)/(1 - \tilde{z}_1 z)$.

Let us show that $|C| = 1$. Since all the points on the boundary of the unit hyperbola in the z plane are mapped into points on the boundary of the unit hyperbola of the w plane, i.e., $|w| = 1$ for $z = \exp[h\,\theta]$, from (7.5.22) we obtain

$$|w| \equiv 1 = \frac{|C|}{\exp[h\,\theta]}\frac{|(\exp[-h\,\theta] - z_1)|}{|(\exp[h\,\theta] - \tilde{z}_1)|} \equiv |C|,$$

hence $C = \exp[h\,\alpha]$ and we obtain (7.5.22). □

We conclude this paragraph by studying the hyperbolic counterpart of the linear-fractional transformations of complex variables used by Poincaré for representing in a half-plane the non-Euclidean geometry which Beltrami and Klein considered in a circle ([25], p. 55) (see Chapter 9).

For a complex variable the linear-fractional transformation $w = (z-i)/(z+i)$ maps the internal points of the unit circle with center at $O \equiv (0, 0)$ into the half-plane $y > 0$, and, in particular, its circumference into the $y = 0$ axis. Considering the same transformation in the hyperbolic plane, we have

Theorem 7.20. *The linear-fractional function*

$$w = \frac{z - h}{z + h} \equiv 1 - \frac{2\,h}{z + h} \Rightarrow w - 1 = -\frac{2\,h}{z + h}, \tag{7.5.23}$$

whose inverse transformation is

$$z \equiv x + h\,y = h\frac{1 + w}{1 - w} \equiv \frac{2v + h(1 - u^2 + v^2)}{1 + u^2 - v^2 - 2u} \tag{7.5.24}$$

maps the left and right arms of unit hyperbolas into the $y = 0$ axis.

The following proof allows us to see how the topology of a hyperbolic plane determines a more intricate mapping with respect to the one of the complex plane, but also has similar properties.

Proof. Divisors of zero in the z plane are given by the straight lines $y \pm x = -1$ and in the w plane by the straight lines $u \pm v = 1$.

Since $J \propto h \cdot (-h) \equiv -1 < 0$, the corresponding equilateral hyperbolas of the w and z planes are of different kinds.

The center of inversion in the z plane is at $C_z \equiv (0, -1)$, in the w plane is at $C_w \equiv (1, 0)$. The unit hyperbola with center at $z_0 \equiv (0, -1)$ is mapped into a hyperbola with semi-diameter $p = 2$ and center at $w_0 \equiv (1, 0)$. Internal and external points, as usual, are inverted.

Equilateral hyperbolas are mapped into equilateral hyperbolas that pass through the transformed domain.

In the complex plane the limiting circle (mapped into the x axis) is given by $w\bar{w} = 1$. Now let us consider in the hyperbolic plane the same equation $w\tilde{w} = 1$, which is mapped on the z plane into $(z - h)(\tilde{z} + h) = (z + h)(\tilde{z} - h)$ and, as in the complex plane, represents the straight line $z - \tilde{z} = 0$, i.e, the x axis.

Let us investigate in more detail the transformation of the two arms of a unit hyperbola. By considering the parametric equations

$$1) \quad u = \cosh\theta, \quad v = \sinh\theta, \quad 2) \quad u = -\cosh\theta, \quad v = -\sinh\theta, \quad (7.5.25)$$

their mapping into the (x, y) plane is obtained by (7.5.24)

$$1)\ x + hy \ = \ \frac{\sinh\theta}{(1 - \cosh\theta)} \equiv -\coth(\theta/2),$$

$$2)\ x + hy \ = \ -\frac{\sinh\theta}{(1 + \cosh\theta)} \equiv -\tanh(\theta/2). \quad (7.5.26)$$

Now we see how the two arms of the unit hyperbola and some points are mapped in agreement with the topology of the hyperbolic plane.

- The arm 1) of (7.5.26) $\in Rs$ is mapped into the exterior part of segment $-1 \leftrightarrow +1$ of the $y = 0$ axis (sectors Ls, Rs with respect to zero divisors of the z plane).

- The arm 2) of (7.5.26) $\in Ls$ is mapped into the segment $-1 \leftrightarrow +1$ of the $y = 0$ axis (sectors Us with respect to zero divisors of the z plane).

- Point $(-1, 0)$ corresponds to point $+\infty$ of both arms of the hyperbola.

- Point $(+1, 0)$ corresponds to point $-\infty$.

- The vertex of the right arm ($w_0 \equiv w - 1 = 0$) corresponds to $x \to \infty$.

- The vertex of the left arm corresponds to point $z \equiv (0, 0)$.

- Points of the hyperbola with $v > 0$ are mapped into the semi-axis $x < 0$.

- Points $v < 0$ are mapped into the semi-axis $x > 0$.

- The axes bisectors $u = \pm v$ are mapped into $y = 1 + \pm x$.

- The center of the "limiting hyperbola" is mapped into the point $(0, h)$.

The first two results are the same as the mapping of the circle into the complex plane. With the same symbol, in the complex plane we have

$$z \equiv x + \mathrm{i}\, y = \mathrm{i}\,\frac{1 + w}{1 - w} \equiv \frac{2v + \mathrm{i}\,(1 - u^2 - v^2)}{1 + u^2 + v^2 - 2u}$$

and for $u = \cos\phi$, $v = \sin\phi$ we have $z = -\cot(\phi/2)$. The half-circle on the right and left sides with respect to the coordinates' origin are mapped as the corresponding arms of the unit hyperbola.

Let us consider the conjugate arms of the limiting hyperbola $\in Us,\ Ds$: $w\tilde{w} = -1$ in the z plane; we have

$$(z - h)(\tilde{z} + h) = -(z + h)(\tilde{z} - h) \Rightarrow z\tilde{z} = 1 \in Ls,\ Rs.$$

We say, as a particular example of a general rule, that conjugate arms are not mapped into conjugate arms, i.e., only conjugate hyperbolas with center at the "inversion center" of the z plane are mapped into conjugate hyperbolas with center at the "inversion center" of the w plane.   □

7.6 Some Introductory Remarks on Commutative Hypercomplex Systems with Three Unities

We have seen in Section 2.2 that the types of two-dimensional systems derive from the kind of solutions of the characteristic equation of degree 2. The same happens for hypercomplex numbers with more than two dimensions and the types of the numbers are linked to the possible solutions of an equation of degree N. Then the number of the systems is always limited. Here we make some introductory notes and give the multiplication tables for systems with unity or in decomposed form.

The classification of these systems is known. In particular for non-separable (commutative and non-commutative) systems the classification is reported in [13], while for separable systems it is reported in [24].

The commutative systems with unity are five. Two of these systems are non-separable and have the following multiplication tables:

e_0	e_1	e_2		e_0	e_1	e_2		
e_1	e_2	0	$(a);$	e_1	0	0	$(b).$	(7.6.1)
e_2	0	0		e_2	0	0		

In the tables for non-separable systems the versor e_0 represents unity. In general, the product $e_i\, e_k$ is in the $(i + 1)^{th}$ row and in the $(k + 1)^{th}$ column. By analogy with two-dimensional systems we call these systems *generalized parabolic systems*.

The other three systems are separable and each of them is composed of a system with two units and a system with one unit (real). By setting $\alpha = 0, \pm 1$ the three separable systems can be summarized in the following table, where e_0 and e_2 are the idempotent sub-unities.

$$
\begin{array}{|c|c|c|}
\hline
e_0 & e_1 & 0 \\
\hline
e_1 & \alpha e_0 & 0 \\
\hline
0 & 0 & e_2 \\
\hline
\end{array}
\tag{7.6.2}
$$

7.6.1 Some Properties of the Three-Units Separable Systems

We obtain systems with a unity versor by means of a linear transformation. If we call the new versors $1, i, j$ we have $1 = e_0 + e_2$. As far as the other versors are concerned, we can put i, j as an arbitrary linear transformation of e_0, e_1, e_2. In particular, by setting $i = e_1$; $j = e_0 - e_2$, we have in vector-matrix form

$$
\begin{pmatrix} 1 \\ i \\ j \end{pmatrix} = \begin{pmatrix} 1 & 0 & 1 \\ 0 & 1 & 0 \\ 1 & 0 & -1 \end{pmatrix} \begin{pmatrix} e_0 \\ e_1 \\ e_2 \end{pmatrix}
\tag{7.6.3}
$$

and the inverse

$$
\begin{pmatrix} e_0 \\ e_1 \\ e_2 \end{pmatrix} = \frac{1}{2} \begin{pmatrix} 1 & 0 & 1 \\ 0 & 2 & 0 \\ 1 & 0 & -1 \end{pmatrix} \begin{pmatrix} 1 \\ i \\ j \end{pmatrix}.
\tag{7.6.4}
$$

The multiplication table for the new versors is easily obtained from the multiplication table (7.6.2)

$$
\begin{array}{|c|c|c|}
\hline
1 & i & j \\
\hline
i & \frac{\alpha}{2}(1+j) & i \\
\hline
j & i & 1 \\
\hline
\end{array}.
\tag{7.6.5}
$$

We say that $\zeta \equiv e_0\,\tau + e_1\,\xi + e_2\,\eta = t + ix + jy \equiv z$. The transformation for the variables is given by (7.6.3) and (7.6.4) by substituting the variables τ, ξ, η for the versors 1, i, j and t, x, y for e_0, e_1, e_2, respectively. The characteristic matrix is obtained by multiplying the hypercomplex number (ζ) by the versors and taking the coefficients of the versors. By using table (7.6.2) we obtain

$$
\begin{pmatrix} \tau & \alpha\,\xi & 0 \\ \xi & \tau & 0 \\ 0 & 0 & \eta \end{pmatrix},
\tag{7.6.6}
$$

from which it follows that the invariant is given by $\rho = \eta\,(\tau^2 - \alpha\,\xi^2)$; this quantity is equal to the product of the invariants of the component systems. As a function of the variables t, x, y, the invariant becomes

$$
\rho = (t - y)[(t + y)^2 - \alpha\,x^2].
\tag{7.6.7}
$$

The plane $t = y$ is a divisor of zero for all three systems. The existence of other divisors of zero depends on the two-dimensional component systems.

Let us see how the three-dimensional systems are related to the possible solutions of a cubic equation. The systems of Table (7.6.2) correspond, depending on the $\alpha = 1,\ 0,\ -1$ value, to three real solutions, three real but two equal solutions, one real and two complex conjugate solutions, respectively. The systems of Table (7.6.1) correspond to three real but equal solutions.

The Exponential and Logarithm Functions

As an example we report these functions obtained by means of the property (7.3.8)

$$\exp[\zeta] = e_0 \exp[\tau + e_1\,\xi] + e_2 \exp[\eta].\tag{7.6.8}$$

For $\alpha = 1$ we have (Section 4.1.1)

$$\exp[\zeta] = e_0 \exp[\tau]\,(\cosh\xi + e_1 \sinh\xi) + e_2 \exp[\eta].$$

For the variable z we obtain

$$\exp[t + i\,x + j\,y] = \frac{1}{2}\{\exp[t + y]\,\cosh x + \exp[t - y]\}$$
$$+\,i\exp[t + y]\,\sinh x + \frac{j}{2}\{\exp[t + y]\,\cosh x - \exp[t - y]\}\,.\tag{7.6.9}$$

By inverting these expressions, we obtain the definition of the logarithm

$$\ln[t + ix + jy] = \frac{1}{2}\ln\left[(t - y)\sqrt{(t + y)^2 - x^2}\right] + i\tanh^{-1}\left[\frac{x}{t + y}\right]$$
$$+\,\frac{j}{2}\ln\left[\frac{\sqrt{(t + y)^2 - x^2}}{t - y}\right].\tag{7.6.10}$$

For $\alpha = -1$ the hyperbolic functions must be replaced by the circular functions and $-x^2$ by x^2.

The Generalized Cauchy–Riemann Conditions

Let us put
$$F(z) = U(t,\,x,\,y) + i\,V(t,\,x,\,y) + j\,W(t,\,x,\,y)\,;\tag{7.6.11}$$

then the function $F(z)$ is called a *function of the hypercomplex variable* $z = t + i\,x + j\,y$, if the real functions U, V, W have partial derivatives with respect to the variables t, x, y and these derivatives satisfy (7.2.1), which in this case become

$$\frac{\partial F}{\partial t} = \frac{1}{i}\frac{\partial F}{\partial x} = \frac{1}{j}\frac{\partial F}{\partial y}.\tag{7.6.12}$$

Substituting in the function F the components U, V, W, we obtain the following partial differential equations:

$$\frac{\partial U}{\partial x} + i\frac{\partial V}{\partial x} + j\frac{\partial W}{\partial x} = i\left(\frac{\partial U}{\partial t} + i\frac{\partial V}{\partial t} + j\frac{\partial W}{\partial t}\right) \equiv \frac{\alpha}{2}\frac{\partial V}{\partial t} + i\left(\frac{\partial U}{\partial t} + \frac{\partial W}{\partial t}\right) + j\frac{\alpha}{2}\frac{\partial V}{\partial t},$$

$$\frac{\partial U}{\partial y} + i\frac{\partial V}{\partial y} + j\frac{\partial W}{\partial y} = j\left(\frac{\partial U}{\partial t} + i\frac{\partial V}{\partial t} + j\frac{\partial W}{\partial t}\right) \equiv \frac{\partial W}{\partial t} + i\frac{\partial V}{\partial t} + j\frac{\partial U}{\partial t}.$$

Equating the coefficients of the versors we obtain the (GCR) conditions. We express these conditions writing the partial derivatives of V, W as functions of the partial derivatives of U,

$$\frac{\partial V}{\partial t} = \frac{\partial V}{\partial y} = \frac{2}{\alpha}\frac{\partial U}{\partial x}, \quad \frac{\partial V}{\partial x} = \frac{\partial U}{\partial t} + \frac{\partial U}{\partial y},$$

$$\frac{\partial W}{\partial t} = \frac{\partial U}{\partial y}, \quad \frac{\partial W}{\partial x} = \frac{\partial U}{\partial x}, \quad \frac{\partial W}{\partial y} = \frac{\partial U}{\partial t}.$$

It is easy to verify that these conditions are satisfied by the functions U, V, W given by (7.6.9) and (7.6.10).

Chapter 8

Hyperbolic Variables on Lorentz Surfaces

8.1 Introduction

In this chapter we start from two fundamental memoirs of Gauss [39] and Beltrami [3] on complex variables on surfaces described by a definite quadratic form. By using the functions of a hyperbolic variable, we extend the results of the classic authors to surfaces in space-times described by non-definite quadratic forms.

To make our exposition self-consistent, we are obliged to begin by summarizing the essential points of the authors' memoirs which can quite rightly be considered two milestones in the history of mathematics, not only for their scientific contents but also for the *perfection of style and neatness of expressions*[1]. Of course our summary can not substitute for reading of the original sources or, at least, an unabridged complete translation, but allows us to stress the points we consider essential for our treatment. Here we briefly recall the basics of Gauss' differential geometry and some remarks taken from Beltrami's introduction to the paper we summarize in Section 8.4.

The starting idea of Gauss was: to study the properties of surfaces in Euclidean space by means of a quadratic differential form which gives the distance between two neighboring points (infinitesimal distance). This quadratic form is obtained by applying Pythagoras' theorem to two points on a surface in Euclidean space, represented by Cartesian coordinates:

$$d\,s^2 = d\,x^2 + d\,y^2 + d\,z^2. \tag{8.1.1}$$

The equation of the surface $f(x,y,z) = 0$ can be replaced by the parametric equations

$$x = x(u,\,v)\,;\; y = y(u,\,v)\,;\; z = z(u,\,v), \tag{8.1.2}$$

where the two parameters u,v can be taken as curvilinear coordinates on the surface. Thus one can write the differentials of x,y,z as functions of the differentials of u,v:

$$d\,x = \frac{\partial x}{\partial u}\,d\,u + \frac{\partial x}{\partial v}\,d\,v\,;\; d\,y = \frac{\partial y}{\partial u}\,d\,u + \frac{\partial y}{\partial v}\,d\,v\,;\; d\,z = \frac{\partial z}{\partial u}\,d\,u + \frac{\partial z}{\partial v}\,d\,v. \tag{8.1.3}$$

---

[1]These last words are taken from the commemoration of Beltrami at the London Mathematical Society [10].

By substituting (8.1.3) into (8.1.1), Gauss obtained the[2] ◁ metric (or line) element ▷

$$ds^2 = E(u,v)\, du^2 + 2\, F(u,v)\, du\, dv + G(u,v)\, dv^2. \tag{8.1.4}$$

Here we report the meaning of the metric element and the required regularity conditions from the introduction of Beltrami's paper (Section 8.4).

◁ We recall that when we study a surface just from its metric element it must be considered regardless of its representation, as an example in a Cartesian space. Actually these representations can generate misleading. What we have to take into account is that different points on the surface correspond to different values of coordinates u, v. The possibility that one point in the space corresponds to different values of u, v, is a consequence of its representation.

The nature of coordinate lines $u = $ const., $v = $ const. remains undefined, while for giving exact formulations, we consider the regions Ω of the surface S in which the following conditions are satisfied:

1. the functions E, F, G in every point are real, monodromic, continue and finite;

2. the functions E, G, $EG - F^2 \equiv H^2$, in every point are positive and different from zero so that the differential form is positive definite.

Some of these conditions are necessary so that all points of Ω are real, the other ones give conditions about the coordinate system. Actually, since in general $\cos\theta = \frac{F}{\sqrt{EG}}$, $\sin\theta = \frac{H}{\sqrt{EG}}$, where θ is the angle between the tangent to coordinate lines in point u, v, the condition $H > 0$ implies that θ is different from 0° or 180°, i.e., that two curves of different coordinate systems are tangent. Summarizing, all small regions of the surface around an internal point of Ω are covered by a network of coordinate curves similar to the ones of two systems of parallel lines on a plane (systems which, in general, are not orthogonal). These regions can be called *ordinary regions*. ▷

Coming back to Gauss, he showed that from line element (8.1.4) it is possible to obtain all the geometrical features of a surface, in particular the angle between two curves, the area of a part of the surface and the finite distance between two faraway points. This last aspect allows us to introduce the geodesic lines which generalize a property of a straight line on a plane: the shortest line between two given points. In Chapter 9 we see how the introduction of geodesic lines has allowed Beltrami to give a Euclidean interpretation of non-Euclidean geometries.

Another fundamental quantity which can be obtained from the metric element is the *Gauss curvature*. We give its geometrical meaning: let us consider a straight line orthogonal to a surface and the sheaf of planes specified by the straight line. The intersection of each of these planes with the surface gives a line. Every one of these lines is better approximated, in the given point, by a circle

---

[2]In this and subsequent chapters, we use ◁ ... ▷ to identify material that is a literal translation of the original author's words.

called an *"osculating circle"*. The circles with the minimum and maximum radius are in two orthogonal planes and Gauss found (*Theorema Egregium*) that *their product is independent of the surface representation*, i.e., *it is an invariant quantity with respect to the transformations of parameters u, v.* Gauss called *curvature* the reciprocal of this quantity which can be calculated by means of the functions E, F, G and of their first and second derivatives. Actually it is given by

$$K = -\frac{1}{2H}\left[\frac{\partial}{\partial u}\left[\frac{1}{H}\left(\frac{\partial G}{\partial u} - \frac{F}{E}\frac{\partial E}{\partial v}\right)\right] + \frac{\partial}{\partial v}\left[\frac{1}{H}\left(\frac{\partial E}{\partial v} - 2\frac{\partial F}{\partial u} + \frac{F}{E}\frac{\partial E}{\partial u}\right)\right]\right]$$

(8.1.5)

with $H^2 = EG - F^2$.

In general the curvature changes from point to point. If it is constant in all points, the curvature radii with respect to the sheaf of planes are also constant. If this property holds true, the surface is called a *constant curvature surface* as we shall see in Chapter 9.

8.2   Gauss: Conformal Mapping of Surfaces

The development of the theory of conformal mappings is directly connected to the involvement of Gauss in geodesic observations [11].

In 1822 the Danish Academy of Science announced a prize for a derivation of all possible projections of the Earth's surface which could be used for the production of maps. The precise problem was to map a given arbitrary area into another area in such a way that "the image is similar to the original in the smallest parts". Gauss won the prize with an essay [39] which laid the formalization of ◁ The theory of conformal mapping for functions of a complex variable. ▷ We give below the salient points of the Gauss essay.

Let us consider two surfaces represented by means of their line elements (8.1.4) as functions of the parameters t, u, and T, U, respectively:

$$\text{I) } ds^2 = e(t,u)\,dt^2 + 2\,f(t,u)\,dt\,du + g(t,u)\,du^2\,; \qquad (8.2.1)$$

$$\text{II) } dS^2 = E(T,U)\,dT^2 + 2\,F(T,U)\,dT\,dU + G(T,U)\,dU^2\,. \qquad (8.2.2)$$

For representing the first surface on the second one we must define a bijective mapping $T(t,u)$, $U(t,u)$. This mapping must satisfy both the conditions imposed by the problem and by the nature of surfaces. After this transformation the line element of the second surface is written as

$$dS^2 = E(t,u)\,dt^2 + 2\,F(t,u)\,dt\,du + G(t,u)\,du^2\,. \qquad (8.2.3)$$

The requirements of the problem can be put in the form: ◁ let us consider two arbitrary infinitesimal arcs on a surface and the transformed ones on the second surface, the latter must be proportional to the former and the angles between them must be equal. ▷

Gauss' solution of the problem. Since the conditions must be true for every dt, du, then e, f, g must be proportional to E, F, G, respectively. Moreover the trigonometric functions of the angles depend on the ratios between e, f, g, then the second condition is satisfied if the first is.

The analytical condition required by the problem becomes

$$\frac{E}{e} = \frac{F}{f} = \frac{G}{g} = m^2, \tag{8.2.4}$$

where m is the magnification ratio of the linear elements of the first surface when they are mapped on the second one. This ratio is, in general, a function of t, u, i.e., of points. If it is constant the map is a homothety between the finite parts of the surfaces too.

Let us now return to quadratic forms and decompose them into linear factors. Since $ds^2 > 0$, the differential form is definite, then it must be $f^2 - eg < 0$ and the factors are complex conjugate

$$ds^2 = \frac{1}{e}\left[e\,dt + \left(f + \mathrm{i}\sqrt{eg - f^2}\right)du\right]\left[e\,dt + \left(f - \mathrm{i}\sqrt{eg - f^2}\right)du\right]; \tag{8.2.5}$$

the problem is reduced to two separate integrations of linear differential equations. The first one is given by

$$e\,dt + (f + \mathrm{i}\sqrt{eg - f^2})du = 0, \tag{8.2.6}$$

the second one is obtained by exchanging $\mathrm{i} \to -\mathrm{i}$. Equation (8.2.6) can be integrated by means of an integrating factor (in general complex) function of t, u, as we shall see in Beltrami's work (Section 8.4).

The integral can be split into a "real" and an "imaginary" part and written as $p + \mathrm{i}q$, where p and q are real functions of t, u. The integral of the second equation is $p - \mathrm{i}q$ and the integrating factor is the complex conjugate of the first one. Differentiating, we have

$$ds^2 = n(dp + \mathrm{i}\,dq)(dp - \mathrm{i}\,dq) \equiv n(dp^2 + dq^2), \tag{8.2.7}$$

where n is the square modulus of the integrating factor and is a real finite function of t, u. This expression of line element is called *isometric*. In the same way the second form can be decomposed into linear terms and we can write $dS^2 = N(dP^2 + dQ^2)$, where P, Q, N are real functions of T, U. These integrations (which, from a practical point of view, can be performed just for particular cases) must be done before solving the problem.

If we express T, U by means of such functions of t, u that satisfy the problem conditions given by (8.2.4), we have $dS^2 = m^2\,ds^2$ and, using the new variables,

$$\frac{(dP + \mathrm{i}\,dQ)(dP - \mathrm{i}\,dQ)}{(dp + \mathrm{i}\,dq)(dp - \mathrm{i}\,dq)} = m^2\frac{n}{N}; \tag{8.2.8}$$

the numerator of the left-hand side is divisible by the denominator if $(dP + i\,dQ)$ is divisible by $(dp + i\,dq)$ and $(dP - i\,dQ)$ by $(dp - i\,dq)$, or vice versa. Then $(dP + i\,dQ)$ is zero when $dp + i\,dq = 0$, i.e., ◁ $P + iQ$ is constant if $p + iq$ is constant. This means that $P + iQ$ is a function just of $p + iq$ and $P - iQ$ just of $p - iq$. ▷ Or, inversely, if $P + iQ$ is a function of $p - iq$ and $P - iQ$ of $p + iq$. In particular, it can be shown that if $P + iQ$ is a function of $p + iq$ and $P - iQ$ of $p - iq$, the angles are the same if $P + iQ$ is a function of $p - iq$ and $P - iQ$ of $p + iq$, the angles change sign. These conditions are also sufficient.

Setting
$$P + iQ = f(p + iq), \quad P - iQ = f'(p - iq),$$

the function f' is determined by f: if the constants in f are real, then $f' \equiv f$, so that to real values of p, q, real values of P, Q correspond. If there are imaginary elements we must exchange i for $-$i. In every case $f' = \bar{f}$, then $f\,f' = |f|^2$, and *a problem described by a differential quadratic form has reduced to a problem for a linear form.* We also obtain

$$P = \frac{1}{2}(f + f'), \quad Q = \frac{1}{2}(f - f').$$

Calling ϕ, ϕ' the derivative of f and f',

$$\phi(p + iq) = \frac{dP + i\,dQ}{dp + i\,dq}, \quad \phi'(p - iq) = \frac{dP - i\,dQ}{dp - i\,dq},$$

we have

$$m = \sqrt{\frac{N}{n}\phi\phi'} \equiv \sqrt{\frac{N}{n}}|\phi|. \tag{8.2.9}$$

Then, using curvilinear coordinates and integrating the differential form (8.2.6), we obtain the ◁ complete solution of the proposed problem by applying the theory of functions of a complex variable. ▷

8.2.1 Mapping of a Spherical Surface on a Plane

The general theory is now applied to the proposed problem: let us consider the sphere with radius R represented by

$$x = R\cos t \sin u, \quad y = R\sin t \sin u, \quad z = R\cos u; \tag{8.2.10}$$

the metric element is

$$ds^2 = R^2(\sin^2 u\,dt^2 + du^2) \equiv R^2\sin^2 u\left(dt + i\frac{d u}{\sin u}\right)\left(dt - i\frac{d u}{\sin u}\right). \tag{8.2.11}$$

We must solve the integrable differential equation $dt \mp i\frac{d u}{\sin u} = 0$, the integral of which is $t \pm i \ln \cot(u/2) = Const.$. Introducing the new variables

$$p = t, \quad q = \ln \cot(u/2), \quad \text{and} \quad dp = dt, \quad dq = \frac{du}{\sin u}, \tag{8.2.12}$$

we have the isometric form

$$ds^2 = R^2 \sin^2 u \, (dp^2 + dq^2) \equiv \frac{R^2}{\cosh^2 q} \, (dp^2 + dq^2). \tag{8.2.13}$$

In the last passage we have used the equality $\sin u \equiv \frac{2\cot(u/2)}{1+\cot^2(u/2)} = \frac{2\exp q}{1+\exp 2q} \equiv 1/\cosh q$. Let us now consider the plane with metric element

$$dS^2 = dT^2 + dU^2; \ P + iQ \equiv T + iU \ \Rightarrow \ E = G = 1, \ F = 0.$$

Then, if f is an arbitrary function, we put

$$P = \Re\{f(t + i \ln \cot u/2)\} \text{ and } Q = \Im\{f(t + i \ln \cot u/2)\}.$$

Let us take a linear function $f(v) = kv$; we have

$$T + iU = k(t + i \ln \cot u/2) \tag{8.2.14}$$

and the amplification ratio is

$$m = \frac{k}{R \sin u}. \tag{8.2.15}$$

This representation is the Mercator projection. Setting $f(v) = k \exp[iv]$, we have

$$T + iU = k \exp[\ln \tan(u/2) + it] = k(\cos t + i \sin t) \tan(u/2) \tag{8.2.16}$$

and the amplification ratio is

$$m = \frac{k \tan(u/2)}{R \sin u}. \tag{8.2.17}$$

This representation corresponds to a polar stereographic projection.

8.2.2   Conclusions

We have seen how the idea of decomposing the differential quadratic form of a line element into two linear complex conjugate factors allows us to obtain the solution of the problem by considering just one of them.

Let us now examine the main passages of Gauss' demonstration and the meaning assumed by i.

1. For obtaining the metric element in isometric form, the quadratic form is decomposed into two linear forms. In this passage the i is introduced as the *square root of* -1.

2. The linear form is considered as a two-component quantity (real and imaginary parts), then i assumes the meaning of a *versor*.

3. The last step is the introduction of functions of a complex variable and i becomes a *symmetry preserving operator*.

Since all these properties are in common with hyperbolic numbers, as shown in Section 2.2.1 for point 1) and in Section 7.2.3 for point 3), in the next section *all the above results are extended to non-definite line elements*.

8.3 Extension of Gauss Theorem: Conformal Mapping of Lorentz Surfaces

Let us consider the same problem solved by Gauss with reference to Lorentz surfaces, i.e., with $f^2 - eg > 0$, represented by their line elements

$$\text{I)} \quad ds^2 \;=\; e(t, u)\, dt^2 + 2\, f(t, u)\, dt\, du + g(t, u)\, du^2\,; \tag{8.3.1}$$

$$\text{II)} \quad dS^2 \;=\; E(T, U)\, dT^2 + 2\, F(T, U)\, dT\, dU + G(T, U)\, dU^2\,. \tag{8.3.2}$$

Thanks to observations in Section 2.2, we can decompose the line elements into the product of two hyperbolic conjugate linear factors

$$ds^2 = \frac{1}{e}\left[e\, dt + \left(f + h\,\sqrt{f^2 - eg}\,\right) du \right]\left[e\, dt + \left(f - h\,\sqrt{f^2 - eg}\,\right) du \right]\,; \tag{8.3.3}$$

the problem, as before, is reduced to the integration of a linear differential equation

$$e\, dt + (f + h\sqrt{f^2 - eg})\, du = 0. \tag{8.3.4}$$

The integral of (8.3.4) can be divided into a "real" and a "hyperbolic" part and written as $p + h\,q$, where p and q are real functions of t, u. The integral of the other factor of (8.3.3) is $p - h\,q$. Differentiating, we have

$$ds^2 = n(dp + h\, dq)(dp - h\, dq) \equiv n(dp^2 - dq^2), \tag{8.3.5}$$

where n is the square modulus of an integrating factor (generally hyperbolic) and is a real finite function of t, u.

In the same way the second form can be decomposed into linear terms and we can write $dS^2 = N(dP^2 - dQ^2)$, where P, Q, N are real functions of T, U.

Let us now introduce the hyperbolic variable $z = p + h\,q$ and the hyperbolic function $w = P + h\,Q$ and the differentials $dz = dp + h\,dq$, $d\tilde{z} = dp - h\,dq$, $dw = dP + h\,dQ$, $d\tilde{w} = dP - h\,dQ$; we have

$$\frac{dS^2}{ds^2} \equiv \frac{N}{n}\frac{dw}{dz}\left(\widetilde{\frac{dw}{dz}}\right). \tag{8.3.6}$$

From Section 7.5 we know that this ratio is independent of direction, and the hyperbolic angle between two lines is preserved, if w is a hyperbolic function of the hyperbolic variable z. Calling J the Jacobian determinant, we have that the constant "amplification" (M) is given by

$$M = \frac{N}{n}\, J\,. \tag{8.3.7}$$

We can extend to the pseudo-Euclidean plane the well-known theorem valid in the Euclidean plane:

Theorem 8.1. *The mappings by functions of a hyperbolic variable preserve the isometric form of the non-definite line elements.*

Proof. If we consider line elements in isometric form, we have

$$
\begin{aligned}
d\,u^2 - d\,v^2 &\equiv (d\,u + h\,d\,v)(d\,u - h\,d\,v) \\
&\equiv (u_x + h\,v_x)(u_x - h\,v_x)(d\,x + h\,d\,y)(d\,x - h\,d\,y) \\
&\equiv (u_x^2 - v_y^2)(d\,x^2 - d\,y^2) = J(d\,x^2 - d\,y^2)\,.
\end{aligned}
\tag{8.3.8}
$$

$\square$

8.4   Beltrami: Complex Variables on a Surface

The starting point of Beltrami's paper is the just exposed Gauss' work. In particular Beltrami applies the results of his own research on differential geometry ([4], **I**, p. 143) and the properties of the quantities called ◁ differential parameters ▷ [34].

Gauss started from a particular problem, solved it in a general way and applied the results to the original problem for which the necessary integrations could be performed in elementary ways. Beltrami pursues the common properties to solutions of Gauss' problem and obtains a result which goes beyond the specific problem and becomes very important for partial differential elliptic equations [75].

Let us consider the line element (8.1.4) and, following Beltrami, summarize some expressions that will be used. Let us consider the line elements ds, δs starting from point u, v, to which the increments du, dv and δu, δv, respectively, correspond and the angle ϵ between them which is considered positive from ds to δs. The following expressions are true for every ϵ:

$$
\begin{aligned}
ds\delta s \cos \epsilon &= Edu\delta u + F(du\delta v + dv\delta u) + Gdv\delta v, \\
ds\delta s \sin \epsilon &= H(du\,\delta v - dv\,\delta u).
\end{aligned}
\tag{8.4.1}
$$

Calling $\Delta\Omega$ the area of the parallelogram with sides ds, δs, we have

$$
\Delta\Omega = \pm ds\,\delta s \sin \epsilon \equiv \pm H(du\,\delta v - dv\,\delta u).
$$

Now we recall the expressions of *differential parameters* ([4], **I**, p. 143)

$$
\Delta_1 \phi = \frac{E\phi_{,v}^2 - 2F\phi_{,v}\phi_{,u} + G\phi_{,u}^2}{H^2},
$$

$$
\Delta_1(\phi,\,\psi) = E\phi_{,v}\psi_{,v} - F(\phi_{,v}\psi_{,u} + \phi_{,u}\psi_{,v}) + G\phi_{,u}\psi_{,u}H^2,
$$

$$
\Delta_2 \phi = \frac{1}{H}\left[\left(\frac{G\phi_{,u} - F\phi_{,v}}{H}\right)_{,u} + \left(\frac{E\phi_{,v} - F\phi_{,u}}{H}\right)_{,v}\right],
\tag{8.4.2}
$$

where ϕ and ψ are two arbitrary functions of u, v.

- $\Delta_1\phi$ is called the differential parameter of the first order,

- $\Delta_1(\phi,\,\psi)$ is called the mixed differential parameter of the first order,

- $\Delta_2\phi$ is called the differential parameter of the second order.

The order is determined by the maximum order of the derivatives of functions ϕ, ψ. The differential parameters are invariant with respect to the transformations of variables u, v. They also have the following properties:

1. Since the equations $\phi = const$, $\psi = const$ represent two surfaces, $\Delta_1\,\phi$ represents the squared length of the gradient of ϕ as well as of a vector orthogonal to the surface $\phi = const$; for the same reason, if $\Delta(\phi,\,\psi) = 0$, then the two surfaces $\phi = const$ and $\psi = const$ are orthogonal.

2. If $\Delta_2\phi = 0$, and we take the curves $\phi{=}Const$, together with the orthogonal curves, as a coordinate system, the metric element is in isometric form.

8.4.1 Beltrami's Equation

Let there be
$$U\,du + V\,dv, \quad U\,du + V'dv,$$
the imaginary conjugate factors, introduced by Gauss, in which the right-hand side of (8.1.4) is decomposed. We have

$$U = \sqrt{E}, \ \ V = \frac{F + \mathrm{i}\,H}{\sqrt{E}}, \ \ V' = \frac{F - \mathrm{i}\,H}{\sqrt{E}}. \tag{8.4.3}$$

Then $V\,V' = G$. With the same symbols of Section 8.2, we consider the ratio

$$\frac{U\delta u + V\delta v}{U\,du + V\,dv} \equiv \frac{E\delta u + F\delta v + \mathrm{i}\,H\delta v}{E\,du + F\,dv + \mathrm{i}\,H\,dv}.$$

The complex term in the right-hand side (obtained by multiplying the numerator and denominator by the conjugate of the denominator) can be put in exponential form $\rho e^{\mathrm{i}\,\lambda}$ and, equating the real and imaginary parts from (8.4.1), we have

$$\rho\cos\lambda = \frac{\delta s}{ds}\cos\epsilon, \quad \rho\sin\lambda = \frac{\delta s}{ds}\sin\epsilon,$$

and $\rho = \frac{\delta s}{ds}$, $\lambda = \epsilon$; then

$$\frac{U\delta u + V\delta v}{\delta s} = \frac{U\,du + V\,dv}{ds}e^{\mathrm{i}\,\epsilon}. \tag{8.4.4}$$

If we take $\delta s = ds$, the element in the left-hand side can be considered as given by a rotation of an angle ϵ of the element in the right-hand side. Then the complex differential binomial $U\delta u + V\delta v$ is obtained from the binomial $U\,du + V\,dv$ multiplied by $e^{\mathrm{i}\,\epsilon}$, where ϵ is the rotation angle. This property is the same as that is satisfied by complex variable $x + iy$ represented in a Gauss plane. This property is true even if we multiply $U du + V dv$ by an integrating factor (in general complex) $\gamma(u, v)$, so that $\gamma(U du + V dv) = dp + i\, dq \equiv dw$. Since $\bar{\gamma}(U du + V' dv) = dp - i\, dq \equiv d\bar{w}$, we see that the variables p, q are the same used by Gauss for expressing ds^2 in isometric coordinates. Actually we have

$$ds^2 \equiv (U du + V dv)(U du + V' dv) \equiv \frac{dp^2 + dq^2}{|\gamma|^2}.$$

From these observations we infer that *if we want to fruitfully apply to surfaces the theory of functions of a complex variable, we do not have to use $u + i\, v$ but rather the variables p, q, introduced by Gauss, after multiplication by a complex integrating factor $\gamma(u, v)$, so that $\gamma(U du + V dv) = dp + i\, dq \equiv dw$ and $\bar{\gamma}(U du + V' dv) = dp - i\, dq \equiv d\bar{w}$.*

The integration can be performed just for particular problems, however some peculiar characteristics of the Gauss functions $f(u, v)$, for transforming a metric element into an isometric form, can be known.

Let us consider its derivative

$$f_{,w} = \frac{f_{,u} du + f_{,v} dv}{\gamma(U du + V dv)} \equiv \frac{f_{,u} + f_{,v} v_{,u}}{\gamma(U + V v_{,u})}. \tag{8.4.5}$$

Following Riemann, this derivative must be independent of direction $v_{,u}$, then it must be

$$U f_{,v} - V f_{,u} = 0.$$

All the characteristic properties of $f(w)$ are contained in this equation.

Actually, by substituting U, V from (8.4.3), we have

$$E f_{,v} - F f_{,u} = i H f_{,u}. \tag{8.4.6}$$

By squaring, we have

$$E f_{,v}^2 - 2 F f_{,u} f_{,v} + G f_{,u}^2 = 0,$$

which means

$$\Delta_1 f = 0. \tag{8.4.7}$$

This is the first property: *the functions $f(w)$, as functions of u, v, have null the first differential parameter and are the only ones with this property.*

Writing the function f by means of its real and imaginary parts, $f = \phi + i\psi$, and substituting in (8.4.7) we obtain $\Delta_1 \phi - \Delta_1 \psi + 2i\Delta_1(\phi, \psi) = 0$, and since ϕ, ψ are real functions it follows that

$$\text{I)} \quad \Delta_1(\phi, \psi) = 0 , \qquad\qquad \text{II)} \quad \Delta_1 \phi = \Delta_1 \psi.$$

The first equation means that the family of curves $\phi = Const$, $\psi = Const$ are orthogonal. The second one that curves $\phi = Const$, $\psi = Const$ are isometric.

Substituting $f = \phi + i\psi$ in (8.4.6) we obtain the real equations

$$\frac{E\phi_{,v} - F\phi_{,u}}{H} = -\psi_{,u}; \qquad \frac{E\psi_{,v} - F\psi_{,u}}{H} = \phi_{,u}, \qquad (8.4.8)$$

and also

$$\frac{G\phi_{,u} - F\phi_{,v}}{H} = \psi_{,v}; \qquad \frac{G\psi_{,u} - F\psi_{,v}}{H} = -\phi_{,v}. \qquad (8.4.9)$$

These equations give necessary and sufficient conditions for characterizing the real and imaginary parts of the function f, which allow us to solve the proposed problem.

Differentiating and equating the second mixed derivatives, we obtain an equation for both ϕ or ψ which is today called *Beltrami's equation*

$$L\phi \equiv \left[\left(\frac{G\phi_{,u} - F\phi_{,v}}{H} \right)_{,u} + \left(\frac{E\phi_{,v} - F\phi_{,u}}{H} \right)_{,v} \right] = 0 \qquad (8.4.10)$$

and the operator L is called *Beltrami's operator* [29]. The same equation is obtained for the function ψ. Thanks to (8.4.2), we can write

$$\Delta_2 \phi = 0, \qquad \Delta_2 \psi = 0. \qquad (8.4.11)$$

The complex functions $f = \phi + i\,\psi$, given by two conjugate solutions of Beltrami's equations (8.4.11) can be called, following Beltrami, *complex variables on surfaces.* Then summarizing we have obtained for functions ϕ, ψ the following properties:

1. The first and the second differential parameters are zero.

2. The real components have the second differential parameters equal to zero.

3. All real functions ϕ, which satisfy Beltrami's equation, can be considered as the real part of a complex function $f(w)$.

Functions of Ordinary Complex Variables

$f(w)$ is a function of $u + iv$ if (7.1.1) are satisfied. By means of these conditions we can eliminate ψ from (8.4.8), and we find, for the unknown $\phi_{,v}$, $\phi_{,u}$, the homogeneous linear system

$$(H - E)\phi_{,v} + F\phi_{,u} = 0, \quad F\phi_{,v} - (H - E)\phi_{,u} = 0;$$

this system is satisfied only if the determinant of the system $(H - E)^2 + F^2 = 0$, equivalent to the conditions $F = 0$, $H = E$. Since $H^2 = EG$, it follows that $E = G$, which are the conditions for isometric expression of the line element.

So we have: *for isometric coordinates on surfaces the same properties which are satisfied on a plane (referred to x, y coordinates) by the function of an ordinary complex variable x + i y hold.* Actually Beltrami's equation allows us to reduce the general partial differential elliptic equations to canonic form [75].

8.5   Beltrami's Integration of Geodesic Equations

In the previous chapters we have seen how hyperbolic numbers and functions of
a hyperbolic variable can be successfully used for solving problems in space-time
as well as complex analysis is used for problems in the Euclidean plane. So it is
natural to try to extend Beltrami's results, just reported, to "Lorentz surfaces",
i.e., to space-time surfaces described by a non-definite quadratic differential form.

   We begin by recalling another Beltrami theorem which refers to integration
of geodesic equations.

8.5.1   Differential Parameter and Geodesic Equations

We recall some concepts of classical differential geometry [6] and [34] which do
not appear in recent books [28] and [57]. In particular we report a method for
obtaining the equations of the geodesics firstly obtained by E. Beltrami ([4], Vol.
I, p. 366) who extended to line elements an integration method used by Jacobi for
integrating the dynamics equations.

   Let us extend the definitions of differential parameters to an N-dimensional
semi-Riemannian space V_N with a line element given by ([34], p. 39)

$$ds^2 = e \sum_{i,\,k=1}^{N} g_{ik}\, dx^i\, dx^k, \tag{8.5.1}$$

where $e = \pm 1$ in a way that $ds^2 \geq 0$. If U, V are any real functions of the x^i
$(i = 1, 2, \ldots, N)$, the invariants defined by

$$\Delta_1 U = g^{ih}\frac{\partial U}{\partial x^i}\frac{\partial U}{\partial x^h} \equiv g^{ih}U_{,i}\,U_{,h}, \tag{8.5.2}$$

$$\Delta_1\,(U,\,V) = g^{ih}\frac{\partial U}{\partial x^i}\frac{\partial V}{\partial x^h} \equiv g^{ih}U_{,i}\,V_{,h} \tag{8.5.3}$$

where g^{ik} are the reciprocal elements of the metric tensor g_{ik}, are called *Beltrami's
differential parameters of the first order*. Since the equations $U = const$, $V = const$
represent $(N - 1)$-dimensional hypersurfaces in V_N, $\Delta_1 U$ represents the squared
length of the gradient of U as well as of a vector orthogonal to the hypersurface
$U = const$; for the same reason, if $\Delta(U, V) = 0$, then the two hypersurfaces
$U = const$ and $V = const$ are orthogonal.

The Equations of the Geodesics

We report an important application of differential parameters, due to Beltrami,
which is an alternative way (with respect to the classical one) for obtaining the
equations for geodesics [34]. This method, recently applied for integrating important fields of General Relativity [7], is used in this section, in Chapter 9 and in
Appendix B.

Let us now consider the non-linear partial differential equation of the first order

$$\Delta_1 \tau = e,$$ (8.5.4)

where e is the same as in (8.5.1). The solution of this equation depends on an additive constant and on $N - 1$ essential constants α_i [34]. Now if we know a complete solution of (8.5.4), we can obtain the equations of the geodesics by the following theorem ([6], p. 299) and ([34], p. 59).

Theorem 8.2. *When a complete solution of (8.5.4) is known, depending on $N - 1$ essential constants α_i, the equations of the geodesics are given by*

$$\partial \tau / \partial \alpha_i = \beta_i,$$ (8.5.5)

where β_i are new arbitrary constants, and the arc of the geodesics is given by the value of τ.

Obviously the above theorem is particularly useful for applications with respect to Euler's differential system

$$\frac{d^2 x^l}{ds^2} + \sum_{i,\,k=1}^{N} \Gamma_{ik}^l \frac{dx^i}{ds} \frac{dx^k}{ds} = 0,$$ (8.5.6)

when we have a complete solution of (8.5.4) at our disposal or this solution is easily obtained.

To be sure, the integration of a partial differential equation is usually considered as a problem more difficult than that of a system of ordinary differential equations. Actually Hamilton and Jacobi achieved a major success by recognizing that this relationship may be reversed. In particular a result due to Bianchi [6] and [34] is the following:

If the fundamental form (8.5.1) can be re-expressed in the generalized Liouville form

$$ds^2 = e\left[X_1(x_1) + X_2(x_2) + \cdots + X_N(x_N)\right] \sum_{i=1}^{N} e_i (dx^i)^2,$$ (8.5.7)

where $e_i = \pm 1$ and X_i is a function of x^i alone, then a complete integral of (8.5.4) is

$$\tau = c + \sum_{i=1}^{N} \int \sqrt{e_i(e\,X_i + \alpha_i)}\, dx^i,$$ (8.5.8)

where c and α_i are constants, the latter being subject to the condition $\sum_{i=1}^{N} \alpha_i = 0$.

In this case the geodesic equations are immediately given by

$$\frac{\partial \tau}{\partial \alpha_l} \equiv \frac{1}{2} \int \frac{e_i\, dx_i}{\sqrt{e_i(e\,X_i + \alpha_l)}} = \beta_l.$$ (8.5.9)

Geodesics Equations in the Hyperbolic Plane

Let us extend to infinitesimal distances the hyperbolic or space-time distance defined in Section 4.1.1,

$$ds^2 = dx^2 - dy^2. \tag{8.5.10}$$

By means of the method just summarized, we can find the geodesics in pseudo-Euclidean plane, described by line element (8.5.10). At first we have to solve the partial differential equation with constant coefficients

$$\Delta_1 \tau \equiv \left(\frac{\partial \tau}{\partial x}\right)^2 - \left(\frac{\partial \tau}{\partial y}\right)^2 = e \equiv \pm 1.$$

The elementary solution is given by $\tau = Ax + By + C$ with the condition $A^2 - B^2 = \pm 1$. Then we can put

$$\text{I)} \quad A = \cosh\theta, \ B = \sinh\theta, \quad \text{if} \quad \Delta_1 \tau \ = \ +1 \ ; \tag{8.5.11}$$

$$\text{II)} \quad B = \cosh\theta, \ A = \sinh\theta, \quad \text{if} \quad \Delta_1 \tau \ = \ -1 \ . \tag{8.5.12}$$

The equations of the geodesics are obtained by equating $\partial \tau / \partial \theta$ to a constant c:

$$\text{I)} \quad \text{if} \quad \Delta_1 \tau = +1 \ \Rightarrow \ x \sinh\theta + y \cosh\theta = c,$$

$$\text{II)} \quad \text{if} \quad \Delta_1 \tau = -1 \ \Rightarrow \ x \cosh\theta + y \sinh\theta = c. \tag{8.5.13}$$

Equations (8.5.13) demonstrate the result anticipated in Chapter 4: *In the pseudo-Euclidean plane it is more appropriate to write the straight line equations by means of hyperbolic trigonometric functions instead of circular trigonometric functions as it is done in Euclidean plane.* We note that integration of the equations of the geodesics in Euler's form (8.5.6) would not give the condition on the integration constants of (8.5.11) and (8.5.12) which allows us to introduce the hyperbolic trigonometric functions in a natural way. With the definitions of Chapter 4, we see that the geodesics (8.5.13) represent lines of the first kind for $\Delta_1 \tau = +1$ and lines of the second kind for $\Delta_1 \tau = -1$.

Equations (8.5.13) in parametric form become

$$\text{I)} \quad \begin{cases} x = \phi \cosh\theta + x_o \\ y = -\phi \sinh\theta + y_o \end{cases} \quad \text{and} \quad \text{II)} \quad \begin{cases} x = \phi \sinh\theta + x_o \\ y = -\phi \cosh\theta + y_o \end{cases}. \tag{8.5.14}$$

By means of this integration method we have obtained at once the two kinds of straight lines anticipated in Section 4.3.

Now let us consider, together with straight line (8.5.13, I), a second one ($x \sinh\theta_1 + y \cosh\theta_1 = \gamma$) and calculate with the methods of differential geometry the trigonometric functions of the crossing angle θ_c. Actually in the Riemannian geometry by means of differential parameters we have $\cos\theta_c = \frac{\Delta_1(u,v)}{\sqrt{\Delta_1(u)\Delta_1(v)}}$. Now performing the calculations with the metric element (8.5.10) we obtain

$\Delta_1(u, v) = \cosh(\theta - \theta_1)$, $\Delta_1(u) = \Delta_1(v) = 1$. The cosine of Riemannian geometry has been substituted by hyperbolic cosine, the same result is obtained for sine.

The same results hold for surface geometry and expressions (8.5.2) become

$$ds\delta s \cosh \epsilon = Edu\delta u + F(du\delta v + dv\delta u) + Gdv\delta v, \qquad (8.5.15)$$

$$ds\delta s \sinh \epsilon = H(du\delta v - dv\delta u). \qquad (8.5.16)$$

To these expressions we can give the same meaning of classical differential geometry, (8.5.15) represents the scalar product and, thanks to results of Chapter 4, (8.5.16) represents an area.

8.6 Extension of Beltrami's Equation to Non-Definite Differential Forms

Using the theory developed in previous chapters by means of hyperbolic numbers, we can extend Beltrami's results to Lorentz surfaces, by the following

Theorem 8.3. *The "Gauss' problem", for both definite and non-definite quadratic forms, is solved by the same Beltrami's equation, taking the absolute value of H.*

Proof. Let us consider a differential quadratic form

$$ds^2 = Edu^2 + 2Fdudv + Gdv^2 \qquad (8.6.1)$$

with $\Delta^2 \equiv F^2 - EG > 0$, which represents a Lorentz surface, and its decomposition into linear elements

$$Udu + Vdv, \quad Udu + V'dv.$$

By analogy with definite forms, we have the decomposition into two hyperbolic conjugate factors

$$U = \sqrt{E}, \ V = \frac{F + h\Delta}{\sqrt{E}}, \ V' = \frac{F - h\Delta}{\sqrt{E}}; \qquad (8.6.2)$$

we proceed as Beltrami and, with the same meaning of symbols, we put the ratio

$$\frac{U\delta u + V\delta v}{Udu + Vdv} \equiv \frac{E\delta u + F\delta v + h\,H\delta v}{Edu + Fdv + h\,Hdv} = \rho e^{h\,\lambda}$$

in exponential form and, equating the real and hyperbolic parts, by means of (8.5.16) we have

$$\rho \cosh \lambda = \frac{\delta s}{ds} \cosh \epsilon, \quad \rho \sinh \lambda = \frac{\delta s}{ds} \sinh \epsilon.$$

Then $\rho = \frac{\delta s}{ds}$, $\lambda = \epsilon$, and

$$\frac{U\delta u + V\delta v}{\delta s} = \frac{U du + V dv}{ds} e^{h\,\epsilon}. \qquad (8.6.3)$$

If $\delta s = ds$, the element on the left-hand side is obtained from the element on the right-hand side by means of a hyperbolic rotation ϵ. This result can be interpreted as a Lorentz transformation between the differential quantities $U\delta u + V\delta v$ and $U du + V dv$. Then, as for a complex variable, if we want to fruitfully apply to Lorentz surfaces the theory of functions of a hyperbolic variable, we do not have to use $u + h\,v$ but rather the variables p, q obtained by Gauss integration, after multiplication by a hyperbolic integrating factor $\gamma(u, v)$, so that $\gamma(U du + V dv) = dp + h\,dq \equiv dw$ and $\bar{\gamma}(U du + V' dv) = dp - h\,dq \equiv d\bar{w}$. So we have

$$ds^2 \equiv (U du + V dv)(U du + V' dv) \equiv \frac{dp^2 - dq^2}{|\gamma|^2}. \qquad (8.6.4)$$

The integration can be performed just in particular cases, in any case some characteristics of the transformation functions $f(u, v)$ can be known. Equation (8.4.6) becomes

$$E f_{,v} - F f_{,u} = h\,\Delta f_{,u.}. \qquad (8.6.5)$$

Squaring, we have $E f_{,v}^2 - 2F f_{,u} f_{,v} + G f_{,u}^2 = 0$, which means

$$\Delta_1 f = 0. \qquad (8.6.6)$$

Then *the first differential parameter of functions $f(w)$ is zero*.

Calling ϕ, ψ the real and hyperbolic parts we can write $f = \phi + h\,\psi$, and substituting in (8.6.6) we find $\Delta_1 \phi + \Delta_1 \psi + 2h\,\Delta_1(\phi,\ \psi) = 0$ and, for the real functions ϕ, ψ, we obtain

$$\Delta_1(\phi,\ \psi) = 0, \quad \Delta_1 \phi = -\Delta_1 \psi.$$

The first equation indicates that curves $\phi = Const$ are orthogonal to curves $\psi = Const$, the second equation indicates that curves $\phi = Const$, $\psi = Const$ are isometric.

Substituting $f = \phi + h\,\psi$ in (8.6.5), we find the real equations

$$\frac{E\phi_{,v} - F\phi_{,u}}{\Delta} = \psi_{,u}\,; \qquad \frac{E\psi_{,v} - F\psi_{,u}}{\Delta} = \phi_{,u} \qquad (8.6.7)$$

and also

$$\frac{G\phi_{,u} - F\phi_{,v}}{\Delta} = -\psi_{,v}\,; \qquad \frac{G\psi_{,u} - F\psi_{,v}}{\Delta} = -\phi_{,v}. \qquad (8.6.8)$$

These equations show the differential properties of real and hyperbolic parts of the functions f which solve the proposed problem.

Actually, differentiating and equating the second mixed derivatives, we obtain an equation for both ϕ or ψ which, as for complex variables, corresponds to Beltrami's equation

$$L\phi \equiv \left[\left(\frac{G\phi_{,u} - F\phi_{,v}}{\Delta} \right)_{,u} + \left(\frac{E\phi_{,v} - F\phi_{,u}}{\Delta} \right)_{,v} \right] = 0, \qquad (8.6.9)$$

which can be written

$$\Delta_2 \phi = 0, \qquad \Delta_2 \psi = 0. \qquad (8.6.10)$$

$\square$

Summarizing we have obtained for functions ϕ, ψ the following properties:

1. The first and the second differential parameters are zero.

2. The real components have the second differential parameters equal to zero.

3. A real function ϕ, which satisfies (8.6.9), can be considered as the real part of a function $f(w)$ of a hyperbolic variable.

Conclusion

In Chapter 4 we have seen that the introduction of hyperbolic numbers has allowed a Euclidean formalization of trigonometry in the "flat" Minkowski space-time. In a similar way, thanks to the introduction of functions of a hyperbolic variable, we have extended the classical results of Gauss and Beltrami to two-dimensional varieties with non-definite metric elements (Lorentz surfaces).

Chapter 9

Constant Curvature Lorentz Surfaces

9.1 Introduction

It is known that, in the XVIIIth century, the growth of a new physics also drove the mathematics into new ways with respect to Euclidean geometry which, following Plato and Galileo, had been considered the "measure and interpretation" of the world. From then on new subjects emerge, súch as differential calculus, complex numbers, analytic, differential and non-Euclidean geometries, functions of a complex variable, partial differential equations and, at the end of the XIXth century, group theory.

This apparent breaking up of a unitary culture was unified again with the constant curvature surfaces studies and, in this way, many subjects, which looked to be abstract, found a concrete application.

The most important aspects of these surfaces derive from the fundamental works of Riemann ([61], p. 304), (1854) and Beltrami [2], (1868). In the former it was shown that the constant curvature spaces are the only ones, among the spaces which today we call *Riemann spaces*, in which the same motions (roto-translations) of a flat Euclidean space are possible. In the latter it was shown that the non-Euclidean geometries of Bolyai–Lobachevski and Riemann himself can be represented in a Euclidean way, on constant curvature surfaces (negative and positive constant curvature, respectively).

Here we summarize the most important aspects of these studies, which have gone beyond mathematical importance and brought about a great change in scientific thought. In the historical description of the principal steps, we refer to the works of Beltrami, who made the most important contribution, developing ideas which were completed, from a formal point of view, by Klein (1872) and Poincaré (1882), using complex variables.

Beltrami, referring to the paper of Gauss summarized in Section 8.2, observed that among the representations of the sphere on a plane, the one in which the geodesics on the sphere (great circles) are transformed into geodesics on a plane (straight lines), is the most important from a practical point of view: this transformation allows a simple measurement of distances. From a geometrical point of view, it is known that this transformation corresponds to the projection of points of the spherical surface from the center of the sphere to a plane, tangent to the sphere at a pole. Beltrami inquired about the existence of other surfaces, represented by line element (8.1.4), for which a transformation of variables may exist, which transforms geodesics on the surface into straight lines on a plane.

The answer to this question is contained in a paper of 1865 ([2], p. 263), in which he stated that ◁ The only surfaces which may be represented on a plane, in such a way that geodesics correspond to straight lines, are just the ones which have constant curvature (null, positive, negative). ▷ The results for the first two cases were known because they correspond to the plane and the sphere. The research of the meaning of the third case led Beltrami to write a paper which is considered a milestone, not only for the history of science, but also for all scientific thought: *Saggio di interpretazione della geometria non-euclidea*, today known as *Saggio* ([2], p. 374).

Let us recall[1], from the introduction of this paper, some important passages in which the author inquires into the basic meaning of Euclidean geometry and extends it: ◁ The fundamental criterion of demonstration of elementary geometry is the *possibility of superimposing equal figures*. This criterion not only applies to plane, but also to the surfaces on which equal figures may exist in different positions. Surfaces with constant spherical curvature have this property unconditionally... The straight line is a fundamental element in figures and constructions of elementary geometry. A particular characteristic of this element is that it is completely determined by just two of its points... This particular feature not only characterizes straight lines on the plane, but also is peculiar to geodesic lines on surfaces... ▷ In particular ◁ an analogous property of straight lines completely applies to constant curvature surfaces, that is: if two surfaces, with the same constant curvature, are given and a geodesic line exists on each surface, superimposing one surface on the other, in such a way that the geodesics are superimposed in two points, they are superimposed along all their extension. The analogies between spherical and plane geometries are based on the aforesaid property. ▷ The same geometrical considerations, which are evident for the sphere, may be extended to negative constant curvature surfaces, which, as it has been recalled before, can be represented on a plane, in such a way that every geodesic corresponds to a straight line.

Beltrami demonstrated that ◁ the theorems of non-Euclidean geometry (of Lobachevsky) for the plane figures made by straight lines, are valid for the analogous figures made by geodesic lines on the Euclidean surfaces with negative constant curvature, which we can call pseudo-spherical. ▷ Then ◁ results that seemed to be inconsistent with the hypothesis of a plane can become consistent with a surface of the above-mentioned kind and, in this way, they can get a simple and satisfactory explanation. ▷ A geometrical representation of this constant curvature surface ◁ is given by the surface obtained by rotating "the constant tangent curve" also called "tractrix", ▷ and the geodesics on this "pseudo-sphere" may be mapped on a plane and become the chords of a circle, in which distances are not measured by means of Euclidean geometry, but by Gauss geometry. The lengths become infinite near the circumference, so the circle represents a *limiting circle*[2]. Chords

---

[1]Many of the arguments, summarized in Beltrami's words, will be more thoroughly explored in the remainder of this chapter.

[2]A representation of the figures in this circle has been given by M.C. Escher in the woodcuts

of infinite length in the circle may be compared with straight lines in the plane. Beltrami concludes: ◁ With the present work we have offered the development of a case in which abstract geometry finds a representation in the concrete (Euclidean) geometry, but we do not want to omit to state that the validity of this new order of concepts is not subordinate to the existence (or not) of such a correspondence (as it surely happens for more than two dimensions). ▷

Now we recall the very important epistemological aspect of this work: the formulation of Euclidean geometry starts from postulates which are in agreement with the experience and all the following statements are deduced from these postulates (axiomatic-deductive method). Lobachevsky uses the same axiomatic-deductive method, but he starts from "abstract" postulates, which do not arise from experience. The work of Beltrami demonstrates that the theorems of Lobachevsky are valid on a surface of Euclidean space, i.e., *also starting from arbitrary axioms, we can obtain results in agreement with our "Euclidean" experience*. These concepts can be generalized and applied to scientific research: arbitrary hypotheses acquire validity if the consequent results are in agreement with experience. The most important discoveries of modern physics (Maxwell equations, special and general relativity, quantum theory, Dirac equation) arose from this "conceptual broad-mindedness".

Coming back to Beltrami's interpretation, a lot of great mathematicians completed his interpretation from many points of view (elementary mathematics, differential and projective geometries, group theory [36]) and the definitive mathematical formalization was performed by means of complex analysis. All these studies for a definite line element could have been extended to non-definite forms but these last ones were considered too far from common sense as Poincaré pointed out in a famous paper [59]. These considerations were surpassed by the theory of special relativity. Actually the most important objection of Poincaré was that the common sense can not accept geometries in which the distance between two distinct points is zero. As is today well known, this is the property of the Minkowski space-time, then the relation between space and time stated by special relativity gives a physical meaning to non-definite quadratic forms; following Beltrami, we can say that *results which seemed to be inconsistent with the hypotheses of Euclidean geometry (symmetry) can become consistent with another important physical symmetry and, in this way, they can get a satisfactory explanation*. This "symmetry" can be extended to surfaces characterized by non-definite line elements which are today named *Lorentz surfaces*. Then as Riemann surfaces with constant curvature (definite line elements) form one of the essential topics of geometry, so Lorentz surfaces with constant curvature (non-definite line elements) are relevant for physics [77].

A milestone in the study of the former is the application of complex numbers theory, which yields the best mathematical formalization [25], [29] and [81].

Circle Limit I–IV (1958–60). In these paintings the correspondence between the representation and the properties peculiar to non-Euclidean geometry are increasing from I to III ("mathematically perfect" [63]).

In the same way, the formalization of hyperbolic trigonometry in Chapter 4, the
introduction of functions of a hyperbolic variable in Chapter 7, the extension of
Gauss' and Beltrami's theorems in Chapter 8 allow us to use "hyperbolic mathematics" for a complete formalization of constant curvature Lorentz surfaces. We
shall see that the functions of a hyperbolic variable and the methods of classical
differential geometry allow us to establish a complete analogy between Riemann's
and Lorentz's constant curvature surfaces. In particular, expressions of the line elements and equations of the geodesics written as functions of hyperbolic variables
are the same as the ones written as functions of complex variables and motions
on both these surfaces may be expressed by complex or hyperbolic bilinear transformations.

   The obtained results also have a physical meaning. Actually, as we show in
Chapter 10, if the variables x, y are interpreted as time and one space variable,
respectively, the geodesics on constant curvature Lorentz surfaces represent relativistic hyperbolic motion. This result has been obtained just as a consequence
of the space-time symmetry stated by the Lorentz group, but it goes beyond special relativity, because the constant acceleration is linked with the curvature of
surfaces, as stated by Einstein's field equations [31], as we show in Chapter 10.

9.2   Constant Curvature Riemann Surfaces

We begin this chapter by recalling the classical studies on constant curvature
Riemann surfaces. This exposition allows us to show how the study of constant
curvature surfaces can be considered as an emblematic example of the development
of the scientific method in the last two centuries. Actually the starting point is
the study of the positive constant curvature Riemann surface represented by the
sphere already considered in Gauss' work (Section 8.2). Its study is performed by
means of Euclidean geometry and also its representation on a plane is obtained as a
geometrical projection of the spherical surface on a Cartesian plane. The following
step is the study of the Euclidean "tractrix" for which the representation on a
Cartesian plane can be obtained by a variable transformation. The same result
can also be obtained [29] by considering a two-sheets hyperboloid represented in a
Minkowski three-dimensional space-time which, for its non-definite metric, is out
of our geometrical intuition. When this analytical way has been taken it becomes
possible to extend the obtained results to a one-sheet hyperboloid represented in
a three-dimensional space with non-definite metric. On this hyperboloid we can
represent the constant curvature Lorentz surfaces [81].

9.2.1   Rotation Surfaces

The parametric equations of the sphere (8.2.10) can be considered as representative
of a rotation surface. These surfaces are very important in the present study
because, among them, there are positive and negative constant curvature surfaces.

Therefore we start by introducing the general form of the line element of rotation surfaces.

Line Element

Let us consider in a three-dimensional space, structured by an orthogonal Cartesian frame (O, X, Y, Z), the curve $Z = F(X)$ on the X, Z plane and the surface obtained by rotating this curve around the Z axis. Calling r, v two parameters, we have the rotation surface

$$X = r \cos v, \quad Y = r \sin v, \quad Z = F(r); \tag{9.2.1}$$

the function $Z = F(r)$ is called the *generating curve* or also *meridian curve*.

The line element (8.1.4), as a function of the parameters r, v, is given by

$$ds^2 = [1 + F'^2(r)]dr^2 + r^2 dv^2, \tag{9.2.2}$$

where $F'(r)$ indicates the derivative of $F(r)$ with respect to r. Introducing the new variable

$$u = \int \sqrt{1 + F'^2(r)} \, dr, \tag{9.2.3}$$

we can express r as a function of u and obtain

$$ds^2 = du^2 + r^2(u)dv^2 \equiv r^2(u) \left(\frac{du^2}{r^2(u)} + dv^2 \right), \tag{9.2.4}$$

which is considered ([6], p. 141) the characteristic form for line elements of rotation surfaces. The isometric form is obtained by introducing the new variable $\rho = \pm \int d u / r(u)$, and we have the line element

$$ds^2 = f(\rho)(d\rho^2 + dv^2). \tag{9.2.5}$$

It is known that, given a general line element, we can obtain a corresponding surface just in some particular case, but this is always possible for line element (9.2.4), since a problem about differential equations is considered as solved if we obtain a solution in integral form. We have ([6], p. 337)

$$Z(u) = \pm \int \sqrt{1 - \left(\frac{dr}{du}\right)^2} \, du, \tag{9.2.6}$$

or

$$Z(r) = \pm \int \sqrt{\left(\frac{du}{dr}\right)^2 - 1} \, dr. \tag{9.2.7}$$

Equations of Geodesics

For the line element expressed by (9.2.5), the equations of geodesics can be easily obtained by means of Theorem 8.2, p. 131.

Actually let us consider a surface represented by the line element

$$ds^2 = f(\rho)(d\rho^2 + d\phi^2).$$

The first step is the solution of the partial differential equation

$$\Delta_1 \tau \equiv \left(\frac{\partial \tau}{\partial \rho}\right)^2 + \left(\frac{\partial \tau}{\partial \phi}\right)^2 = f(\rho).$$

Here we follow a direct procedure, without using the general solution given in Section 8.5.1. By the substitution $\tau = A\phi + \tau_1(\rho) + C$, where A, C are constants, we obtain the solution in integral form,

$$\tau = A\phi + \int \sqrt{f(\rho) - A^2}\, d\rho + C. \tag{9.2.8}$$

The equations of geodesics are given, in implicit form, by

$$\frac{\partial \tau}{\partial A} \equiv \phi - \int \frac{A\, d\rho}{\sqrt{f(\rho) - A^2}} = B. \tag{9.2.9}$$

Now we show

Theorem 9.1. *The coordinates ρ and ϕ can be expressed as functions of the line parameter τ.*

Proof. From the identity

$$\tau - A\frac{\partial \tau}{\partial A} \equiv \tau - AB, \tag{9.2.10}$$

in which the terms on the left-hand side are evaluated by (9.2.8) and (9.2.9), we obtain a relation between τ and ρ,

$$\tau = AB + \int \frac{f(\rho)\, d\rho}{\sqrt{f(\rho) - A^2}}, \tag{9.2.11}$$

and, by using (9.2.9), we obtain a relation between τ and ϕ. $\qquad\square$

This result allows one to write the geodesics in the same parametric form as the Euler equations given by (8.5.6).

9.2.2 Positive Constant Curvature Surface

Line Element

For metric elements (9.2.4), the curvature (8.1.5) becomes

$$K = -\frac{1}{r(u)}\frac{d^2 r(u)}{du^2}. \tag{9.2.12}$$

If we look for a surface with $K = cost = \frac{1}{R^2}$ we obtain $r(u)$ for a positive constant curvature surface

$$r(u) = a \sin\left(\frac{u}{R} + b\right), \tag{9.2.13}$$

where a and b are integration constants. We do not lose in generality by setting $b = 0$, $a = R$ and, from (9.2.6), we have

$$Z(u) = \pm \int \sqrt{1 - \cos^2(u/R)}\, du = \pm R\cos(u/R). \tag{9.2.14}$$

Inverting (9.2.13), from (9.2.7) we have

$$Z(r) = \pm\sqrt{R^2 - r^2}. \tag{9.2.15}$$

By substituting (9.2.14) to X and Y in (9.2.1), we obtain the parametric equations of the sphere (8.2.10):

$$X = R\sin\left(\frac{u}{R}\right)\cos v, \quad Y = R\sin\left(\frac{u}{R}\right)\sin v, \quad Z = R\cos\left(\frac{u}{R}\right), \tag{9.2.16}$$

where $0 \leq u \leq R\pi$, $0 \leq v < 2\pi$. With respect to Gauss' expressions (8.2.10) in which u is the central angle, in (9.2.16), u is the arc on the surface. The metric element is given by

$$ds^2 = du^2 + R^2 \sin^2(u/R)\, dv^2 \equiv R^2 \sin^2(u/R)\left(\frac{du^2}{R^2\sin^2(u/R)} + dv^2\right). \tag{9.2.17}$$

Isometric Form of the Line Element, Representation of a Surface on the Plane

Using the Gauss procedure described in Section 8.2, by means of the transformation $u, v \to \rho, \phi$ where

$$\begin{aligned}
\rho &= -\int \frac{du}{R\sin(u/R)} \equiv \ln\cot(u/2R),\\
\phi &= v,
\end{aligned} \tag{9.2.18}$$

we obtain the isometric form of a line element

$$ds^2 = R^2\frac{d\rho^2 + d\phi^2}{\cosh^2\rho}. \tag{9.2.19}$$

This isometric form is preserved (Section 8.2) in a complex exponential transformation

$$x + \mathrm{i}\,y = \exp[\rho + \mathrm{i}\,\phi] \equiv \exp[\rho](\cos\phi + \mathrm{i}\sin\phi), \qquad (9.2.20)$$

which gives

$$ds^2 = 4R^2 \frac{dx^2 + dy^2}{(1 + x^2 + y^2)^2}. \qquad (9.2.21)$$

This transformation has a simple geometrical interpretation.

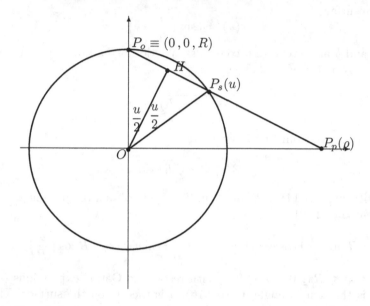

Figure 9.1: Projection of a plane section of a sphere from its north pole into its equatorial plane. From the similar triangles $P_p O P_o$ and $O H P_o$, we obtain $\overline{OP_o} = R\,\frac{\overline{OH}}{\overline{OP_o}} \equiv R\tan u/2$ and the relation $P_s(u) \to P_p(\rho)$.

Referring to Fig. 9.1, let us consider the projection of a point P_s, of the sphere $X^2 + Y^2 + Z^2 = R^2$, from the pole $P_0\,(0, 0, R)$ into the point P_p of the equatorial plane $Z = 0$. By considering the similar triangles OP_pP_0 and OHP_0, we obtain the polar coordinates [29]

$$\text{radial coordinate} \;\; : \;\; OP_p = R\,\frac{\cos(u/(2\,R))}{\sin(u/(2\,R))} = R\exp[\rho],$$

$$\text{angular coordinate} \;\; : \;\; v = \phi; \qquad\qquad (9.2.22)$$

then x, y can be considered as normalized (divided by R) Cartesian coordinates of a polar projection into the equatorial plane.

If on this plane we introduce the complex variable $z = x + \mathrm{i}\,y$ and the conjugate $\bar{z} = x - \mathrm{i}\,y$, we can write

$$ds^2 = 4R^2 \frac{dx^2 + dy^2}{(1 + x^2 + y^2)^2} \equiv 4R^2 \frac{dz\,d\bar{z}}{(1 + z\bar{z})^2}. \tag{9.2.23}$$

Relations Between Coordinates on the Plane and on the Spherical Surface

Here we see how a complex variable allows one to obtain useful results in a simple way. We begin by stating the link between the coordinates of the points on the sphere and the coordinates of the points on the plane. From (9.2.18) and (9.2.20) we have

$$z\bar{z} + 1 = \frac{1}{\sin^2(u/2R)} \; ; \; z + \bar{z} = 2\cot(u/2R)\cos v,$$

$$z\bar{z} - 1 = \frac{\cos(u/R)}{\sin^2(u/2R)} \; ; \; z - \bar{z} = 2\,\mathrm{i}\,\cot(u/2R)\sin v$$

and (9.2.16) become

$$X = R\frac{z + \bar{z}}{z\,\bar{z} + 1}, \quad Y = -\mathrm{i}\,R\frac{z - \bar{z}}{z\,\bar{z} + 1}, \quad Z = R\frac{z\bar{z} - 1}{z\,\bar{z} + 1}. \tag{9.2.24}$$

The inverse transformation gives the complex coordinate as a function of the coordinates of a point on the sphere

$$z = \frac{X + \mathrm{i}\,Y}{R - Z}.$$

As a first application we show: *in the polar projection all the circles on the sphere are mapped into circles on the equatorial plane and vice-versa.*

Proof. Actually the circles on the sphere are given by the intersections of the planes

$$aX + bY + cZ + d = 0 \tag{9.2.25}$$

with the sphere. By means of (9.2.24) we obtain the lines on the (x, y) plane

$$a(z + \bar{z}) - \mathrm{i}\,b(z - \bar{z}) + c(z\bar{z} - 1) + d(z\bar{z} + 1) = 0. \tag{9.2.26}$$

If $c + d \neq 0$, these lines are circles, if $c + d = 0$, they are straight lines. Equation (9.2.26) depends on three parameters, therefore represent all the circles on the (x, y) plane. □

In particular if $d = 0$, the planes (9.2.25) pass through the center of the sphere and their intersections are great circles, i.e., geodesic lines which depend on two parameters. From (9.2.26) we obtain the geodesic on the (x, y) plane

$$a(z + \bar{z}) - \mathrm{i}\,b(z - \bar{z}) + c(z\bar{z} - 1) = 0.$$

Geodesic lines can be also obtained by the integration method exposed in Section 9.2.1, as we shall see for the other constant curvature surfaces. The results for this and the following cases are summarized in Table 9.1.

Polar isometric forms		
	ds^2	Equations of the geodesics
(1)	$R^2\dfrac{d\rho^2+d\phi^2}{\cosh^2\rho}$	$\sin(\phi-\sigma)=\tan\epsilon\,\sinh\rho$ $\tanh\rho=\cos\epsilon\,\sin\left(\frac{\tau-\tau_0}{R}\right)$
(2)	$R^2\dfrac{d\rho^2+d\phi^2}{\sinh^2\rho}$	$\sin(\phi-\sigma)=\tanh\epsilon\,\cosh\rho$ $\coth\rho=\cosh\epsilon\,\cosh\left(\frac{\tau-\tau_0}{R}\right)$
(3)	$R^2\dfrac{d\rho^2-d\phi^2}{\cosh^2\rho}$	$\sinh(\sigma-\phi)=\tanh\epsilon\,\sinh\rho$ $\tanh\rho=\cosh\epsilon\,\sin\left(\frac{\tau-\tau_0}{R}\right)$
(4)	$R^2\dfrac{d\rho^2-d\phi^2}{\sinh^2\rho}$	$\sinh(\sigma-\phi)=\tan\epsilon\,\cosh\rho$ $\coth\rho=\cos\epsilon\,\cosh\left(\frac{\tau-\tau_0}{R}\right)$
Cartesian isometric forms		
	ds^2	Equations of the geodesics
(1)	$4R^2\dfrac{dx^2+dy^2}{(1+x^2+y^2)^2}$ $4R^2\dfrac{dz\,d\bar z}{(1+z\bar z)^2}$	$x^2+y^2+2\dfrac{\sin\sigma\,x-\cos\sigma\,y}{\tan\epsilon}-1=0$ $z\bar z+\mathrm{i}\dfrac{z\exp[-\mathrm{i}\,\sigma]-\bar z\exp[\mathrm{i}\,\sigma]}{\tan\epsilon}-1=0$
(2)	$4R^2dx^2+dy^2(1-x^2-y^2)^2$ $4R^2\dfrac{dz\,d\bar z}{(1-z\bar z)^2}$	$x^2+y^2+2\dfrac{\sin\sigma\,x-\cos\sigma\,y}{\tanh\epsilon}+1=0$ $z\bar z+\mathrm{i}\dfrac{z\exp[-\mathrm{i}\,\sigma]-\bar z\exp[\mathrm{i}\,\sigma]}{\tanh\epsilon}+1=0$
(3)	$4R^2\dfrac{dx^2-dy^2}{(1+x^2-y^2)^2}$ $4R^2\dfrac{dz\,d\bar z}{(1+z\bar z)^2}$	$x^2-y^2-2\dfrac{\sinh\sigma\,x-\cosh\sigma\,y}{\tanh\epsilon}-1=0$ $z\bar z+h\dfrac{z\exp[-h\sigma]-\bar z\exp[h\sigma]}{\tanh\epsilon}-1=0$
(4)	$4R^2\dfrac{dx^2-dy^2}{(1-x^2+y^2)^2}$ $4R^2\dfrac{dz\,d\bar z}{(1-z\bar z)^2}$	$x^2-y^2-2\dfrac{\sinh\sigma\,x-\cosh\sigma\,y}{\tan\epsilon}+1=0$ $z\bar z+h\dfrac{z\exp[-h\sigma]-\bar z\exp[h\sigma]}{\tan\epsilon}+1=0$

Table 9.1: *Line elements and equations of the geodesics for constant curvature surfaces with definite (1, 2) and non-definite (3, 4) line elements.* In rows (1) and (2) $z = x + \mathrm{i}\,y$ is a complex variable and in rows (3) and (4) $z = x + hy$ is a hyperbolic variable. x, y are the coordinates in the Cartesian representation; ϵ, σ are constants connected with the integration constants (A and B) of the equations of the geodesics as follows: $\sigma = B$ and $\epsilon = \sin^{-1}(A/R)$ in rows (1), (4) and $\epsilon = \sinh^{-1}(A/R)$ in rows (2), (3). The constants are determined by setting two conditions that fix the geodesic, as the passage in two points (see Table 9.2) or one point and the tangent in this point. τ is the line element on the arc of geodesics. $\tau_0 = AB$.

Motions

Now we see another important application of the complex variable formalism. From a geometrical point of view, given a surface and a generic figure represented on it, the allowed displacements of the figure on the surface, are called "motions". From an analytical point of view, we call motions the transformations of variables which leave the line element unchanged. Following Klein (see Chapter 3), the motions are a group and this gives motions a physical relevance and the name itself.

Actually the invariants of the physical conservation laws can be considered as the invariants of a group. In particular the conservation laws of momentum and angular momentum derive from the invariance of Newton dynamics equations with respect to translations and rotations, respectively. We have also seen (Section 7.1), how from the invariance with respect to rotations the Green function for two-dimensional Laplace equation can be obtained. In a similar way in Chapter 10, the relativistic hyperbolic motion shall be obtained, from the invariance with respect to hyperbolic rotations (Lorentz's transformations).

Now we observe a kind of "unification" of the constant curvature surfaces of the translation and rotation groups on the Euclidean plane. Actually the roto-translation group is "composed" by two independent groups characterized by a two-parameters group (translations) and a one-parameter group (rotations). On the other hand, referring to the most evident Euclidean example of a spherical surface, the motions are represented by just one three-parameters (as an example the Euler angles) group which gives the rotations around the center.

This Euclidean vision of motions is lost when we represent the spherical surface on a plane, since the induced geometry (distances, geodesics) is defined by a Gauss metric, i.e., just in an analytical way. In any case the use of complex numbers allows a very simple analytical description of motions.

Actually let us take a representation of the sphere on the complex z plane. From (9.2.26) we see that the circles on the sphere are mapped into circles on the plane, then since the motions on the sphere preserve the geometric figures, the same must happen for the circles on the z plane. We know from the theory of complex variables [69] that this property is held by bilinear mappings

$$w = \frac{\alpha z + \beta}{\gamma z + \delta}, \qquad (9.2.27)$$

where α, β, γ, δ are complex constants, three of which are independent. Among these general transformations we must take the ones which, from a geometrical point of view, transform, on the sphere, great circles into great circles and, on the plane, geodesic circles into geodesic circles. From an analytical point of view these transformations must leave the line element unchanged. By means of the geometrical approach these transformations have been obtained by Cayley and are named after him: *Cayley transformations* ([6], p. 152). The analytical approach can be found in ([29], p. 106). Actually the general mapping (9.2.27), is particularized to a transformation depending on two complex constants (α, β) and their conjugates

$$w = \frac{\alpha z + \beta}{-\bar{\beta} z + \bar{\alpha}} \qquad (9.2.28)$$

with the unimodularity condition $\alpha\bar{\alpha} + \beta\bar{\beta} = 1$. Since this condition is real it reduces the four parameters to three, and (9.2.28) depends on three real parameters instead of the six parameters of (9.2.27). The inverse transformation is

$$z = \frac{\bar{\alpha} w - \beta}{\bar{\beta} w + \alpha}. \qquad (9.2.29)$$

9.2.3 Negative Constant Curvature Surface

Line Element

As well as for the positive constant curvature surface with line element (9.2.4), setting $K = -\frac{1}{r(u)}\frac{d^2 r(u)}{du^2} \equiv cost \equiv -\frac{1}{R^2}$, we obtain, depending on the integration constants, three typical forms representative of negative constant curvature surfaces ([6], p. 337)

$$ds^2 = du^2 + \gamma^2 \exp\left[\frac{2u}{R}\right] dv^2, \tag{9.2.30}$$

$$ds^2 = du^2 + \lambda^2 \cosh^2\left[\frac{u}{R}\right] dv^2, \tag{9.2.31}$$

$$ds^2 = du^2 + \lambda^2 \sinh^2\left[\frac{u}{R}\right] dv^2. \tag{9.2.32}$$

In particular, for the form (9.2.30) we obtain by means of (9.2.7) the generating function expressed by elementary functions

$$Z(r) = \pm\left\{ \sqrt{R^2 - r^2} - R\ln\left[\frac{R + \sqrt{R^2 - r^2}}{r}\right] \right\}. \tag{9.2.33}$$

Introducing (9.2.33) into (9.2.1), parametric equations of the surface in three-dimensional Euclidean space are obtained, as functions of parameters r, v, with $0 < r \le R$, $0 \le v < 2\pi$.

The curve given by (9.2.33) in the reference frame (O, r, Z), is a "tractrix" with asymptote coincident with the Z axis. This curve is also called "the constant tangent curve" for the following property: if we consider a tangent straight line to the curve, the length of the segment between the contact point (T) and the point in which the tangent line in T crosses the asymptote of the curve, is constant. This property can be demonstrated by an analytical procedure starting from (9.2.33). As we have noted in Section 9.1, this rotation surface was called *pseudosfera* (pseudo-sphere) by Beltrami.

In the cases of line elements (9.2.31) and (9.2.32), $Z(r)$ can be expressed by means of elliptic integrals ([6], p. 337). In the present study we consider the line element (9.2.32), similar to line element (9.2.17) obtained for the sphere and, in a similar way ([29], p. 98), we obtain the polar isometric form

$$ds^2 = R^2 \frac{d\rho^2 + d\phi^2}{\sinh^2 \rho}. \tag{9.2.34}$$

By means of exponential transformation (9.2.20) we obtain the Cartesian isometric form, which like (9.2.23), can be also expressed by means of the complex variable $z = x + iy$ as

$$ds^2 = 4R^2 \frac{dx^2 + dy^2}{(1 - x^2 - y^2)^2} \equiv 4R^2 \frac{dz d\bar{z}}{(1 - z\bar{z})^2}. \tag{9.2.35}$$

This expression holds in the internal part of the circle $x^2 + y^2 < 1$ ([6], p. 614), ([29], p. 100) and, as $x^2 + y^2 \to 1$, we have $ds^2 \to \infty$.

Equations of Geodesics

Geodesic lines can be obtained by the integration method exposed in Section 9.2.1, substituting the general function $f(\rho)$ of (9.2.9) and (9.2.11) with $R^2/\sinh^2 \rho$. It is known that if we express the line element with new variables, by means of the same transformations we obtain the geodesics in a new reference frame. In particular by means of the transformation (9.2.20), we obtain the geodesics on an (x, y) plane. The results are given in the second row of Table 9.1.

9.2.4 Motions

As we have seen for positive constant curvature surfaces, the expression of line element (9.2.35) as a function of a complex variable allows us to obtain the motions. Actually the transformations which leave unchanged the expression of the line element are given again by bilinear complex transformations, similar to Cayley's formula, which leave also unchanged the limiting circle ([6], p. 618), ([29], p. 122)

$$w = \frac{\alpha z + \beta}{\bar{\beta} z + \bar{\alpha}}, \quad \text{with the inverse} \quad z = \frac{-\bar{\alpha} w + \beta}{\bar{\beta} w - \alpha}, \tag{9.2.36}$$

with $\alpha\bar{\alpha} - \beta\bar{\beta} = 1$.

Representation of Negative Constant Curvature Surfaces in a Half-Plane

This representation, due to Poincaré ([6], p. 614), ([25], p. 55) and ([29], p. 101), is epoch-making from historical and practical points of view. Actually, from the historical point of view, it has allowed the representation of the non-Euclidean geometry of Lobachevsky on a half-plane, i.e., out of a limited zone as it is the interior of a circle. From a practical point of view this representation allowed one to obtain the geodesic distance between two points as a cross ratio which is the quantity preserved in projective geometry with respect to the group of bilinear transformations. Moreover it represents another example of the applications of complex formalism for studying the extension of Euclidean geometry.

We add the personal note that just this use of the complex variable stimulates us to use the hyperbolic formalism for extending the Minkowski (or Lorentz) geometry to constant curvature surfaces shown in this chapter and, for doing this, to formalize the arguments as exposed in previous chapters.

From complex analysis we know ([6], p. 615 and [69]), that the transformation

$$\zeta \equiv \xi + i\eta = i\frac{1+z}{1-z} \equiv \frac{-2y}{(1-x)^2 + y^2} + i\frac{1 - (x^2 + y^2)}{(1-x)^2 + y^2}, \tag{9.2.37}$$

whose inverse is

$$z = \frac{\zeta - i}{\zeta + i},$$

(9.2.38)

maps the interior part of the unit circle with center at the coordinate axes origin of the z plane into the half-plane $\eta > 0$ of the ζ plane. In particular, the center of the circle is mapped into the point $(0, i)$, and its circumference into the $\eta = 0$ axis. The metric element (9.2.35) becomes ([6], p. 615)

$$ds^2 = R^2 \frac{d\xi^2 + d\eta^2}{\eta^2}.$$

(9.2.39)

Equations of Geodesics in the Half-Plane

Applying the method exposed in Section 9.2.1, in particular (9.2.9), to line element (9.2.39), $(f(\rho) = R^2/\eta^2)$, we obtain the equations of the geodesics ([6], p. 609), ([29], p. 101)

$$(\xi - B)^2 + \eta^2 = C,$$

(9.2.40)

where B and C are constants. Equation (9.2.40) represents circles orthogonal to the limiting straight line $\eta = 0$. Their center is on the ξ axis, and the straight line orthogonal to the ξ axis are also (particular) geodesics. We note that by means of the transformation (9.2.38), we can obtain the geodesics on the (x, y) plane. In agreement with the properties of conformal mappings and, in particular, of bilinear transformations, the geodesics are circles orthogonal to the limiting circle.

Motions in the Half-Plane

For this representation the motions are given by the unimodular bilinear transformations with real coefficients α, β, γ, δ ([6], p. 609),

$$\omega \equiv \mu + i\nu = \frac{\alpha \zeta + \beta}{\gamma \zeta + \delta}.$$

(9.2.41)

These transformations do not change the limiting line $\eta = 0$.

Actually we have ([29], p. 122)

Theorem 9.2. *The bilinear mappings represent the motions in the representations of negative constant curvature surfaces if they preserve the domains in which the Lobachevsky metric is represented.*

Geodesic Distance Between Two Points in the Half-Plane

Referring to line element (9.2.39) we begin by taking two points on the geodesic line represented by the $\xi = 0$ axis. In particular $A \equiv (0, a)$, $B \equiv (0, b)$ with

$0 < a < b$. The linear line element becomes $ds = d\eta/\eta$, and the (non-Euclidean) distance (indicated by d), is given by

$$d = R \int_a^b \frac{d\eta}{\eta} \equiv R \ln \frac{b}{a}. \tag{9.2.42}$$

This expression can be written so that it is invariant for bilinear mapping, i.e., as the cross ratio defined in the following way: given, on a line, four points (A, B, C, D), the quantity $\frac{\overline{AC}/\overline{AD}}{\overline{BC}/\overline{BD}}$, which is indicated by $(A, B; C, D)$, is called *cross ratio* ([81], p. 215). This relation also holds for the limiting case $A \to \infty$, $B \to 0$ ([6], p. 612), which represents the present one.

    Actually if we consider, together with points a, b, the points "∞" and "0" of the y axis, (9.2.42) can be written as

$$d = R \ln(b/a) \equiv R \ln(\infty, 0; a, b).$$

Taking into account that this expression is invariant with respect to bilinear mappings [30], we can extend the measure of distance to every point on the half-plane. Actually by means of (9.2.41) we can map the points A, B into every other two points (A', B') of the half-plane $\eta > 0$. By means of the same transformation the line $\xi = 0$ becomes the geodesic circle which passes through A', B', and the points ∞, 0 become the points in which the geodesic circle crosses the ξ axis. Indicating by M_1, M_2, these two last points, the geodesic distance can be written as

$$d = R \ln(M_1, M_2; A', B'). \tag{9.2.43}$$

This expression shows that, as it happens in Euclidean geometry, distance on constant curvature surfaces can be expressed as a function of the coordinates of the points. In some texts [25] and [30], in which the methods of differential geometry are not used, (9.2.43) is taken as an axiomatic definition of the distance since it satisfies the necessary requirements.

    From (9.2.42) or (9.2.43) we can see that $d(AB) = -d(BA)$. We can also see that if the point A is fixed and $B \to M_1$ or $\to M_2$, the cross ratio becomes ∞ or 0, respectively and, in both cases $d \to \infty$. *The points on the limiting line are at infinite non-Euclidean distance from all the finite points.*

    In Section 9.4 we shall see how the distance between two points can be expressed as a function of complex (or hyperbolic) variables in points A, B. And a similar result is also true for positive constant curvature surfaces for both definite and non-definite line elements. These results are summarized in Table 9.2.

9.2.5   Two-Sheets Hyperboloid in a Semi-Riemannian Space

Now we give another formalization of the geometry on a negative constant curvature Riemann surface, widely considered in modern books ([29], p. 98). In particular this treatment derives from the concept introduced by special relativity which, giving a physical meaning to non-definite quadratic forms, allows one to see in a similar frame both the two constant curvature Riemann surfaces and the two constant curvature Lorentz surfaces. Practically, by considering three-dimensional spaces with non-definite metrics [34], we can apply the same procedure developed for the sphere to other quadric surfaces, obtaining similar results [81].

Let us consider in three-dimensional space, structured by an orthogonal Cartesian frame (O, X, Y, Z), a two-sheets hyperboloid defined by the equation

$$X^2 + Y^2 - Z^2 = -R^2 \tag{9.2.44}$$

or, as a function of the parameters u, v,

$$X = R \sinh{(u/R)} \cos v; \quad Y = R \sinh{(u/R)} \sin v; \quad Z = \pm R \cosh{(u/R)}. \tag{9.2.45}$$

It is a rotation surface generated by an up-down hyperbola in the (X, Z) plane rotating around the Z axis. We suppose this figure in a space with non-definite metric (three-dimensional Minkowski space-time)

$$ds^2 = dX^2 + dY^2 - dZ^2. \tag{9.2.46}$$

The metric element for the two-sheets hyperboloid (9.2.45) is given by

$$ds^2 = du^2 + R^2 \sinh^2(u/R) \, dv^2 \equiv R^2 \sinh^2(u/R) \left(\frac{du^2}{R^2 \sinh^2(u/R)} + dv^2 \right), \tag{9.2.47}$$

which is the same as (9.2.32). This result is equivalent to the one obtained in Section 9.2.2 for the sphere in Euclidean space. Then the sphere in three-dimensional Euclidean space and the two-sheets hyperboloid in three-dimensional Minkowski space-time are positive and negative constant curvature Riemann surfaces [29]. This results in an example of a more general theorem ([34], p. 201).

Before introducing this theorem, let us point out that in a flat semi-Riemannian space with line element

$$ds^2 = \sum_{\alpha=0}^{N} c_\alpha (dx^\alpha)^2, \quad \text{with } c_\alpha \text{ positive or negative constants,} \tag{9.2.48}$$

the surfaces

$$\sum_{\alpha=0}^{N} c_\alpha (x^\alpha)^2 = \pm R^2, \quad \text{with the same } c_\alpha \text{ of (9.2.48),}$$

are called *fundamental hyperquadrics* [34] and [57]. For these surfaces we have ([34] p. 203)

Theorem 9.3. *The fundamental hyperquadrics are N-dimensional spaces of constant Riemannian curvature and are the only surfaces of constant Riemannian curvature of an $(N + 1)$-dimensional flat semi-Riemannian space.*

In particular, the constant curvature is positive if in the equation of the fundamental hyperquadric there is $+R^2$ and negative if there is $-R^2$.

9.3 Constant Curvature Lorentz Surfaces

9.3.1 Line Element

The theorem just exposed can be used for finding the constant curvature Lorentz surfaces, but here we follow a direct approach. Let us consider a Lorentz surface with the line element given by

$$ds^2 = du^2 - r^2(u)\, dv^2. \tag{9.3.1}$$

For this line element the Gauss curvature K is given by [29]

$$K = -\frac{1}{r(u)}\frac{d^2 r(u)}{du^2}.$$

In particular, if we put $K = \pm R^{-2}$ and with a suitable choice of the initial conditions, we obtain

- for positive constant curvature surfaces (PCC)

$$ds^2 = du^2 - R^2 \sin^2(u/R)\, dv^2, \tag{9.3.2}$$

- for negative constant curvature surfaces (NCC)

$$ds^2 = du^2 - R^2 \sinh^2(u/R)\, dv^2. \tag{9.3.3}$$

9.3.2 Isometric Forms of the Line Elements

As shown in Section 8.3, the non-definite line elements can be transformed into the isometric form [29]

$$ds^2 = f(\rho, \phi)(d\rho^2 - d\phi^2). \tag{9.3.4}$$

In particular, (9.3.2) is reduced into isometric form by the transformation

$$\rho = -\int \frac{du}{R\sin(u/R)} \equiv \ln \cot(u/2R); \qquad \phi = v$$

and (9.3.3) by the transformation

$$\rho = -\int \frac{du}{R\sinh(u/R)} \equiv \ln \tanh(u/2R); \qquad \phi = v.$$

We obtain

$$ds^2 = R^2\frac{d\rho^2 - d\phi^2}{\cosh^2 \rho} \quad \text{for PCC}, \qquad ds^2 = R^2\frac{d\rho^2 - d\phi^2}{\sinh^2 \rho} \quad \text{for NCC}. \tag{9.3.5}$$

As stated by Theorem 8.3 these isometric forms are preserved from transformations with functions of the hyperbolic variable. Then by means of the *hyperbolic polar transformation* (4.1.6)

$$x + h\,y = \exp[\rho + h\,\phi] \equiv \exp[\rho](\cosh\phi + h\sinh\phi),\qquad(9.3.6)$$

(9.3.5) can be rewritten in the form

$$ds^2 = 4R^2\frac{dx^2 - dy^2}{(1 + x^2 - y^2)^2}\quad\text{for PCC,}\qquad ds^2 = 4R^2\frac{dx^2 - dy^2}{(1 - x^2 + y^2)^2}\quad\text{for NCC,}$$

$$(9.3.7)$$

where x, y can be considered the coordinates in a Cartesian representation.

9.3.3   Equations of the Geodesics

With the method summarized in Section 9.2.1, we can find the equations of the geodesics for a surface with the particular line element

$$ds^2 = f(\rho)(d\rho^2 - d\phi^2)\,.$$

We obtain

$$\tau = A\,\phi + \int\sqrt{f(\rho) + A^2}\,d\rho + C\,.\qquad(9.3.8)$$

The equations of the geodesics are given by

$$\frac{\partial\tau}{\partial A} \equiv \phi + \int\frac{A\,d\rho}{\sqrt{f(\rho) + A^2}} = B.\qquad(9.3.9)$$

As in Section 9.2.1, from (9.3.8) and (9.3.9) we obtain a relation between ρ and the line parameter τ:

$$\tau - A\frac{\partial\tau}{\partial A} \equiv \int\frac{f(\rho)\,d\rho}{\sqrt{f(\rho) + A^2}} = \tau - A\,B.\qquad(9.3.10)$$

Application to Constant Curvature Surfaces

Let us consider a positive constant curvature surface with the line element given by (9.3.5). For this case we have $f(\rho) = R^2\cosh^{-2}\rho$ and (9.3.8)–(9.3.10) become, respectively,

$$\tau = A\,\phi + \int\frac{\sqrt{R^2 + A^2\cosh^2\rho}}{\cosh\rho}d\rho + C,\qquad(9.3.11)$$

$$\sqrt{(R/A)^2 + 1}\,\sinh(B - \phi) = \sinh\rho,\qquad(9.3.12)$$

$$\tau - AB = R\,\sin^{-1}\left[\frac{\tanh\rho}{\sqrt{1 + (A/R)^2}}\right].\qquad(9.3.13)$$

In order to obtain simplified final expressions we put in (9.3.12) and (9.3.13) $A = R \sinh \epsilon$, $B = \sigma$. In a similar way we obtain the equations of the geodesics for the negative constant curvature surfaces represented by (9.3.5). In this case we put $A = R \sin \epsilon$, $B = \sigma$.

Since we know from differential geometry [57] that, if we transform a line element, the same transformation holds for the equations of the geodesics, by substituting (9.3.6) into (9.3.12) we obtain the equations of the geodesics in a Cartesian (x, y) plane. Equation (9.3.12) becomes

$$2 \exp[\rho](\sinh \phi \cosh \sigma - \cosh \phi \sinh \sigma) = \tanh \epsilon (\exp[2\rho] - 1).$$

Substituting the x, y variables from (9.3.6) the expressions reported in Table 9.1 are obtained.

In the Cartesian representation of this table, we report the line elements and the equations of the geodesics as functions of complex (definite line elements 1 and 2) or hyperbolic (non-definite line elements 3 and 4) variables. These expressions for definite and non-definite line elements are the same if they are written as functions of the z variable (complex or hyperbolic, respectively) [81].

The "Limiting Curves"

The results reported in Table 9.1 for constant curvature surfaces with definite line element are well known and the geodesics are represented in the (x, y) plane by circles limited by the *"limiting circle"* [6] and [29]. The equation of the limiting circle is obtained by equating to zero the denominator of the line element in Cartesian coordinates. For negative constant curvature surfaces this equation is given by $x^2 + y^2 = R^2$, while for positive constant curvature surfaces it is given by $x^2 + y^2 = -R^2$, representing a circle with imaginary points. Then for positive constant curvature surfaces the geodesics are complete circles. Now we show that the same situation applies to constant curvature surfaces with non-definite line elements.

For positive as well as for negative constant curvature surfaces, the geodesics are hyperbolas of the form

$$(y - y_0)^2 - (x - x_0)^2 = d^2, \tag{9.3.14}$$

where x_0, y_0, d depend on two parameters which can be obtained as functions of A and B from the equations reported in Table 9.1. In particular, the half-diameter d of the geodesic hyperbolas, as well as the radius of the geodesic circles on the constant curvature Riemann surfaces, are

$$d = \frac{R}{A} \equiv \frac{1}{A\sqrt{|K|}}. \tag{9.3.15}$$

From the same equations we see that for $\epsilon \to 0$ ($A \to 0$) the geodesics are given by straight lines through the coordinate axes origin, as it happens for constant curvature Riemann surfaces. The limiting hyperbolas are:

$$y^2 - x^2 = 1 \quad \text{for PCC}; \qquad x^2 - y^2 = 1 \quad \text{for NCC}.$$

The former limiting hyperbola does not intersect the geodesic hyperbolas.

For the latter we find the intersecting points by subtracting the equation of the limiting hyperbola ($\Phi_2 \equiv [x^2 - y^2 - 1 = 0]$), multiplied by $\tan \epsilon$, from the geodesic hyperbolas ($\Phi_1 \equiv [(x^2 - y^2 + 1) \tan \epsilon - 2(x \sinh \sigma - y \cosh \sigma) = 0]$). In this way we have the system

$$x^2 - y^2 - 1 = 0,$$
$$\tan \epsilon - x \sinh \sigma + y \cosh \sigma = 0. \tag{9.3.16}$$

Now we show that as for Riemann negative constant curvature surfaces, we have

Theorem 9.4. *The geodesics cross orthogonally the limiting hyperbola.*

Proof. Let us calculate the scalar product (in the metric of the hyperbolic plane) of the gradients to Φ_1 and Φ_2 in their crossing points:

$$\frac{\partial \Phi_1}{\partial x} \frac{\partial \Phi_2}{\partial x} - \frac{\partial \Phi_1}{\partial y} \frac{\partial \Phi_2}{\partial y} = 4 \left[\tan \epsilon \, (x^2 - y^2) - x \sinh \sigma + y \cosh \sigma \right].$$

Since in the crossing points $x^2 - y^2 = 1$, due to (9.3.16), this product is zero. Then, as for definite line elements, Φ_1 and Φ_2 are pseudo-orthogonal [34].   □

We note that this property, as many others, can be considered a direct consequence of the fact that the equilateral hyperbolas represented in the pseudo-Euclidean plane satisfy the same theorems as the circles in the Euclidean plane, as shown in Chapter 4.

From an algebraic point of view this also follows from the fact that the equations of circles and equilateral hyperbolas are the same, if they are expressed in terms of complex and hyperbolic variables, respectively.

9.3.4   Motions

We have

Theorem 9.5. *The motions are given by a bilinear hyperbolic transformation.*

Proof. Actually, as reported in Table 9.1, the expressions of the line elements of Euclidean and pseudo-Euclidean constant curvature surfaces, written in terms of complex or hyperbolic variables, respectively, are the same. Then, also the transformations which leave unchanged the line elements are the same, if they are written in terms of complex or hyperbolic variables.

In particular, calling $z = x + h\,y$ and $w = p + h\,q$ two hyperbolic variables and α, β two hyperbolic constants, we obtain the following expressions for the motions:

$$w = \frac{\alpha\,z + \beta}{-\tilde{\beta}\,z + \tilde{\alpha}} \; ; \quad (\alpha\,\tilde{\alpha} + \beta\,\tilde{\beta} = 1) \qquad \text{for PCC;} \tag{9.3.17}$$

$$w = \frac{\alpha\,z + \beta}{\tilde{\beta}\,z + \tilde{\alpha}} \; ; \quad (\alpha\,\tilde{\alpha} - \beta\,\tilde{\beta} = 1) \qquad \text{for NCC.} \tag{9.3.18}$$

These transformations are also reported, without demonstration, in ([81], p. 288). As for constant curvature Riemann surfaces, they depend on three real constants.

□

9.4 Geodesics and Geodesic Distances on Riemann and Lorentz Surfaces

Let us recall from differential geometry that two points on a surface generally determine a geodesic and the distance between these points can be calculated by a line integral of the linear line element. It is known that, for the constant curvature surfaces, the equation of the geodesic as well as the geodesic distance can be determined in an algebraic way as functions of the point coordinates. In what follows we determine these expressions for all the four constant curvature Lorentz and Riemann surfaces, in the representation on the (x, y) plane; then with line elements given by (9.2.23), (9.2.35) and (9.3.7). For Lorentz surfaces, only points for which the geodesic exists ([57], p. 150), are considered. The method we propose is based on the obtained results that the motions which transform geodesic lines into geodesic lines are given by bilinear transformations.

We proceed in the following way: we take the points P_1, P_2 in the complex (hyperbolic) representative plane $z = x + i\,y$ ($z = x + h\,y$) and look for the parameters of the bilinear transformation (9.3.17) or (9.3.18) which maps these points on the geodesic straight line $q = 0$ of the complex (hyperbolic) plane $w = p + i\,q$ ($w = p + h\,q$). The inverse mapping of this straight line will give the equation of the geodesic determined by the given points. Moreover this approach allows us to obtain the distance between two points as a function of the point coordinates, and we show that the proposed method works for positive constant curvature surfaces as well as for the negative ones.

9.4.1 The Equation of the Geodesic

Theorem 9.6. *For Riemann and Lorentz positive and negative constant curvature surfaces, the geodesics between two points are obtained as functions of the points' coordinates as it is summarized in Table 9.2.*

Proof. Here we consider positive constant curvature Lorentz surfaces for which the motions are given by the bilinear transformations (9.3.17). Let us consider the two points

$$P_1 : (z_1 = x_1 + h\,y_1 \equiv \rho_1 \exp[h\theta_1])\,, \ \ P_2 : (z_2 = x_2 + h\,y_2 \equiv \rho_2 \exp[h\theta_2])$$

and look for the parameters (α, β) of the bilinear transformation (9.3.17), which maps P_1, P_2 into points $P_O \equiv (0, 0)$ and $P_l \equiv (l, 0)$, respectively, of the plane $w = p + h\,q$. The α, β parameters are obtained by solving the system

$$w_1 = 0 \Rightarrow \alpha z_1 = -\beta; \qquad w_2 = l \Rightarrow \alpha\,z_2 + \beta = l(-\tilde{\beta}\,z_2 + \tilde{\alpha}).$$

By setting the constants in hyperbolic polar form: $\alpha = \rho_\alpha \exp[h\theta_\alpha]$, $\beta = \rho_\beta \exp[h\theta_\beta]$, we obtain

$$\rho_\beta \exp[h\theta_\beta] \ = \ -\rho_\alpha\rho_1 \exp[h(\theta_\alpha + \theta_1)], \tag{9.4.1}$$

$$(z_2 - z_1) \exp[2h\theta_\alpha] \ = \ l(1 + \tilde{z}_1 z_2). \tag{9.4.2}$$

Equation (9.4.2) can be rewritten as

$$l = \frac{|z_2 - z_1|}{|1 + \tilde{z}_1 z_2|} \exp\left[h\left(2\theta_\alpha + \arg\left(\frac{z_2 - z_1}{1 + \tilde{z}_1 z_2}\right)\right)\right].$$

Since l is real, we have

$$2\theta_\alpha + \arg\left(\frac{z_2 - z_1}{1 + \tilde{z}_1 z_2}\right) = 0, \qquad l = \frac{|z_2 - z_1|}{|1 + \tilde{z}_1 z_2|}. \tag{9.4.3}$$

The constants of the bilinear transformations are given but for a multiplicative constant; therefore setting $\rho_\alpha = 1$, we obtain from (9.4.1)

$$\theta_\beta = \theta_\alpha + \theta_1, \ \ \rho_\beta = -\rho_1. \tag{9.4.4}$$

Now the equation of the geodesic between points P_1, P_2 is derived from the transformation of the geodesic straight line $w - \tilde{w} = 0$ by means of (9.3.17):

$$(x^2 - y^2 - 1)\rho_1 \sinh(\theta_\alpha + \theta_\beta) \ + \ x(\sinh 2\theta_\alpha - \rho_1^2 \sinh 2\theta_\beta) \tag{9.4.5}$$
$$+ \ y(\cosh 2\theta_\alpha + \rho_1^2 \cosh 2\theta_\beta) = 0.$$

Similarly, for the positive constant curvature surfaces with definite line element, by setting $\alpha = \rho_\alpha \exp[i\,\psi_\alpha]$, $\beta = \rho_\beta \exp[i\,\psi_\beta]$, (9.4.1) becomes $\rho_\beta \exp[i\,\psi_\beta] = -\rho_\alpha\rho_1 \exp[i\,(\psi_\alpha + \psi_1)]$. By substituting, in the right-hand side $-1 = \exp[i\pi]$, we obtain

$$\psi_\beta = \pi + \psi_\alpha + \psi_1, \ \ \rho_\beta = \rho_1. \qquad \qquad \square$$

9.4.2 Geodesic Distance

The transformations discussed above allow us to find the geodesic distance $\delta(z_1, z_2)$ between two points P_1 and P_2 as a function of the point coordinates. In particular we have

Theorem 9.7. *For Riemann and Lorentz negative constant curvature surfaces, the distance between two points can be expressed by a cross ratio, which is a function only of the complex or hyperbolic coordinates of the points.*

Proof. Let us take a negative constant curvature Lorentz surface. By using the line element (4) in Table 9.1, in Cartesian isometric form, the distance between $P_0 \equiv (0, 0)$ and $P_l \equiv (l, 0)$, on the $q = 0$ geodesic line, is given by

$$\delta(0, l) = 2R \int_0^l \frac{dp}{1 - p^2} \equiv 2R \cdot \tanh^{-1} l \equiv R \ln \frac{1 + l}{1 - l}. \qquad (9.4.6)$$

This equation can be written as a cross ratio ([30], p. 182), i.e., in a form which is invariant with respect to bilinear transformations ([81], p. 263) and ([25], p. 57). Actually we have

$$\frac{1 + l}{1 - l} \equiv (1, -1; 0, l). \qquad (9.4.7)$$

The points 1 and -1 are the intersection points between the geodesic straight line $q = 0$ and the limiting hyperbola $x^2 - y^2 = 1$, then (9.4.6) and (9.4.7) represent the same result expressed by (9.2.43).

By means of the same arguments of the last paragraph in Section 9.2.4 we can obtain the distance between any two points in the plane, by replacing l with the expression as a function of the point coordinates (9.4.3)

$$\delta(z_1, z_2) = 2R \tanh^{-1} \frac{|z_1 - z_2|}{|1 - \bar{z}_1 z_2|}. \qquad (9.4.8)$$

Equation (9.4.8) is also valid for negative constant curvature Riemann surfaces, with the value of l given in Table 9.2. $\qquad \square$

For negative constant curvature Riemann surfaces the result of (9.4.8) is obtained in a different way in ([25], p. 57).

Theorem 9.8. *Also for the positive constant curvature surfaces the geodesic distance between two points can be written as a function of a cross ratio and then as functions of the point coordinates.*

Proof. Let us consider the line element (1) of Table 9.1, in Cartesian isometric form, and calculate the distance between points P_0 and P_l on the geodesic line $q = 0$,

$$\delta(0, l) = 2R \int_0^l \frac{dp}{1 + p^2} \equiv 2R \tan^{-1} l \equiv \frac{R}{i} \ln \frac{i - l}{i + l} \equiv \frac{R}{i} \ln(-i, i; 0, l). \qquad (9.4.9)$$

Now $-i$, i are the intersection points between the geodesic straight line $q = 0$ and the limiting circle $x^2 + y^2 = -1$. Since the properties of the cross ratio are also valid for imaginary elements, we can again substitute for l its expression reported in Table 9.2, obtaining

$$\delta(z_1, z_2) = 2R \tan^{-1} \frac{|z_1 - z_2|}{|1 + \bar{z}_1 z_2|}. \tag{9.4.10}$$

If we apply the same procedure for non-definite line elements, we obtain the same expression (9.4.10) where z_1 and z_2 are hyperbolic variables. $\qquad\square$

	Motions		l	$\psi_\alpha, \theta_\alpha$	ψ_β, θ_β	ρ_β
(1)	$w = \frac{\alpha z + \beta}{-\beta z + \bar{\alpha}}$	$z = \frac{\bar{\alpha} w - \beta}{\beta w + \alpha}$	$\frac{|z_2 - z_1|}{|1 + \bar{z}_1 z_2|}$	$\frac{1}{2} \arg\left(\frac{1 + \bar{z}_1 z_2}{z_2 - z_1}\right)$	$\pi + \psi_\alpha + \psi_1$	ρ_1
(2)	$w = \frac{\alpha z + \beta}{\beta z + \bar{\alpha}}$	$z = \frac{-\bar{\alpha} w + \beta}{\beta w - \alpha}$	$\frac{|z_2 - z_1|}{|1 - \bar{z}_1 z_2|}$	$\frac{1}{2} \arg\left(\frac{1 - \bar{z}_1 z_2}{z_2 - z_1}\right)$	$\pi + \psi_\alpha + \psi_1$	ρ_1
(3)	$w = \frac{\alpha z + \beta}{-\beta z + \bar{\alpha}}$	$z = \frac{\bar{\alpha} w - \beta}{\beta w + \alpha}$	$\frac{|z_2 - z_1|}{|1 + \bar{z}_1 z_2|}$	$\frac{1}{2} \arg\left(\frac{1 + \bar{z}_1 z_2}{z_2 - z_1}\right)$	$\theta_\alpha + \theta_1$	$-\rho_1$
(4)	$w = \frac{\alpha z + \beta}{\beta z + \bar{\alpha}}$	$z = \frac{-\bar{\alpha} w + \beta}{\beta w - \alpha}$	$\frac{|z_2 - z_1|}{|1 - \bar{z}_1 z_2|}$	$\frac{1}{2} \arg\left(\frac{1 - \bar{z}_1 z_2}{z_2 - z_1}\right)$	$\theta_\alpha + \theta_1$	$-\rho_1$

	The equations of the geodesics
(1)	$(x^2 + y^2 - 1)\rho_1 \sin(\psi_\alpha + \psi_\beta)$ $-x(\sin 2\psi_\alpha - \rho_1^2 \sin 2\psi_\beta) - y(\cos 2\psi_\alpha + \rho_1^2 \cos 2\psi_\beta) = 0$
(2)	$(x^2 + y^2 + 1)\rho_1 \sin(\psi_\alpha + \psi_\beta)$ $+x(\sin 2\psi_\alpha + \rho_1^2 \sin 2\psi_\beta) + y(\cos 2\psi_\alpha - \rho_1^2 \cos 2\psi_\beta) = 0$
(3)	$(x^2 - y^2 - 1)\rho_1 \sinh(\theta_\alpha + \theta_\beta)$ $+x(\sinh 2\theta_\alpha - \rho_1^2 \sinh 2\theta_\beta) + y(\cosh 2\theta_\alpha + \rho_1^2 \cosh 2\theta_\beta) = 0$
(4)	$(x^2 - y^2 + 1)\rho_1 \sinh(\theta_\alpha + \theta_\beta)$ $-x(\sinh 2\theta_\alpha + \rho_1^2 \sinh 2\theta_\beta) - y(\cosh 2\theta_\alpha - \rho_1^2 \cosh 2\theta_\beta) = 0$

Table 9.2: *Equations of the geodesics for the four constant curvature surfaces,* with definite (rows (1) and (2)) and non-definite (rows (3) and (4)) line elements, obtained by the method exposed in Section 9.4.1. In particular, given two points in the z plane, $P_1 \equiv (z_1 = \rho_1 \exp[i\,\psi_1])$, $P_2 \equiv (z_2 = \rho_2 \exp[i\,\psi_2])$, we call $\alpha = \rho_\alpha \exp[i\,\psi_\alpha]$, $\beta = \rho_\beta \exp[i\,\psi_\beta]$ the constants of the bilinear mapping, reported in the second column of the upper table, which maps these points into points $P_0 \equiv (0, 0)$ and $P_l \equiv (l, 0)$ of the w plane. In column 5th, 6th and 7th, these parameters are reported as functions of the coordinates of P_1, P_2.

With these parameters we obtain the geodesics in the z plane by transforming the straight line $w - \widetilde{w} \equiv q = 0$, by means of the same mapping. The equations for the geodesics are reported in the lower part of the table as functions of the parameters.

The value of l (given in the 4th column) allows us to calculate, by means of (9.4.8) or (9.4.10) the geodesic distance between points P_1, P_2.

Chapter 10

Generalization of Two-Dimensional Special Relativity (Hyperbolic Transformations and the Equivalence Principle)

In Chapter 5 we have seen how hyperbolic trigonometry, introduced in the flat pseudo-Euclidean plane, has allowed us a complete treatment of accelerated motions and a consequent formalization of the twin paradox. In this second chapter concerning physical applications we shall see how the expansion from algebraic properties to the introduction of functions of hyperbolic variable allows an intriguing extension to general relativity of the symmetry of hyperbolic numbers, just introduced through special relativity.

This extension can be interpreted in the light of Einstein's words as we briefly summarize in Section 10.3.

10.1 The Physical Meaning of Transformations by Hyperbolic Functions

In Chapter 7 we have seen that the wave equation is invariant with respect to h-conformal mappings. Now we stress the physical relevance of this property, due to the fact that it is equivalent to say that the speed of light does not change in the coordinate systems obtained by these functional transformations, and this infinite group of functional transformations can be considered as a *generalization of the Lorentz–Poincaré two-dimensional group*. We briefly recall how Einstein established the Lorentz transformations of Special Relativity.

As is well known, the starting point was to look for invariance of Maxwell's equations (wave equations in general) with respect to uniform motions of reference frames, in the same way as dynamics equations are invariant with respect to Galileo's group.

Einstein was able to obtain in a straightforward way and by means of elementary mathematics the today named Lorentz transformations, starting from the two postulates

1. all inertial reference frames must be equivalent,

2. light's velocity is constant in inertial reference frames.

Now we note that the second postulate can be replaced by a more general formulation. Actually, as the Michelson experiments pointed out and Einstein himself observed in the introduction of the General Relativity paper (recalled in Section 10.3), the constance of light velocity is not limited to inertial reference frames. Thus we can say that while, from one point of view, Lorentz transformations obtained by Einstein are a right solution of the first requirement, from a more general point of view the second postulate could be formulated in a more general way. Actually, as we have stressed in this book, the link stated between space and time can be interpreted as *The symmetry or group or geometry* between these variables, and this relation holds true for every reference frame and, as a consequence, it must be extended to them. The physical relevance of hyperbolic transformations derives from this fact: *all the coordinate systems they introduce satisfy the required space-time symmetry and, in this way, they represent a generalization of the second postulate, i.e., they allow us to generalize the Lorentz–Poincaré two-dimensional group also to the accelerated reference frames which preserve the right symmetry.*

Following the positions in Section 4.1.2, p. 30 we give to the x variable the physical meaning of a normalized (speed of light $c = 1$) time variable and to y the physical meaning of a space variable, and we write the hyperbolic variable $w = t + h\,x$ and its function $f(w) = u(t,x) + h\,v(t,x)$. The system of partial differential equations (7.3.3) is written

$$u_{,x} = v_{,t}\,; \qquad u_{,t} = v_{,x}. \tag{10.1.1}$$

Now we emphasize the physical meaning of the functional transformations.

It is known that linear Lorentz transformations of special relativity represent a change of inertial frame. Relating the space and time variables to Cartesian coordinates, the inertial motion is represented by straight lines and, applying the linear Lorentz transformations, these lines remain straight lines.

The functional transformations change the straight lines in one reference frame into curved lines in another frame. From a physical point of view, a curved line represents a non-inertial motion, i.e., a motion in a field. The question which arises is whether these functional transformations represent any physical field. Now we show that by extending to hyperbolic functions the procedure reported at the end of Section 7.1 for complex variables, we obtain

Theorem 10.1. *The relativistic hyperbolic motion (Section 5.2)*

$$t = g^{-1}\sinh(g\,\tau)\,; \qquad x = g^{-1}\cosh(g\,\tau) \tag{10.1.2}$$

derives from the solution of a wave equation independent of hyperbolic rotations (Lorentz transformations).

Proof. For the t, x variables, the experimental evidence for symmetry is provided by the invariance under Lorentz transformations or, by (4.1.8), the independence of hyperbolic angle θ. Thus let us find a solution $U(t, x)$ of the wave equation $U_{,tt} - U_{,xx} = 0$, independent of the hyperbolic angle. We can proceed as in Section 7.1, and write the hyperbolic variable $t + hx$ as an exponential function of the variable $\rho + h\theta$ defined in Section 4.1, multiplied by h. The factor h allows us to obtain a function $x(t)$ with $|dx/dt| < 1$ (i.e., the hyperbolic variable $t + hx$ in sectors Us or Ds (see Section 4.1) as it must be from physical considerations. By applying this transformation, we have

$$t = \exp \rho \sinh \theta ; \quad x = \exp \rho \cosh \theta \qquad (10.1.3)$$

with the inverse

$$\rho = \ln \sqrt{x^2 - t^2} ; \quad \theta = \tanh^{-1}(t/x). \qquad (10.1.4)$$

We must solve the equation

$$U_{,\rho\rho} - U_{,\theta\theta} = 0. \qquad (10.1.5)$$

The U invariance under "hyperbolic rotation" means independence of the θ variable. Therefore $U_{,\theta\theta} \equiv 0$, and U depends only on the variable ρ. The partial differential equation (10.1.5) becomes an ordinary differential equation $d^2U/d\rho^2 = 0$, with the elementary solution

$$U = A\rho + B \equiv A \ln \sqrt{x^2 - t^2} + B. \qquad (10.1.6)$$

If we consider, as for the Laplace equation, the "equipotentials" $U = $ const, in the ρ, θ plane, these lines are straight lines which can be expressed in parametric form and, in agreement with (5.2.1) and (5.2.3), written as $\rho =$const.$\equiv \ln(g^{-1})$, $\theta = g\tau$. Going over, by (4.1.4) to t, x variables these straight lines become the hyperbolas (10.1.2), which represent the hyperbolic motion (5.2.1). From (5.2.3), τ is the proper time also in agreement with the fact that it represents the t variable in a reference frame in which x is constant.   □

Equation (10.1.2) is the law of motion of a body in the field of a constant force, as is the case of a uni-dimensional central field, calculated by the relativistic Newton's dynamic law. In this example the motion has been obtained as a transformation through the exponential function of a particular straight line. This transformation belongs to a group which preserves the proper symmetry of space-time. So we can say that, *as well as the space-time symmetry is described by hyperbolic numbers, in the same way space-time fields are formalized by means of functions of hyperbolic variables.* Practically, what we have done is to use suitable mathematics with the symmetries of the problem. In the next sections this principle is applied to the results obtained in Chapter 9.

From a physical point of view these results represent *a mathematical formalization of the equivalence principle* thanks to which, following Einstein ([32], p. 150), ◁ we are able to "produce" a gravitational field merely by changing the system of co-ordinates. ▷ Actually "A theory based on the Poincaré four-dimensional linear group, even being the symmetry group of space-time, provides no natural explanation for the equality of the inertial and gravitational mass, and it can not alone prescribe the form of the dynamical law ([9], p. 4)". On the contrary, the possibility of introducing hyperbolic functions with the space-time symmetry satisfies this requirement.

10.2   Physical Interpretation of Geodesics on Riemann and Lorentz Surfaces with Positive Constant Curvature

Let us now summarize and interpret, from this last point of view, the results of Chapter 9.

In Section 9.2.4 we have briefly recalled the link between groups and physical conservation laws. In particular we have recalled that the independent groups of rotations and translations in the Euclidean plane are unified on constant curvature surfaces.

In Section 7.1 we have also seen how the functions of a complex variable allowed us to obtain the Green function for Laplace's equation from the invariance with respect to rotations.

In a similar way in Section 10.1, from invariance of the wave equation with respect to Lorentz transformations, the hyperbolic motion (corresponding to the uniformly accelerated motion of Special Relativity) can be obtained.

The invariance with respect to translation group does not give similar results, whereas now we see how the "unification" on the constant curvature surfaces of rotation and translation groups (Section 9.2.4) gives some intriguing results.

Let us start by considering the well-studied example of the sphere in a Euclidean space. On this surface the group of roto-translations (which from a geometrical point of view, represents the rotations around the center) is represented by three parameters (such as Euler's angles). If this surface is projected on a plane, for instance by means of a stereographic projection, we lose the intuitive geometrical representation since, on this representative plane, distances and geodesic lines are obtained in the frame of a Gauss metric. Moreover, as we have seen in Section 9.4, in spite of the lack of an intuitive vision, motions, geodesics and geodesic distance are easily expressed, by means of complex variables, as functions of the coordinates of the points.

The Euclidean representation disappears when we study the other constant curvature surfaces such as the pseudo-sphere and the Lorentz constant curvature surfaces which represent the simplest extension of geometry in the pseudo-Euclidean plane. Otherwise, for a physical interpretation of these last surfaces, we assign to one of the variables the meaning of time coordinate, so the geodesic lines now represent the motion of a particle under the action of a particular field. Before going into their physical meaning we come back and look for a physical meaning of the geodesics on the sphere.

10.2.1 The Sphere

We have repeatedly seen how the unimodular multiplicative groups of complex and hyperbolic numbers leave invariable the circles and equilateral hyperbolas centered in the origin of the coordinates, respectively. We know that the functions of a complex variable are related to Laplace's equation, so we give to the rotation symmetry a physical meaning in the light of this equation, saying that the rotation group preserves the equipotential of a central field with a source in the coordinate origin. A similar, but more general, physical interpretation can be given to results obtained by studying the sphere surface and, in this way, we find an *unexpected relation with General Relativity.*

A geodesic circle in the x, y representation (Table 9.1, row (1)) can be considered as an equipotential curve generated by a point source in its center. On the other hand the geodesic circles have the geometrical meaning of stereographic projections, from the north pole to the equatorial plane, of the geodesic great circles on the sphere ([29],p. 96). This projection induces on the (x, y) plane a Gauss metric, so that the radius of the geodesic circle on the plane depends both on the radius of the sphere and on the position (connected with the constant A) of the great circle on the sphere. In fact from data in Table 9.1 we obtain that the radii of these circles are inversely proportional to the constant A times the square root of the curvature of the starting surface: $r = 1/(A\sqrt{K})$. Then the parametric equations of these circles are given by

$$x = x_0 + (A\sqrt{K})^{-1}\cos[A\sqrt{K}\,s], \qquad y = y_0 + (A\sqrt{K})^{-1}\sin[A\sqrt{K}\,s], \quad (10.2.1)$$

where s indicates the line element and x_0, y_0 the center coordinates. Comparing (10.2.1) with (7.1.4) we see that the field intensity is related to the curvature.

10.2.2 The Lorentz Surfaces

Now we extend the above considerations to Lorentz Surfaces. By giving to the variables the meaning of Section 10.1, the geodesics (9.3.14), taking into account (9.3.15), are given in the t, x plane by the hyperbolas: $(x - x_0)^2 - (t - t_0)^2 = 1/(A\sqrt{K})^2$ where K is the Gauss curvature. In parametric form as functions of the line element (proper time), we have

$$t = t_0 + (A\sqrt{K})^{-1}\sinh[A\sqrt{K}\,s], \qquad x = x_0 + (A\sqrt{K})^{-1}\cosh[A\sqrt{K}\,s]. \quad (10.2.2)$$

Comparing (10.2.2) with those of the hyperbolic motion ([52], p. 166) and 10.1.2, we have

$$t = t_0 + (g)^{-1}\sinh g\,\tau; \quad x = x_0 + (g)^{-1}\cosh g\,\tau.$$

Then *the geodesics on constant curvature surfaces, represented on a plane, correspond to a motion with constant acceleration and $g \propto \sqrt{K}$.*

Here we note that the result of (10.2.2) can be considered as a generalization of (10.1.2), obtained again by using space-time symmetry as stated by Lorentz transformations.

Moreover (10.2.2) go over to (10.1.2) since Einstein, when formalizing the general relativity, started from the equivalence principle and postulated that the gravitational field would be described by the curvature tensor in a non-flat space. Here (10.2.2) states a relation between (gravitational) field and space curvature without the need for this postulate.

In addition to the exposed "geometrical description" of a physical effect, we can make sense of the obtained results by recalling some of Einstein's epistemological considerations in the introduction of his fundamental work [32]. In the next section we briefly recall the Einstein way to general relativity, for clarifying these assertions.

10.3   Einstein's Way to General Relativity

For the mathematical formalization of the theory of General Relativity, Einstein, following the approach to Special Relativity, made a practical start from the two postulates ([32], p. 149, p.154)

1. ◁ The laws of physics must be of such a nature that they apply to systems of reference in any kind of motion;

2. In the special case of absence of a gravitational field and for infinitely small four-dimensional regions the theory of Special Relativity is appropriate, if the coordinates are suitably chosen. ▷

The second postulate is equivalent to saying that the space-time symmetry, stated by Lorentz transformation is *a general law of nature*. Otherwise, unlike special relativity for which all the results have been obtained by means of the axiomatic deductive method and an elementary mathematics, now the steps from the starting intuitions to the conclusive formalization were very hard [58], but the astonishing experimental confirmation of the results gives credit to the starting postulates and to intermediate steps.

General Relativity is considered one of the most important scientific works of all times, and all the hypotheses, even if they are of very different nature, agree in a harmonious way.

Einstein, notwithstanding this success, in the description of his effort in looking for the best formalization, seems not completely satisfied as can be noted by the following words ([32], p. 153):

◁ There seems to be no other way which would allow us to adapt systems of co-ordinates to the four-dimensional universe so that we might expect from their application a particularly simple formulation of the laws of nature. So there is nothing for it but to regard all the imaginable systems of co-ordinates, on principle, as equally suitable for the description of nature. ▷

In the first part we find today's well-known principle, following which the description and the solution of a physical problem are simplified if the problem is formalized by means of a suitable mathematics with its symmetries.

The second part concerns the use of differential geometry. This use, following E. Cartan ([13], articles 71, 73, 105), detaches Einstein from group theory, the basis of special relativity and of the starting ideas of general relativity. Actually looking for physical laws which do not change when we change the reference frame is a characteristic property of groups.

We now recall another Einstein's assertion [33] which can be related with the previous one:

◁ Our experience teaches us that the nature represents the realization of what we can imagine of most mathematically simple. I believe that a purely mathematical construction allows us to reveal the concepts which can give us the key for understanding the natural phenomena and the principles that link them together. Obviously the experimental confirmation is the only way for verifying a mathematical construction describing physical phenomena; but just in the mathematics we can find the creative principle. ▷

From these concepts we could presume that Einstein was looking for a transformation group which would preserve the four-dimensional space-time symmetries and by means of "the mathematics linked to this group", obtain the simplest formalization of the law of nature.

We add that since this group must describe fields it must be an infinite-dimensional group represented by functional transformations. Up to now this group has not been found. But we have seen how these requirements are satisfied by the functions of hyperbolic variable for two-dimensional space-time.

10.4 Conclusions

The exposed results have been obtained thanks to the "mathematics generated by hyperbolic numbers" and, in particular, to the introduction of the infinite-dimensional group of functional transformations which preserve the space-time symmetry stated by the Lorentz group. As we have repeatedly shown, this possibility derives from the coincidence between the quadratic invariants for both the multiplicative group of hyperbolic numbers and the two-dimensional Lorentz group.

This coincidence does not hold in more than two dimensions since the physical laws are represented in a Euclidean space (or Minkowski space-time) with a quadratic invariant, while the commutative hypercomplex numbers which allow us to introduce the infinite-dimensional groups of functional transformations, generate geometries different from the multidimensional Euclidean ones.

From these observations we can not yet derive conclusions about the possibility of extending to more than two dimensions the results obtained in two dimensions. In any case if we want to consider the obtained results as a starting point they could stimulate the investigation of multidimensional hypercomplex systems, both for a more complete mathematical formalization and for inspecting physical applications.

The appendices of this book are devoted to the introduction of a four-dimensional hypercomplex numbers system and to a newly formalized theory of commutative multidimensional numbers.

Appendix A

Commutative Segre's Quaternions

Up to now, we have formalized the application of hyperbolic numbers for studying the space-time symmetry stated by special relativity by means of the analogy with complex numbers. This formalization has provided a mathematical tool which allows us an automatic solution of problems in two-dimensional space-time, in the same way as we do in Euclidean plane geometry. As we have repeatedly recalled, this was possible thanks to the coincidence between the multiplicative group of hyperbolic numbers with the two-dimensional Lorentz group.

A relevant characteristic of these group is to possess divisors of zero, a property which was considered, for a long time, very far from the common geometrical intuition [59] and limited the studies of multidimensional commutative systems with this property. Actually, as we have reported in Section 2.1.2, p. 6, the extension of real and complex numbers can be done only by releasing

- the commutative property of the product;

- to be a division algebra, i.e., not to have divisors of zero.

From a geometrical point of view, relevant results [42] have been obtained by means of non-commutative systems without divisors of zero. In particular by the Hamilton quaternions for representing vectors in the three-dimensional Euclidean space and, more generally, by Clifford algebra [45].

On the other hand we have shown in Chapters 7–10, the relevance of the introduction of functions of hyperbolic variables and we know that the functions of complex variables are very important for the determination of two-dimensional physical fields (Section 7.1), described by Cauchy–Riemann partial differential equations, as pointed out in ([54], p. 1252): "The extraordinary apt way in which the properties of functions of a complex variable fit our need for solutions of differential equations in two dimensions, does not apply to equations in more dimensions for which our task will be, in general, more difficult".

Actually we have

- the Hamilton quaternions can be considered as an extension of real and complex numbers as regards the property of being a division algebra;

- the commutative systems can be considered as an extension of real and complex numbers as regards the commutative property of products and the possibility of introducing their functions, which could represent physical fields [71].

Also if many problems must be tackled and solved, the goal, in studying the multi-dimensional commutative systems, is in the possibility of giving a positive answer to [54] for problems in more than two dimensions. Then, since before looking for applications, we need a well-formalized mathematical theory, in these appendices we begin this task from a four-dimensional commutative number system, which has been called [20] *Segre's quaternions* for the reasons summarized in Section A.1.1.

Actually there are two peculiarities of this system which stimulate its study:

1. four coordinates are usually considered for describing the physical world;

2. these numbers can be considered as composed by two systems of complex numbers, which have a well-known physical relevance.

We shall see that, thanks to this last property, many characteristics can be obtained by an extension of complex analysis.

A.1　Hypercomplex Systems with Four Units

The four-units commutative hypercomplex systems are fourteen in number; we recall, from [13] and [24], their multiplication tables.

1. Six of them are non-decomposable.

(a)
e_0	e_1	e_2	e_3
e_1	e_2	e_3	0
e_2	e_3	0	0
e_3	0	0	0

(b)
e_0	e_1	e_2	e_3
e_1	$-e_0$	e_3	$-e_2$
e_2	e_3	0	0
e_3	$-e_2$	0	0

(c)
e_0	e_1	e_2	e_3
e_1	0	0	0
e_2	0	0	0
e_3	0	0	0

$$(\text{A.1.1})$$

(a)
e_0	e_1	e_2	e_3
e_1	e_3	0	0
e_2	0	e_3	0
e_3	0	0	0

(b)
e_0	e_1	e_2	e_3
e_1	e_3	0	0
e_2	0	$-e_3$	0
e_3	0	0	0

(c)
e_0	e_1	e_2	e_3
e_1	e_2	0	0
e_2	0	0	0
e_3	0	0	0

$$(\text{A.1.2})$$

2. Two systems (not reported) are obtained by adding a unit to the three-units systems of Tables (7.6.1).

3. Six systems are obtained by the three two-dimensional systems. Actually calling α', α'' two quantities that can be $0, \pm 1$, we summarize these systems in the following table (three systems are repeated):

e_0	e_1	0	0
e_1	$\alpha'e_0$	0	0
0	0	e_2	e_3
0	0	e_3	$\alpha''e_2$

$$(\text{A.1.3})$$

A.1.1 Historical Introduction of Segre's Quaternions

Out of the fourteen systems of four-units commutative hypercomplex numbers mentioned above, five of them can be considered as a generalization of a historical method due to Corrado Segre [67] who studied a geometrical representation of complex numbers. He proposed the following *complexification*: let there be given the complex number $z = x + i\,y$ and let us substitute the real variables x, y for the complex variables $x \Rightarrow z^1 \equiv x^1 + k\,y^1$ and $y \Rightarrow z^2 \equiv x^2 + k\,y^2$, where k is a unit vector with the same properties of i, i.e., $k^2 = -1$, but different from i since, in a geometrical representation, it represents the unit vector of another axis. Then we have $z = x_1 + k\,y_1 + i\,x_2 + i\,k\,y_2$, and calling $j = i\,k$ the unit vector of a new axis, we obtain a four-dimensional commutative hypercomplex number with the following multiplication table:

$$
\begin{array}{|c|c|c|c|}
\hline
1 & k & i & j \\
\hline
k & -1 & j & -i \\
\hline
i & j & -1 & -k \\
\hline
j & -i & -k & 1 \\
\hline
\end{array}
\qquad (A.1.4)
$$

This multiplication table is similar to the one of Hamilton quaternions; then we can call these numbers *Segre's (commutative) quaternions*.

A.1.2 Generalized Segre's Quaternions

From an algebraic point of view this process of complexification can be generalized to 2^n dimensions ([60], Chap. 5) as proposed by Segre, as well as to other two-dimensional systems. In this case we can take for z and z_m, $(m = 1, 2)$ the hyperbolic numbers (h), parabolic numbers (p) and elliptic numbers (e), as reported in Table A.1. It has been proposed [20] to call these numbers *generalized Segre's quaternions*.

Some years after their geometrical introduction these numbers were reconsidered by F. Severi [68] in association with functions of two complex variables. After this association G. Scorza-Dragoni [66] and U. Morin [53] began to study the existence and differential properties of quaternion functions and, in recent time, has been also reconsidered by Davenport [27] and Price [60]. From a physical point of view, there is an important difference between the functions of two complex variables and functions of commutative quaternions. Actually, while for the former, two real functions of four real variables are defined, for the latter four real functions of four real variables are introduced. These functions, as we have shown in Section 7.2.3, can represent vector fields, which automatically have the same symmetries (groups, geometries) and then the same invariants of the quaternions, since the structure constants act as "symmetry preserving operator" between vectors and the corresponding vector fields. These symmetries are the ones of the geometries "generated" by hypercomplex numbers (Chapter 3). We shall see that when commutative quaternions are represented in their geometry, they keep the same

z	z_m	TABLE	Ref.
h	h	$(A.1.3), \alpha' = \alpha'' = +1$	
h	e	$(A.1.3), \alpha' = \alpha'' = -1$	[66]
e	h	$(A.1.3), \alpha' = \alpha'' = -1$	[66]
e	e	$(A.1.3), \alpha' = \alpha'' = -1$	[60], [66]
h	p	$(A.1.3), \alpha' = \alpha'' = 0$	
p	h	$(A.1.3), \alpha' = \alpha'' = 0$	
e	p	$(A.1.1, b)$	[71], [26]
p	e	$(A.1.1, b)$	[71], [26]
p	p	$(A.1.1, c)$	

Table A.1: *Generalized Segre's quaternions.* The first and second column indicate the two-dimensional numbers (h, p, e) considered. In the third column we report the table in Section A.1, to which the number system corresponds, and in the fourth column the references in which they have been studied or applied.

Note that the three systems from the second to the fourth rows represent the Segre's commutative quaternions which has been recently reconsidered [27] and [60].

differential properties of complex numbers, in particular for conformal mappings. Therefore if a physical field can be associated with their functions, the problem we consider can be simplified, as it can if we use polar coordinates for problems with spherical symmetry or hyperbolic numbers for space-time symmetry [71].

A.2   Algebraic Properties

Here we study three systems among the *generalized Segre's quaternions*, reported in Table A.1: in particular the systems that are decomposable into the two-dimensional hypercomplex systems considered in Section 2.2 and Chapter 6. Here these systems are indicated by $\mathbf{H}_2$. We also indicate by $\mathbf{Q} \subset \mathbf{H}_4$ the set of quaternions defined as

$$\{q = t + i\,x + j\,y + k\,z; \quad t, x, y, z \in \mathbf{R}; \quad i, j, k \notin \mathbf{R}\}, \qquad (A.2.1)$$

where the versors i, j, k satisfy the following multiplication rules.

1	i	j	k
i	α	k	αj
j	k	1	i
k	αj	i	α

$$(A.2.2)$$

In particular, according to the value of α, the $\mathbf{Q}$ systems shall be named after the two-dimensional component systems:

1. for $\alpha < 0$: *Elliptic quaternions*, in particular *canonical (Segre's)* if $\alpha = -1$;

2. for $\alpha = 0$: *Parabolic quaternions*;

3. for $\alpha > 0$: *Hyperbolic quaternions*, in particular *canonical* if $\alpha = 1$.

Here we consider just canonical systems ($\alpha = 0$, ± 1). We observe that the first system is isomorphic with the classical Segre's quaternion (corresponding to the fourth row of Table A.1), with the multiplication table (A.1.4). The second system corresponds to the fifth and the sixth rows, the third system to the first row of Table (A.1.4). The three systems can generally be studied together only by taking into account the different values of α, as was done in Chapter 6 for the two-dimensional component systems.

A characteristic matrix (Section 2.1.3, p. 7) is obtained by taking the coefficients of units of the four numbers q, $i\,q$, $j\,q$, $k\,q$. It is given by

$$\mathcal{M} = \begin{pmatrix} t & \alpha x & y & \alpha z \\ x & t & z & y \\ y & \alpha z & t & \alpha x \\ z & y & x & t \end{pmatrix} \equiv \left(\begin{array}{c|c} A & B \\ \hline B & A \end{array} \right) \qquad (A.2.3)$$

where A and B represent 2×2 matrices. From (A.2.3) we obtain the matrix expressions of versors, written in boldface (Section 2.1)

$$\mathbf{1} = \begin{pmatrix} 1 & 0 & 0 & 0 \\ 0 & 1 & 0 & 0 \\ 0 & 0 & 1 & 0 \\ 0 & 0 & 0 & 1 \end{pmatrix} ; \, \boldsymbol{i} = \begin{pmatrix} 0 & \alpha & 0 & 0 \\ 1 & 0 & 0 & 0 \\ 0 & 0 & 0 & \alpha \\ 0 & 0 & 1 & 0 \end{pmatrix}$$

$$\boldsymbol{j} = \begin{pmatrix} 0 & 0 & 1 & 0 \\ 0 & 0 & 0 & 1 \\ 1 & 0 & 0 & 0 \\ 0 & 1 & 0 & 0 \end{pmatrix} ; \, \boldsymbol{k} = \begin{pmatrix} 0 & 0 & 0 & \alpha \\ 0 & 0 & 1 & 0 \\ 0 & \alpha & 0 & 0 \\ 1 & 0 & 0 & 0 \end{pmatrix} . \qquad (A.2.4)$$

The determinant of matrix (A.2.3) is given by

$$\|\mathcal{M}\| = \begin{vmatrix} t & \alpha x & y & \alpha z \\ x & t & z & y \\ y & \alpha z & t & \alpha x \\ z & y & x & t \end{vmatrix} \Rightarrow \begin{vmatrix} t+y & \alpha(x+z) & 0 & 0 \\ x+z & t+y & 0 & 0 \\ 0 & 0 & t-y & \alpha(x-z) \\ 0 & 0 & x-z & t-y \end{vmatrix}$$

$$\equiv \left[(t+y)^2 - \alpha(x+z)^2 \right] \left[(t-y)^2 - \alpha(x-z)^2 \right] . \qquad (A.2.5)$$

This result can also be obtained from matrix theory, taking into account commutativity between matrices A and B. Actually we have $\|\mathcal{M}\| = \|A^2 - B^2\| \equiv \|A+B\| \times \|A-B\|$; as usual we have indicated the determinants by $\| \cdot \|$. We put $\rho \equiv |q| = \sqrt[4]{|\,\|\mathcal{M}\|\,|}$ and call ρ the *modulus* of q.

For an elliptic system too, determinant (A.2.5) is zero on the planes

$$I \begin{cases} t + y = 0 \\ x + z = 0 \end{cases} \qquad II \begin{cases} t - y = 0 \\ x - z = 0. \end{cases} \tag{A.2.6}$$

The planes of (A.2.6) are called *characteristic planes* or *planes of zero divisors* [53]. We can verify that the product of numbers with coordinates satisfying (A.2.6, I) times numbers with coordinates satisfying (A.2.6, II) is zero. As shown in Section 2.1.5, for these numbers division is not defined.

From the last expression (A.2.5) we obtain at once the characteristic equation (*minimal equation* Section 2.1.6)

$$\left[(t + y - q)^2 - \alpha(x + z)^2 \right] \left[(t - y - q)^2 - \alpha(x - z)^2 \right] = 0. \tag{A.2.7}$$

In Section A.2.1 we find that q and the numbers

$$\bar{q} = t - i\,x + j\,y - k\,z, \quad \bar{\bar{q}} = t + i\,x - j\,y - k\,z, \quad \tilde{q} = t - i\,x - j\,y + k\,z \tag{A.2.8}$$

are solutions of equation (A.2.7), then (A.2.8) are (see p. 10) the *principal conjugations*. We have used, for mnemonic reason, " ¯ " to indicate conjugations with respect to i and k and " ˜ " for conjugations with respect to j. The position of these symbols, from bottom to top, indicates the order of conjugations in the quaternion, from left to right.

The principal conjugations have the same properties as the conjugate of complex and hyperbolic numbers. In particular,

1. the product of q times conjugations (A.2.8) gives the determinant $\|\mathcal{M}\|$;

2. q and conjugations (A.2.8) are a group, in the sense that repeated conjugations are principal conjugations;

3. (A.2.1) and (A.2.8) can be considered as a bijective mapping between t, x, y, z and q, $\bar{\bar{q}}$, $\bar{q}$, $\tilde{q}$. We have for $\alpha \neq 0$,

$$t = \frac{q + \bar{\bar{q}} + \tilde{q} + \bar{q}}{4}, \quad x = i\frac{q - \bar{\bar{q}} + \tilde{q} - \bar{q}}{4\,\alpha},$$
$$y = j\frac{q + \bar{\bar{q}} - \tilde{q} - \bar{q}}{4}, \quad z = k\frac{q - \bar{\bar{q}} - \tilde{q} + \bar{q}}{4\,\alpha}. \tag{A.2.9}$$

The property of point 2 is better seen by considering the conjugations

$$^2\bar{q} = t - i\,x + j\,y + k\,z, \quad ^3\bar{q} = t + i\,x - j\,y + k\,z, \quad ^4\bar{q} = t + i\,x + j\,y - k\,z, \tag{A.2.10}$$

which can be obtained from q by changing a sign to the variable x, y or z, or applying the principal conjugations to

$$^1\bar{q} = t - i\,x - j\,y - k\,z. \tag{A.2.11}$$

The product between (A.2.10) and (A.2.11) is real and gives (A.2.5) changing the sign of one of the variables. It is straightforward to verify that repeating one conjugation (A.2.10) we have a conjugation (A.2.8) while two repeated conjugations give a conjugation of the same kind. Moreover while the components of the functions of principal conjugations, but for their sign, are the same as the quaternion, the components of the functions of these conjugations are, generally, different.

Now we consider two numbers q and $q_1 = t_1 + i\,x_1 + j\,y_1 + k\,z_1$ and calculate their product and products between two conjugations:

$$q\,q_1 = t\,t_1 + \alpha\,x\,x_1 + y\,y_1 + \alpha\,z\,z_1 + i\,(x\,t_1 + t\,x_1 + z\,y_1 + y\,z_1) \qquad (A.2.12)$$
$$+ j\,(t\,y_1 + y\,t_1 + \alpha\,x\,z_1 + \alpha\,z\,x_1) + k\,(z\,t_1 + t\,z_1 + x\,y_1 + y\,x_1)$$
$$\stackrel{q=q_1}{\longrightarrow} t^2 + \alpha\,x^2 + y^2 + \alpha\,z^2 + 2\,i\,(t\,x + y\,z) + 2\,j\,(t\,y + \alpha\,x\,z) + 2\,k\,(t\,z + x\,y),$$

$$q\,\bar{\bar{q}}_1 \equiv \bar{\bar{q}}\,\bar{\bar{q}}_1 = t\,t_1 - \alpha\,x\,x_1 + y\,y_1 - \alpha\,z\,z_1 + i\,(x\,t_1 - t\,x_1 + z\,y_1 - y\,z_1)$$
$$+ j\,(t\,y_1 + y\,t_1 - \alpha\,x\,z_1 - \alpha\,z\,x_1) + k\,(z\,t_1 - t\,z_1 + x\,y_1 - y\,x_1)$$
$$\stackrel{q=q_1}{\longrightarrow} t^2 - \alpha\,x^2 + y^2 - \alpha\,z^2 + 2\,j\,(t\,y - \alpha\,x\,z), \qquad (A.2.13)$$

$$q\,\bar{\bar{q}}_1 \equiv \bar{\bar{q}}\,\bar{\bar{q}}_1 = t\,t_1 + \alpha\,x\,x_1 - y\,y_1 - \alpha\,z\,z_1 + i\,(x\,t_1 + t\,x_1 - z\,y_1 - y\,z_1)$$
$$+ j\,(-t\,y_1 + y\,t_1 - \alpha\,x\,z_1 + \alpha\,z\,x_1) + k\,(z\,t_1 - t\,z_1 - x\,y_1 + y\,x_1)$$
$$\stackrel{q=q_1}{\longrightarrow} t^2 + \alpha\,x^2 - y^2 - \alpha\,z^2 + 2\,i\,(t\,x - y\,z), \qquad (A.2.14)$$

$$q\,\bar{\bar{q}}_1 \equiv \bar{\bar{q}}\,\bar{\bar{q}}_1 = t\,t_1 - \alpha\,x\,x_1 - y\,y_1 + \alpha\,z\,z_1 + i\,(x\,t_1 - t\,x_1 - z\,y_1 + y\,z_1)$$
$$+ j\,(-t\,y_1 + y\,t_1 + \alpha\,x\,z_1 - \alpha\,z\,x_1) + k\,(z\,t_1 + t\,z_1 - x\,y_1 - y\,x_1)$$
$$\stackrel{q=q_1}{\longrightarrow} t^2 - \alpha\,x^2 - y^2 + \alpha\,z^2 + 2\,k\,(t\,z - x\,y). \qquad (A.2.15)$$

We observe from (A.2.13), (A.2.14) and (A.2.15) for $q_1 = q$, that the product of two principal conjugations is a two-dimensional sub-algebra $(1, j)$; $(1, i)$; $(1, k)$, respectively.

In general, quaternions we are studying have the following sub-algebras:

1. real algebra (1),

2. hyperbolic algebra $(1, j)$,

3. general two-dimensional algebras $(1, i)$ and $(1, k)$,

4. three-units algebras $(1, h, j)$ and $(1, l, j)$ with $h = i + k$ and $l = i - k$. The multiplication tables of versors of these last sub-algebras are

$$\text{algebra } (1, h, j) = \begin{array}{|c|c|c|} \hline 1 & h & j \\ \hline h & 2\,\alpha(1+j) & h \\ \hline j & h & 1 \\ \hline \end{array} \; ;$$

$$\text{algebra } (1, l, j) = \begin{array}{|c|c|c|} \hline 1 & l & j \\ \hline l & 2\,\alpha(1-j) & -l \\ \hline j & -l & 1 \\ \hline \end{array} \; . \qquad (A.2.16)$$

Algebras (A.2.16) are the same and, according to α value, they represent a system "composed" by one among the three two-dimensional algebras plus an one unit algebra, i.e., they are the same considered in Table 7.6.5.

A.2.1   Quaternions as a Composed System

Following [53] we see that many properties of commutative quaternions can be obtained in an easy way by using the decomposed form. Let us introduce the idempotent basis

$$e_1 = \frac{1}{2}(1+j); \ \ e_2 = \frac{1}{2}(1-j) \ \ \Rightarrow \ \ e_1^2 = e_1; \ \ e_2^2 = e_2; \ \ e_1 e_2 = 0, \qquad \text{(A.2.17)}$$

with inverse transformations

$$1 = e_1 + e_2; \ \ \ \ j = e_1 - e_2 \qquad\qquad\qquad \text{(A.2.18)}$$

and variables

$$\tau = t + y; \ \ \ \eta = t - y; \ \ \ \xi = x + z; \ \ \ \zeta = x - z, \qquad\qquad \text{(A.2.19)}$$

with inverse transformations

$$t = \frac{\tau + \eta}{2}; \ \ \ y = \frac{\tau - \eta}{2}; \ \ \ x = \frac{\xi + \zeta}{2}; \ \ \ z = \frac{\xi - \zeta}{2}. \qquad \text{(A.2.20)}$$

We also put

$$z^1 = \tau + \mathrm{i}\,\xi; \ \ \ z^2 = \eta + \mathrm{i}\,\zeta. \qquad\qquad\qquad \text{(A.2.21)}$$

z^1, z^2 are two independent variables in the same two-dimensional algebra and each of them is represented in a characteristic plane.

By means of substitutions (A.2.18), (A.2.19) and (A.2.21), quaternion (A.2.1) can be written

$$q = e_1 z^1 + e_2 z^2. \qquad\qquad\qquad\qquad \text{(A.2.22)}$$

The element q is the sum of elements $e_1 z^1$ and $e_2 z^2$ whose product is zero. The algebra of q is said to be *reducible*. From mappings (A.2.19) and (A.2.20) we can obtain z^1, z^2 from q and vice versa.

In Section 2.2.2 (p. 16), we have seen the expressions of the sum, product and (if it exists) division for systems in decomposed form. Now for quaternions (A.2.22) and $q_1 = e_1 z_1^1 + e_2 z_1^2$, we have

$$q\,q_1 = e_1 z^1 z_1^1 + e_2 z^2 z_1^2; \ \ \ q^n = e_1 (z^1)^n + e_2 (z^2)^n; \ \ \ \frac{q}{q_1} = e_1 \frac{z^1}{z_1^1} + e_2 \frac{z^2}{z_1^2}.$$
$$\text{(A.2.23)}$$

We can verify that (A.2.8) become

$$\bar{\bar{q}} = e_1 \bar{z}^1 + e_2 \bar{z}^2; \ \ \ \tilde{q} = e_1 \bar{z}^2 + e_2 \bar{z}^1; \ \ \ \bar{\tilde{q}} = e_1 z^2 + e_2 z^1. \qquad \text{(A.2.24)}$$

Then from (A.2.23) and (A.2.24), we have ([53], p. 1249)

Theorem A.1. *If we carry out rational operations on conjugate numbers we obtain conjugate results.*

We also have

$$q\,\bar{q} = e_1\,z^1\,\bar{z}^1 + e_2\,z^2\,\bar{z}^2; \quad \bar{\bar{q}}\bar{q} = e_1\,z^2\,\bar{z}^2 + e_2\,z^1\,\bar{z}^1; \quad \|\mathcal{M}\| = z^1\,\bar{z}^1 \cdot z^2\,\bar{z}^2. \quad \text{(A.2.25)}$$

If $\alpha = -1$ (elliptic quaternions), the moduli of component systems are positive, and (A.2.25) can be written

$$q\,\bar{q} = e_1\,|z^1|^2 + e_2\,|z^2|^2; \quad \bar{\bar{q}}\bar{q} = e_1\,|z^2|^2 + e_2\,|z^1|^2; \quad |q|^4 = |z^1|^2 \cdot |z^2|^2. \quad \text{(A.2.26)}$$

It has been pointed out that the modulus of hyperbolic numbers can be expressed as the product of two one-dimensional invariants (distances) (p. 16), the same relation (product of invariants of component algebras) holds for quaternions.

Solutions of the Characteristic Equation

For obtaining the solutions of characteristic equation (A.2.7), we use a decomposed form and the following theorem ([65], p. 243).

Theorem A.2. *If algebra A is the sum of algebras B and C, then the minimal equation in A is the product of minimal equations in B and C.*

A minimal equation for z^1, z^2, i.e., of a two-dimensional number in a real field, is an equation of degree 2, which has z^1 or z^2 as root. Then if numbers z^1, z^2 are different numbers, we have

$$\left[(\tau - z')^2 - \alpha\,\xi^2\right]\left[(\eta - z")^2 - \alpha\,\zeta^2\right] = 0, \quad \text{(A.2.27)}$$

then

$$z' = \tau \pm \mathrm{i}\,\xi \rightarrow \left\{ \begin{matrix} z^1 \\ \bar{z}^1 \end{matrix} \right. ; \qquad z" = \eta \pm \mathrm{i}\,\zeta \rightarrow \left\{ \begin{matrix} z^2 \\ \bar{z}^2 \end{matrix} \right. , \quad \text{(A.2.28)}$$

where i indicates a versor of two-dimensional algebra, i.e., we have to consider $\mathrm{i}^2 = \alpha$ and the solutions are complex, coincident (parabolic system) or hyperbolic, depending on the type of component system.

In the idempotent basis defined by (A.2.17), conjugations are obtained with a shift of versors (Section C.3.5), then we have the solutions $q_1 = e_1\,z' + e_2\,z"$; $q_2 = e_1\,z" + e_2\,z'$. If we take the same signs for z', $z"$ we obtain q and principal conjugations (A.2.8). If we take different signs for z', $z"$ we obtain (A.2.10) and (A.2.11).

A.3   Functions of a Quaternion Variable

Curves in Quaternion Space

Definition. Let $\sigma \in \mathbf{R}$ be a parameter and $[\sigma', \sigma"]$ an interval; a curve in quaternion space

$$\lambda : [q(\sigma) = t(\sigma) + ix(\sigma) + jy(\sigma) + kz(\sigma)] \quad \text{(A.3.1)}$$

is a regular mapping for $\sigma' < \sigma < \sigma''$ if $q(\sigma)$ is $\neq 0$ and is not a zero divisor, and the components have a continuous derivative ([53], p. 1252) and ([60], Sect. 31).

From a geometrical point of view the regularity condition means that the curve $q(\sigma)$ does not have more than a point in common with characteristic planes. We can also say that *The modulus of $dq/d\sigma$ must be $\neq 0$.*

A.3.1   Holomorphic Functions

We indicate a quaternion function by

$$F \equiv F_1(t, x, y, z) + i\, F_i(t, x, y, z) + j\, F_j(t, x, y, z) + k\, F_k(t, x, y, z), \quad \text{(A.3.2)}$$

where F_n, with $n = 1, i, j, k$, are real functions with partial derivatives with respect to the variables t, x, y, z. We have

Theorem A.3. *F is said to be a holomorphic function of quaternions if* ([60], *Chap.* 3)

1. *it is differentiable with derivatives $\neq 0$ and not a zero divisor,*

2. *partial derivatives of components satisfy the following partial differential equations called Generalized Cauchy–Riemann conditions (GCR):*

$$(GCR) \quad \begin{cases} F_{1,t} = F_{i,x} = F_{j,y} = F_{k,z} & (a) \\ F_{1,x} = \alpha F_{i,t} = F_{j,z} = \alpha F_{k,y} & (b) \\ F_{1,y} = F_{i,z} = F_{j,t} = F_{k,x} & (c) \\ F_{1,z} = \alpha F_{i,y} = F_{j,x} = \alpha F_{k,t} & (d) \end{cases} \quad \text{(A.3.3)}$$

While we proof Theorem A.3 just for elliptic quaternions, the results can be obtained similarly for parabolic and hyperbolic quaternions.

Proof. GCR conditions can be obtained in many ways and with the same rigor as for CR conditions for functions of a complex variable ([60], Chap. 3); here they are obtained by means of the Hamilton differential operator ∇ (nabla) and its conjugations

$$\nabla = \frac{\partial}{\partial t} + i\frac{\partial}{\partial x} + j\frac{\partial}{\partial y} + k\frac{\partial}{\partial z}; \quad \overline{\overline{\nabla}} = \frac{\partial}{\partial t} - i\frac{\partial}{\partial x} + j\frac{\partial}{\partial y} - k\frac{\partial}{\partial z};$$

$$\overline{\overline{\nabla}} = \frac{\partial}{\partial t} + i\frac{\partial}{\partial x} - j\frac{\partial}{\partial y} - k\frac{\partial}{\partial z}; \quad \widetilde{\nabla} = \frac{\partial}{\partial t} - i\frac{\partial}{\partial x} - j\frac{\partial}{\partial y} + k\frac{\partial}{\partial z}. \quad \text{(A.3.4)}$$

As is known for complex variables ([25], p. 32), the operators (A.3.4) can be formally related to the derivatives with respect to q and its conjugations. For elliptic quaternions, by considering F as a function of q and its conjugations, via the variables t, x, y, z, by the derivative chain rule, and from (A.2.9), we have

$$\frac{\partial F}{\partial q} \equiv \frac{\partial F}{\partial t}\frac{\partial t}{\partial q} + \frac{\partial F}{\partial x}\frac{\partial x}{\partial q} + \frac{\partial F}{\partial y}\frac{\partial y}{\partial q} + \frac{\partial F}{\partial z}\frac{\partial z}{\partial q} = \frac{1}{4}\overline{\overline{\nabla}}F. \quad \text{(A.3.5)}$$

In a similar way we also obtain

$$\nabla = 4\frac{\partial}{\partial\bar{q}}, \quad \overline{\overline{\nabla}} = 4\frac{\partial}{\partial\overline{\bar{q}}}, \quad \widetilde{\overline{\nabla}} = 4\frac{\partial}{\partial\widetilde{\bar{q}}}.$$ (A.3.6)

Then, if we impose that a function given by (A.3.2) is just a function of q (and not of the conjugations of q), we must have

$$\nabla F = 0; \quad \overline{\overline{\nabla}} F = 0; \quad \widetilde{\overline{\nabla}} F = 0$$ (A.3.7)

and, performing the formal multiplications, we obtain (A.3.3). □

We also have

$$\frac{1}{4}\overline{\overline{\nabla}} F = \frac{d\,F(q)}{d\,q} = \frac{\partial F}{\partial t}.$$ (A.3.8)

From relations (A.3.3) or by applying two operators, among (A.3.6), we obtain the following relations among second derivatives:

$$
\begin{aligned}
F_{1,ty} &= \alpha F_{1,xz} = F_{i,xy} = F_{i,tz} = F_{k,yz} \\
&= F_{k,tx} = F_{j,tt} = F_{j,yy} = \alpha F_{j,xx} = \alpha F_{j,zz}, \\
F_{1,tz} &= F_{1,xy} = F_{i,xz} = \alpha F_{i,ty} = F_{j,yz} \\
&= F_{j,tx} = \alpha F_{k,tt} = \alpha F_{k,yy} = F_{k,xx} = F_{k,zz}, \\
F_{1,tx} &= F_{1,yz} = F_{j,xy} = F_{j,tz} = \alpha F_{k,ty} \\
&= F_{k,xz} = \alpha F_{i,tt} = \alpha F_{i,yy} = F_{i,xx} = F_{i,zz}, \\
F_{i,tx} &= F_{i,yz} = F_{j,ty} = \alpha F_{j,xz} = F_{k,tz} \\
&= F_{k,xy} = F_{1,tt} = F_{1,yy} = \alpha F_{1,xx} = \alpha F_{1,zz}.
\end{aligned}
$$ (A.3.9)

From the last four terms of each (A.3.9) we see that second derivatives of all components satisfy the following equations:

1. for elliptic quaternions, Laplace and wave-like equations

$$F_{n,tt} + F_{n,xx} \pm (F_{n,yy} + F_{n,zz}) = 0; \quad F_{n,tt} + F_{n,zz} \pm (F_{n,yy} + F_{n,xx}) = 0,$$

2. for hyperbolic quaternions, wave-like equations

$$F_{n,tt} - F_{n,xx} + F_{n,yy} - F_{n,zz} = 0; \quad F_{n,tt} + F_{n,xx} - F_{n,yy} - F_{n,zz} = 0$$

$$F_{n,tt} - F_{n,xx} - F_{n,yy} + F_{n,zz} = 0.$$

If we multiply the four differential operators of (A.3.4) we obtain a real partial differential operator of degree 4. Actually the product of the four differential operators of (A.3.4) is formally equal to the product of the principal conjugations, thus it is real. Therefore by applying this operator to a holomorphic function,

an equation, called (see App. C) a *characteristic differential equation* is obtained
from (A.3.7) for the components of quaternion functions. This equation represents
a generalization of the Laplace equation for functions of a complex variable and
of a wave equation for functions of a hyperbolic variable.

From a characteristic differential equation or from (A.3.9) we can determine
if a real function of four real variables can be considered as a component of a
holomorphic quaternion function. Now we see that if we know, in a simply connected domain, a component $F_n(t, x, y, z)$, with $n = 1, i, j, k$, of a holomorphic
function, we can calculate, but for a constant term, the other three components
by means of line integrals ([60], Chap. 4). This property is similar to a well-known
property of functions of a complex variable.

Proof. Actually for a linear differential form we can write the line integral

$$F_n(q) = \int_{q_0}^{q} \left(F_{n,\,t}\, dt + F_{n,\,x}\, dx + F_{n,\,y}\, dy + F_{n,\,z}\, dz \right), \tag{A.3.10}$$

where all functions and differentials must be considered as functions of a line
parameter σ.

Now let us assume that we know F_1. From (A.3.3) we can obtain the partial
derivative of the other components and write the differential forms for the unknown
components by means of partial derivatives of F_1. Then

$$F_i(q) = \int_{q_0}^{q} \left(\alpha\, F_{1,\,x}\, dt + F_{1,\,t}\, dx + \alpha\, F_{1,\,z}\, dy + F_{1,\,y}\, dz \right),$$

$$F_j(q) = \int_{q_0}^{q} \left(F_{1,\,y}\, dt + F_{1,\,z}\, dx + F_{1,\,t}\, dy + F_{1,\,x}\, dz \right),$$

$$F_k(q) = \int_{q_0}^{q} \left(\alpha\, F_{1,\,z}\, dt + F_{1,\,y}\, dx + \alpha\, F_{1,\,x}\, dy + F_{1,\,t}\, dz \right). \tag{A.3.11}$$

Relations (A.3.9) represent the integrability conditions. $\square$

We shall see in Section A.3.2 that, given a component, the other ones can
also be obtained by means of an algebraic method.

Analytic Functions

Let us consider a quaternion holomorphic function $F(q)$, which we can express as
a power series in the variable q about q_0,

$$F(q) = \sum_{r=0}^{\infty} c_r\, (q - q_0)^r, \quad \text{where} \quad c_r \in \boldsymbol{Q}. \tag{A.3.12}$$

In particular, if $q_0 = 0$ we have

$$F(q) = \sum_{r=0}^{\infty} c_r\, q^r. \tag{A.3.13}$$

In a similar way as in (A.2.22) we put

$$c_r = e_1 a_r + e_2 b_r , \quad \text{where} \quad a_r, b_r \in \boldsymbol{H_2}. \tag{A.3.14}$$

Substituting (A.2.22), (A.2.23) and (A.3.14) in (A.3.13), we have

$$F(q) = \sum_{r=0}^{\infty} (e_1 a_r + e_2 b_r)(e_1 z^1 + e_2 z^2)^r \equiv \sum_{r=0}^{\infty} (e_1 a_r + e_2 b_r)[e_1 (z^1)^r + e_2 (z^2)^r] \tag{A.3.15}$$

and, from (A.2.17), we obtain

$$F(q) = \sum_{r=0}^{\infty} [e_1 a_r (z^1)^r + e_2 b_r (z^2)^r] \equiv e_1 \sum_{r=0}^{\infty} a_r (z_1)^r + e_2 \sum_{r=0}^{\infty} b_r (z_2)^r$$
$$\equiv e_1 F^1(z^1) + e_2 F^2(z^2). \tag{A.3.16}$$

In particular, if $c_r \in \boldsymbol{R}$, as is the case of elementary functions, it results that $a_r = b_r$ and we have

$$F(q) = e_1 F(z^1) + e_2 F(z^2). \tag{A.3.17}$$

From (A.2.23) and (A.2.24) it follows that *Analytic functions, defined by series with real coefficients, of conjugate variables are conjugate analytic functions*

$$\overline{\overline{F(q)}} = F(\overline{\overline{q}}); \quad \overline{\widetilde{F(q)}} = F(\overline{\widetilde{q}}); \quad \widetilde{\overline{F(q)}} = F(\widetilde{\overline{q}}). \tag{A.3.18}$$

Following the notation of complex analysis we put

$$F(z^m) = u^m + i v^m \quad \text{with} \quad m = 1, 2. \tag{A.3.19}$$

Functions in a Decomposed Algebra

Here we show some differential properties of functions in a decomposed algebra obtained as a consequence of GCR conditions.

If we sum or subtract the first and third terms in (A.3.3, a) and (A.3.3, c), we obtain

$$F_{1,t} \pm F_{1,y} = F_{j,y} \pm F_{j,t} \tag{A.3.20}$$

and if we take t, y as functions of τ, η given by (A.2.20) the left-hand side becomes

$$F_{1,\tau}(\tau_{,t} \pm \tau_{,y}) + F_{1,\eta}(\eta_{,t} \pm \eta_{,y}). \tag{A.3.21}$$

The same expression is obtained for the right-hand side, by substituting $1 \to j$. If we substitute the value (∓ 1) for the τ, η derivatives with respect to t, y and separate the (+) and (-) cases, we have

$$(F_1 - F_j)_{,\tau} = 0, \quad (F_1 + F_j)_{,\eta} = 0. \tag{A.3.22}$$

And, in the same way, we obtain

$$(F_i - F_k)_{,\xi} = 0, \quad (F_i + F_k)_{,\zeta} = 0, \quad (F_1 - F_j)_{,\xi} = 0,$$
$$(F_1 + F_j)_{,\zeta} = 0, \quad (F_i - F_k)_{,\tau} = 0, \quad (F_i + F_k)_{,\eta} = 0. \qquad (\text{A.3.23})$$

With the definitions (A.3.19) we have

$$u^1 = F_1 + F_j, \quad u^2 = F_1 - F_j, \quad v^1 = F_i + F_k, \quad v^2 = F_i - F_k; \qquad (\text{A.3.24})$$

we see that u^1, v^1 are just functions of τ, ξ and u^2, v^2 just of η, ζ. It follows that u^1, v^1 *and* u^2, v^2 *are constant if* τ, ξ *or* η, ζ, *respectively, are constant.*

It is easy to verify that

$$u^1_{,\xi} \equiv (F_1 + F_j)_{,\xi} = \alpha \, (F_i + F_k)_{,\tau} \equiv \alpha \, v^1_{,\tau}; \quad u^1_{,\tau} \equiv (F_1 + F_j)_{,\tau} = (F_i + F_k)_{,\xi} \equiv v^1_{,\xi}$$
$$(\text{A.3.25})$$

satisfy the Cauchy–Riemann conditions for the two-dimensional component algebras (complex, parabolic, hyperbolic). Analogous relations can be obtained for u^2 and v^2.

A.3.2 Algebraic Reconstruction of Quaternion Functions Given a Component

In ([47], p. 205) it is shown that a function of a complex variable can be obtained from one component by means of an algebraic method. This same method can be extended to functions of elliptic and hyperbolic quaternions.

Theorem A.4. *Let the component* $F_1(t, x, y, z)$ *be given. We obtain the quaternion function* $F(q)$, *by means of a suitable substitution of a quaternion variable for real variables* t, x, y, z. *In particular we have*

$$F(q) = 4 \, F_1 \left(\frac{q + \bar{\bar{q}}_0 + \bar{\tilde{q}}_0 + \tilde{q}_0}{4}, \, i \frac{q - \bar{\bar{q}}_0 + \bar{\tilde{q}}_0 - \tilde{q}_0}{4 \, \alpha}, \, j \frac{q + \bar{\bar{q}}_0 - \bar{\tilde{q}}_0 - \tilde{q}_0}{4}, \right.$$
$$\left. k \frac{q - \bar{\bar{q}}_0 - \bar{\tilde{q}}_0 + \tilde{q}_0}{4\alpha} \right) - \overline{\overline{F_0}} - \overline{\widetilde{F_0}} - \widetilde{F_0}, \qquad (\text{A.3.26})$$

where $F_0 \equiv F(q_0)$ *represents the initial value for* $F(q)$ *in* q_0.

In the proof we follow the same procedure of ([47], p. 205):

Proof. If in q_0 the function is holomorphic, from (A.3.12) we can write a series expansion about q_0,

$$F(q) = \sum_{r=0}^{\infty} c_r \, (q - q_0)^r \equiv F_0 + \sum_{r=1}^{\infty} c_r \, (q - q_0)^r, \qquad (\text{A.3.27})$$

where c_r, q, $q_0 \in \mathbf{Q}$. From (A.2.9), written for the functions, we have

$$4 F_1(t, x, y, z) = F(q) + \overline{\overline{F(q)}} + \widetilde{\overline{F(q)}} + \overline{\widetilde{F(q)}}$$

$$\equiv \sum_{r=0}^{\infty} \{ c_r \left[(t - t_0) + i(x - x_0) + j(y - y_0) + k(z - z_0) \right]^r$$

$$+ \bar{c}_r \left[(t - t_0) - i(x - x_0) + j(y - y_0) - k(z - z_0) \right]^r$$

$$+ \bar{\tilde{c}}_r \left[(t - t_0) + i(x - x_0) - j(y - y_0) - k(z - z_0) \right]^r$$

$$+ \tilde{c}_r \left[(t - t_0) - i(x - x_0) - j(y - y_0) + k(z - z_0) \right]^r \}. \quad \text{(A.3.28)}$$

In a quaternion domain in which the series converge we can "complexify" them by means of the substitutions

$$t - t_0 \to \frac{q - q_0}{4}; \quad x - x_0 \to \frac{q - q_0}{4\,i}; \quad y - y_0 \to \frac{q - q_0}{4\,j}; \quad z - z_0 \to \frac{q - q_0}{4\,k}. \quad \text{(A.3.29)}$$

With these substitutions the last three terms in square brackets of (A.3.28), are zero except for the known term. If we call γ_0 the coefficient of unity in c_0 we have

$$4 F_1(t, x, y, z) = 4\gamma_0 + \sum_{r=1}^{\infty} c_r \left[(t - t_0) + i(x - x_0) + j(y - y_0) + k(z - z_0) \right]^r$$

$$\equiv 4\gamma_0 + F(q) - F_0. \quad \text{(A.3.30)}$$

Now by means of (A.3.29) we substitute the variables in F_1, in the following way shown for the variable t,

$$t = \frac{q + 4t_0 - q_0}{4} \equiv \frac{q + 3t_0 - i\,x_0 - j\,y_0 - k\,z_0}{4} \equiv \frac{q + \bar{q}_0 + \bar{\tilde{q}}_0 + \tilde{q}_0}{4}.$$

In the last passage we substitute the variables in q_0, with their expressions given by (A.2.9). In the same way we obtain the expressions for x, y, z. Then

$$t \to \frac{q + \bar{q}_0 + \bar{\tilde{q}}_0 + \tilde{q}_0}{4}; \quad x \to \frac{q - \bar{q}_0 + \bar{\tilde{q}}_0 - \tilde{q}_0}{4\,i};$$

$$y \to \frac{q + \bar{q}_0 - \bar{\tilde{q}}_0 - \tilde{q}_0}{4\,j}; \quad z \to \frac{q - \bar{q}_0 - \bar{\tilde{q}}_0 + \tilde{q}_0}{4\,k} \quad \text{(A.3.31)}$$

and the function $F(q)$ given by (A.3.26) is obtained. $\qquad \square$

A.4 Mapping by Means of Quaternion Functions

A.4.1 The "Polar" Representation of Elliptic and Hyperbolic Quaternions

Quaternions q can be represented in a four-dimensional space by means of an association of components q^{α} with Cartesian coordinates of a point $P \equiv (q^{\alpha})$. Now we note that the most appropriate geometrical representation of these numbers is in a space with the geometry they generate. This geometry (Section 3.2) is characterized by a metric fixed by the characteristic determinant, an algebraic form of degree 4, which is invariant for the four-parameters group of the translations and a three-parameters group which, as an extension of the terminology for complex and hyperbolic numbers, we call *rotations*.

It is well known that for complex and hyperbolic numbers, the exponential function allows us to introduce the circular or hyperbolic angles that represent the "rotations" of the related geometries. The same is true for quaternion geometries for which the existence of differential calculus allows us to introduce functions (see Section A.5). Moreover, from the definition of exponential function, for $q \neq 0$ and non-zero divisors, as well as for complex and hyperbolic variables (4.1.6), the *polar quaternion* representation can be introduced,

$$q \equiv t + i\,x + j\,y + k\,z = \rho \exp[i\,\phi + j\,\theta + k\,\psi], \qquad (A.4.1)$$

where ρ is the modulus and ϕ, θ, ψ represent three circular and/or hyperbolic angles. By analogy with complex numbers we can call them **arguments**. Modulus and principal arguments for elliptic quaternions are given as functions of components of q, by

$$
\begin{aligned}
\rho &= \sqrt[4]{[(t+y)^2 + (x+z)^2]\,[(t-y)^2 + (x-z)^2]}, \\
\phi &= \frac{1}{2}\tan^{-1}\left[\frac{2\,(t\,x - y\,z)}{t^2 - x^2 - y^2 + z^2}\right], \\
\theta &= \frac{1}{2}\tanh^{-1}\left[\frac{2\,(t\,y + x\,z)}{t^2 + x^2 + y^2 + z^2}\right], \\
\psi &= \frac{1}{2}\tan^{-1}\left[\frac{2\,(t\,z - x\,y)}{t^2 + x^2 - y^2 - z^2}\right].
\end{aligned}
\qquad (A.4.2)
$$

For hyperbolic quaternions, corresponding expressions can be obtained from components of the logarithm function (Section A.5).

As well as for Euclidean and hyperbolic geometries (or trigonometries), exponential mapping can give the most suitable formalization for the quaternion geometries (and trigonometries). Actually if we put a quaternion constant in polar form

$$b \equiv b_1 + i\,b_i + j\,b_j + k\,b_k \rightarrow \rho_\beta \exp[i\,\beta_i + j\,\beta_j + k\,\beta_k] \qquad (A.4.3)$$

and we take constant b so that $\rho_\beta = 1$, the motions of the geometry are given by the three parameters $\beta_i, \beta_j, \beta_k$. We can check at once that

$$q\,\bar{\bar{q}}_1\,\tilde{\bar{q}}_2\,\tilde{q}_3; \quad q\,\bar{\bar{q}}\,\tilde{\bar{q}}_1\,\tilde{q}_1; \quad q\,\tilde{\bar{q}}\,\bar{\bar{q}}_1\,\tilde{q}_1; \quad q\,\tilde{\bar{q}}\,\bar{\bar{q}}_1\,\bar{q}_1 \qquad (A.4.4)$$

are invariant quantities with respect to the mapping $q' = b\,q$.

Quantities (A.4.4) allow an operative definition of arguments and a link between "distances" and angles as well as in the formalization of hyperbolic trigonometry (Chapter 4, 6). Although the way for formalizing quaternion "trigonometry" may be considered straightforward, the practical realization is not immediate and goes beyond the scope of this book. The reason for these very short considerations is that the introduction of the three arguments allows us to show in Section A.4 that mappings by means of quaternion functions have the same properties of conformal mapping of a complex variable.

A.4.2 Conformal Mapping

Let us consider a bijective mapping $W = F(q)$ where $F(q)$ is a holomorphic quaternion function, i.e., a bijective mapping between two real four-dimensional spaces given by a correspondence of the components of $W \equiv (w_1, w_i, w_j, w_k)$ and $F(q)$. A necessary condition for bijective mappings is that Jacobian determinant J must be $\neq 0$.

For functions of hypercomplex variables the Jacobian determinant is equal to the characteristic determinant of the derivative of $F(q)$ (Section 7.2.1). For commutative quaternions here considered, thanks to representations of q with unity and in decomposed form, we have

$$J[F(q)] \equiv \|F'(q)\| = J[F^1(z^1)] \cdot J[F^2(z^2)], \qquad (A.4.5)$$

which is zero when the derivatives with respect to complex variables z^1, z^2 are zero, i.e., if z^1 or z^2 are constant or $F^{m\prime}(z^m)$ are zero divisors. Then, since holomorphic functions are defined if $F'(q) \neq 0$ and is not a zero divisor, it results that *If $F(q)$ is derivable the mapping is bijective.*

Equations (A.2.19) and (A.2.21) state that z^1 or z^2 is constant if they lie in planes parallel to characteristic planes

$$I \begin{cases} \tau \equiv t + y = \text{const}_1 \\ \xi \equiv x + z = \text{const}_2 \end{cases} \qquad II \begin{cases} \eta \equiv t - y = \text{const}_3 \\ \zeta \equiv x - z = \text{const}_4 \end{cases}. \qquad (A.4.6)$$

Moreover if t, x, y, z satisfy (A.4.6), from (A.3.24) and the considerations of Section A.3.1 it follows that *Planes parallel to characteristic planes are mapped into planes parallel to characteristic planes*: The topologies of starting and transformed spaces are the same.

Now we see that, if we represent quaternions in a four-dimensional space in which the geometry is the one they generate (Section 3.2), for every q_0 in which $F(q_0)$ is holomorphic we have

Theorem A.5. *The quaternion conformal mappings have exactly the same properties as the conformal mappings of a complex variable, in particular:*

1. *The stretching is constant in quaternion geometry and is given by $\|F'(q)\|$ which is equal to the Jacobian determinant of the mapping.*

2. *The three angles (arguments) of quaternion geometry, between any two curves passing through q_0, are preserved.*

Proof.

1. From the definition of the derivative of quaternion functions we have

$$\lim_{\substack{\Delta q \to 0 \\ |\Delta q| \neq 0}} \frac{\Delta W}{\Delta q} \equiv \frac{dW}{dq} = F'(q) \tag{A.4.7}$$

and for conjugate functions (A.4.7)

$$\frac{d\overline{W}}{d\overline{q}} = \overline{\overline{F'(q)}}, \qquad \frac{d\widetilde{W}}{d\widetilde{q}} = \widetilde{\widetilde{F'(q)}}, \qquad \frac{d\overline{\widetilde{W}}}{d\overline{\widetilde{q}}} = \overline{\widetilde{F'(q)}}. \tag{A.4.8}$$

If we multiply the sides of (A.4.7) for the corresponding ones of (A.4.8) we have

$$\|dW\| = \|F'(q)\| \cdot \|dq\| \, ; \tag{A.4.9}$$

then in quaternion geometry the stretching does not depend on direction, moreover $\|F'(q)\|$ is the characteristic determinant of the derivative of $F(q)$, and thus it is equal to the Jacobian determinant of $F(q)$ (Section 7.2.1).

2. Let us consider, in the representative space of quaternion q, a regular curve λ given by (A.3.1) and passing through $q_0 \equiv q(\sigma_0)$. This curve is mapped into a curve Λ of W space passing through $W_0 \equiv F(q_0) = F[q(\sigma_0)]$. For the derivative chain rule we have

$$W'_0 = F'(q_0) \, q'(\sigma_0), \tag{A.4.10}$$

since derivability requires that $F'(q_0) \neq 0$ and for regular curves $q'(\sigma_0) \neq 0$ we can put all terms in (A.4.10) in exponential form and by calling ϑ_l, ϕ_l, γ_l (with $l = i, j, k$) the arguments of W', $F'(q)$, $q'(\sigma)$, respectively, we have

$$\begin{aligned}
\exp[i\,\vartheta_i + j\,\vartheta_j + k\,\vartheta_k] &= \exp[i\,\phi_i + j\,\phi_j + k\,\phi_k]\exp[i\,\gamma_i + j\,\gamma_j + k\,\gamma_k] \\
&\Rightarrow \quad \vartheta_l - \gamma_l = \phi_l \, ; \tag{A.4.11}
\end{aligned}$$

then *The mapping $W = F(q)$ "rotates" all curves passing through q_0 by the same quantities given by the arguments of the derivative at q_0.* □

A.4.3 Some Considerations About Scalar and Vector Potentials

It is well known that fields can be obtained by means of a derivation process from scalar and/or vector potentials. The most important field obtained in this way is the electromagnetic one defined in relativistic four-dimensional space-time. We also know that for a two-dimensional Laplace field we can define both a scalar and a "vector potential" [74, p. 199]. In fact if we have a scalar potential U for a two-dimensional Laplace equation, we can find the function V that, together with U define a function of a complex variable $F(z) = U + iV$. If we consider in complex-vector form the field given by $\operatorname{grad} U \equiv (U_{,x}, U_{,y})$, we can write

$$\vec{E} = U_{,x} + i\,U_{,y} \xrightarrow{\text{from CR}} U_{,x} - i\,V_{,x} = \overline{F'(z)} \equiv \frac{d\,\overline{F(z)}}{d\,\bar{z}}, \qquad (A.4.12)$$

i.e., as Euler first stated [25, p. 37], the conjugate functions of a complex variable have a physical relevance. In modern language, we know that the CR conditions for $f(\bar{z})$ correspond to $\operatorname{curl} E = 0$; $\operatorname{div} E = 0$ [74]. Summing up, in the considered two-dimensional example, we can obtain the field by means of a partial derivative of a scalar potential U or as a derivative of a "vector potential" $F(\bar{z})$.

Now we see that *the same result holds for quaternion functions*.

Proof. In fact let us consider a four-dimensional potential U_1, that is a component of an elliptic quaternion function $F(q)$. From Eqs. (A.3.11) or Eq. (A.3.26), we can determine the other components U_i, U_j, U_k. If we consider a field given by $\operatorname{grad} U_1$ we have

$$\vec{E} = \quad U_{1,t} + i\,U_{1,x} + j\,U_{1,y} + k\,U_{1,z}, \qquad (A.4.13)$$

$$\xrightarrow{\text{GCR}} \quad U_{1,t} - i\,U_{i,t} + j\,U_{j,t} - k\,U_{k,t} = \overline{\overline{F'(q)}} \equiv \frac{d\,\overline{\overline{F(q)}}}{d\,\bar{\bar{q}}}.$$

$\qquad\qquad\qquad\qquad\qquad\qquad\qquad\qquad\qquad\qquad\qquad\qquad\qquad\qquad\Box$

A.5 Elementary Functions of Quaternions

An easy way for obtaining the components of a quaternion function is to use (A.3.17), but we show other approaches here as well. We can verify that *the derivatives are formally equal to the ones for real and complex variables*.

Elliptic Quaternions

The exponential function

The components of an exponential function can be obtained in many ways, in particular we can use the property shown in App. C for which the exponential and logarithm functions have the same properties of the corresponding functions of a real or complex variable. Then we have $\exp[t + i\,x + j\,y + k\,z] \equiv \exp[t] \cdot \exp[i\,x] \cdot \exp[j\,y] \cdot \exp[k\,z]$ and by writing the exponential of two-dimensional algebras in terms of trigonometric functions (circular and/or hyperbolic), we obtain

$$
\begin{aligned}
F_1 &= \exp[t](\cos x\,\cosh y\,\cos z - \sin x\,\sinh y\,\sin z)\\
F_i &= \exp[t](\sin x\,\cosh y\,\cos z + \cos x\,\sinh y\,\sin z)\\
F_j &= \exp[t](\cos x\,\sinh y\,\cos z - \sin x\,\cosh y\,\sin z)\\
F_k &= \exp[t](\cos x\,\cosh y\,\sin z + \sin x\,\sinh y\,\cos z).
\end{aligned}
\qquad (A.5.1)
$$

The logarithm function

It can be obtained as the inverse of the exponential function.

$$F_1 = \frac{1}{4} \ln\left\{ \left[(t+y)^2 + (x+z)^2 \right] \left[(t-y)^2 + (x-z)^2 \right] \right\} \tag{A.5.2}$$

$$F_i = \frac{1}{2} \left\{ \tan^{-1}\left[\frac{x+z}{t+y} \right] + \tan^{-1}\left[\frac{x-z}{t-y} \right] \right\} \equiv \frac{1}{2} \tan^{-1}\left[\frac{2\,(t\,x - y\,z)}{t^2 - x^2 - y^2 + z^2} \right]$$

$$F_j = \frac{1}{4} \ln\left[\frac{(t+y)^2 + (x+z)^2}{(t-y)^2 + (x-z)^2} \right] \equiv \frac{1}{2} \tanh^{-1}\left[\frac{2\,(t\,y + x\,z)}{t^2 + x^2 + y^2 + z^2} \right]$$

$$F_k = \frac{1}{2} \left\{ \tan^{-1}\left[\frac{x+z}{t+y} \right] - \tan^{-1}\left[\frac{x-z}{t-y} \right] \right\} \equiv \frac{1}{2} \tan^{-1}\left[\frac{2\,(t\,z - x\,y)}{t^2 + x^2 - y^2 - z^2} \right].$$

The circular trigonometric functions

Cosine: $\cos q \equiv (\exp[i\,q] + \exp[-i\,q])/2$

$$F_1 = \cos t \cosh x \, \cos y \, \cosh z - \sin t \sinh x \, \sin y \, \sinh z$$
$$F_i = -\sin t \sinh x \, \cos y \, \cosh z - \cos t \cosh x \, \sin y \, \sinh z$$
$$F_j = -\sin t \cosh x \, \sin y \, \cosh z + \cos t \sinh x \, \cos y \, \sinh z$$
$$F_k = -\sin t \cosh x \, \cos y \, \sinh z - \cos t \sinh x \, \sin y \, \cosh z. \tag{A.5.3}$$

Sine: $\sin q \equiv (\exp[i\,q] - \exp[-i\,q])/(2\,i)$

$$F_1 = \sin t \cosh x \, \cos y \, \cosh z + \cos t \sinh x \, \sin y \, \sinh z$$
$$F_i = \cos t \sinh x \, \cos y \, \cosh z - \sin t \cosh x \, \sin y \, \sinh z$$
$$F_j = \cos t \cosh x \, \sin y \, \cosh z + \sin t \sinh x \, \cos y \, \sinh z$$
$$F_k = \cos t \cosh x \, \cos y \, \sinh z - \sin t \sinh x \, \sin y \, \cosh z. \tag{A.5.4}$$

The hyperbolic trigonometric functions

Hyperbolic cosine: $\cosh q \equiv (\exp[q] + \exp[-q])/2$

$$F_1 = \cosh t \cos x \, \cosh y \, \cos z - \sinh t \sin x \, \sinh y \, \sin z$$
$$F_i = \sinh t \sin x \, \cosh y \, \cos z + \cosh t \cos x \, \sinh y \, \sin z$$
$$F_j = \sinh t \cos x \, \sinh y \, \cos z - \cosh t \sin x \, \cosh y \, \sin z$$
$$F_k = \sinh t \cos x \, \cosh y \, \sin z + \cosh t \sin x \, \sinh y \, \cos z. \tag{A.5.5}$$

Hyperbolic sine: $\sinh q \equiv (\exp[q] - \exp[-q])/2$

$$F_1 = \sinh t \cos x \, \cosh y \, \cos z - \cosh t \sin x \, \sinh y \, \sin z$$
$$F_i = \cosh t \sin x \, \cosh y \, \cos z + \sinh t \cos x \, \sinh y \, \sin z$$
$$F_j = \cosh t \cos x \, \sinh y \, \cos z - \sinh t \sin x \, \cosh y \, \sin z$$
$$F_k = \cosh t \cos x \, \cosh y \, \sin z + \sinh t \sin x \, \sinh y \, \cos z. \tag{A.5.6}$$

Hyperbolic Quaternions

The exponential function

$$F_1 = \exp[t](\cosh x \, \cosh y \, \cosh z + \sinh x \, \sinh y \, \sinh z)$$
$$F_i = \exp[t](\sinh x \, \cosh y \, \cosh z + \cosh x \, \sinh y \, \sinh z)$$
$$F_j = \exp[t](\cosh x \, \sinh y \, \cosh z + \sinh x \, \cosh y \, \sinh z)$$
$$F_k = \exp[t](\cosh x \, \cosh y \, \sinh z + \sinh x \, \sinh y \, \cosh z). \qquad (A.5.7)$$

The logarithm function

It can be obtained in many ways. In particular we utilize the identity

$$\ln q \equiv \frac{1}{4} \left\{ \ln \left[q \, \bar{\bar{q}} \, \tilde{\bar{q}} \, \tilde{\bar{q}} \right] + \ln \left[\frac{q \, \bar{\tilde{q}}}{\bar{\tilde{q}} \, \tilde{q}} \right] + \ln \left[\frac{q \, \tilde{\tilde{q}}}{\bar{\bar{q}} \, \tilde{q}} \right] + \ln \left[\frac{q \, \tilde{q}}{\bar{\bar{q}} \, \tilde{q}} \right] \right\}. \qquad (A.5.8)$$

From (A.2.13)–(A.2.15) for $q = q_1$ we observe that the arguments of the second, third and fourth terms in the right-hand side of (A.5.8), are ln of ratios between conjugate two-dimensional numbers in the algebras $(1, i)$, $(1, j)$, $(1, k)$, respectively. Then they are the arguments of these numbers multiplied by the corresponding versors[1]. In particular we have the following.

$$F_1 = \frac{1}{4} \ln \left\{ \left[(t + y)^2 - (x + z)^2 \right] \left[(t - y)^2 - (x - z)^2 \right] \right\}$$
$$F_i = \frac{1}{4} \ln \left[\frac{(t + x)^2 - (y + z)^2}{(t - x)^2 - (y - z)^2} \right] \equiv \frac{1}{2} \tanh^{-1} \left[\frac{2(t\,x - y\,z)}{t^2 + x^2 - y^2 - z^2} \right]$$
$$F_j = \frac{1}{4} \ln \left[\frac{(t + y)^2 - (x + z)^2}{(t - y)^2 - (x - z)^2} \right] \equiv \frac{1}{2} \tanh^{-1} \left[\frac{2(t\,y - x\,z)}{t^2 - x^2 + y^2 - z^2} \right]$$
$$F_k = \frac{1}{4} \ln \left[\frac{(t + z)^2 - (x + y)^2}{(t - z)^2 - (x - y)^2} \right] \equiv \frac{1}{2} \tanh^{-1} \left[\frac{2(t\,z - x\,y)}{t^2 - x^2 - y^2 + z^2} \right]. \qquad (A.5.9)$$

The circular trigonometric functions

In this case we can not relate these functions to the exponential function of an imaginary variable. Here these functions are defined by means of the coincidence between absolutely convergent series: if we indicate the generic versor by ($l = i, j, k$), we have $\cos lx \equiv \cos x$; $\sin lx \equiv l \sin x$; by means of this position and using the usual (real and complex analysis) rules for the sums of arguments of trigonometric functions, we obtain:

---

[1] It is known that for complex numbers we have $\ln[z/\bar{z}] = 2\,i \arg z$.

Cosine:

$$F_1 = \cos t \cos x \, \cos y \, \cos z + \sin t \sin x \, \sin y \, \sin z$$
$$F_i = -(\sin t \sin x \, \cos y \, \cos z + \cos t \cos x \, \sin y \, \sin z)$$
$$F_j = -(\sin t \cos x \, \sin y \, \cos z + \cos t \sin x \, \cos y \, \sin z)$$
$$F_k = -(\sin t \cos x \, \cos y \, \sin z + \cos t \sin x \, \sin y \, \cos z). \qquad (A.5.10)$$

Sine:

$$F_1 = \sin t \cos x \, \cos y \, \cos z - \cos t \sin x \, \sin y \, \sin z$$
$$F_i = \cos t \sin x \, \cos y \, \cos z - \sin t \cos x \, \sin y \, \sin z$$
$$F_j = \cos t \cos x \, \sin y \, \cos z - \sin t \sin x \, \cos y \, \sin z$$
$$F_k = \cos t \cos x \, \cos y \, \sin z - \sin t \sin x \, \sin y \, \cos z. \qquad (A.5.11)$$

The hyperbolic trigonometric functions

Hyperbolic cosine: $\cosh q \equiv (\exp[q] + \exp[-q])/2$

$$F_1 = \cosh t \cosh x \cosh y \cosh z + \sinh t \sinh x \sinh y \sinh z$$
$$F_i = \sinh t \sinh x \cosh y \cosh z + \cosh t \cosh x \sinh y \sinh z$$
$$F_j = \sinh t \cosh x \sinh y \cosh z + \cosh t \sinh x \cosh y \sinh z$$
$$F_k = \sinh t \cosh x \cosh y \sinh z + \cosh t \sinh x \sinh y \cosh z. \qquad (A.5.12)$$

Hyperbolic sine: $\sinh q \equiv (\exp[q] - \exp[-q])/2$

$$F_1 = \sinh t \cosh x \cosh y \cosh z + \cosh t \sinh x \sinh y \sinh z$$
$$F_i = \cosh t \sinh x \cosh y \cosh z + \sinh t \cosh x \sinh y \sinh z$$
$$F_j = \cosh t \cosh x \sinh y \cosh z + \sinh t \sinh x \cosh y \sinh z$$
$$F_k = \cosh t \cosh x \cosh y \sinh z + \sinh t \sinh x \sinh y \cosh z. \qquad (A.5.13)$$

Parabolic Quaternions

The exponential function

$$F_1 = \exp[t] \, \cosh y \, ; \qquad\qquad F_i = \exp[t] \, (x \cosh y + z \, \sinh y)$$
$$F_j = \exp[t] \, \sinh y \, ; \qquad\qquad F_k = \exp[t] \, (z \cosh y + x \, \sinh y). \qquad (A.5.14)$$

The logarithm function

It can be obtained as the inverse of the exponential function.

$$F_1 = \frac{1}{2} \ln[t^2 - y^2] \, ; \qquad\qquad F_i = \frac{t\,x - y\,z}{t^2 - y^2}$$
$$F_j = \tanh^{-1}\left[\frac{y}{t}\right] \, ; \qquad\qquad F_k = \frac{t\,z - x\,y}{t^2 - y^2}. \qquad (A.5.15)$$

The hyperbolic trigonometric functions

Hyperbolic cosine: $\cosh q \equiv (\exp[q] + \exp[-q])/2$

$$F_1 = \cosh t \cosh y \; ; \qquad F_i = x \sinh t \cosh y + z \cosh t \sinh y$$
$$F_j = \sinh t \sinh y \; ; \qquad F_k = x \cosh t \sinh y + z \sinh t \cosh y. \qquad (A.5.16)$$

Hyperbolic sine: $\sinh q \equiv (\exp[q] - \exp[-q])/2$

$$F_1 = \sinh t \cosh y \; ; \qquad F_i = x \cosh t \cosh y + z \sinh t \sinh y$$
$$F_j = \cosh t \sinh y \; ; \qquad F_k = x \sinh t \sinh y + z \cosh t \cosh y. \qquad (A.5.17)$$

The circular trigonometric functions

The circular trigonometric functions can be calculated taking into account that in two-dimensional parabolic algebra we have $\cos ix = \cos kz = 1; \quad \sin ix = i x; \quad \sin kz = k z$ (see Section C.5.2, p. 233). Directly or from (A.5.3) and (A.5.4) we obtain:

Cosine:

$$F_1 = \cos t \cos y \; ; \qquad F_i = -x \sin t \cos y - z \cos t \sin y$$
$$F_j = -\sin t \sin y \; ; \qquad F_k = -x \cos t \sin y - z \sin t \cos y. \qquad (A.5.18)$$

Sine:

$$F_1 = \sin t \cos y \; ; \qquad F_i = x \cos t \cos y - z \sin t \sin y$$
$$F_j = \cos t \sin y \; ; \qquad F_k = z \cos t \cos y - x \sin t \sin y. \qquad (A.5.19)$$

A.6   Elliptic-Hyperbolic Quaternions

The elliptic-hyperbolic quaternions make up the four-dimensional system which corresponds to the values $\alpha' = -1; \alpha'' = 1$ in Table (A.1.3), i.e., a system composed of a two-dimensional elliptic and a two-dimensional hyperbolic algebra.

We start from the table

$$
\begin{array}{|c|c|c|c|}
\hline
e_1 & e_2 & 0 & 0 \\
\hline
e_2 & -e_1 & 0 & 0 \\
\hline
0 & 0 & e_3 & e_4 \\
\hline
0 & 0 & e_4 & e_3 \\
\hline
\end{array}
\qquad (A.6.1)
$$

and we introduce the unity and the other versors by means of the transformation

$$
\begin{pmatrix} 1 \\ i \\ j \\ k \end{pmatrix} =
\begin{pmatrix}
1 & 0 & 1 & 0 \\
0 & 1 & 0 & 1 \\
1 & 0 & -1 & 0 \\
0 & 1 & 0 & -1
\end{pmatrix}
\begin{pmatrix} e_1 \\ e_2 \\ e_3 \\ e_4 \end{pmatrix}
\qquad (A.6.2)
$$

and the inverse

$$
\begin{pmatrix} e_1 \\ e_2 \\ e_3 \\ e_4 \end{pmatrix} = \frac{1}{2} \begin{pmatrix} 1 & 0 & 1 & 0 \\ 0 & 1 & 0 & 1 \\ 1 & 0 & -1 & 0 \\ 0 & 1 & 0 & -1 \end{pmatrix} \begin{pmatrix} 1 \\ i \\ j \\ k \end{pmatrix}.
\tag{A.6.3}
$$

We obtain, from (A.6.2), the *multiplication table for new versors*

$$
\begin{array}{|c|c|c|c|}
\hline
1 & i & j & k \\
\hline
i & -j & k & -1 \\
\hline
j & k & 1 & i \\
\hline
k & -1 & i & -j \\
\hline
\end{array}
\tag{A.6.4}
$$

We note that in this algebra $(1, i)$ and $(1, k)$, are not subalgebras and their powers are given by all four versors.

We choose again $q = t + i\,x + j\,y + k\,z$ and the characteristic matrix is given by

$$
\mathcal{M} = \begin{pmatrix} t & -z & y & -x \\ x & t & z & y \\ y & -x & t & -z \\ z & y & x & t \end{pmatrix} \equiv \left(\begin{array}{c|c} A & B \\ \hline B & A \end{array} \right).
\tag{A.6.5}
$$

From (A.6.5) we obtain the matrix form of the versors (Section 2.1)

$$
i = \begin{pmatrix} 0 & 0 & 0 & -1 \\ 1 & 0 & 0 & 0 \\ 0 & -1 & 0 & 0 \\ 0 & 0 & 1 & 0 \end{pmatrix} ; \quad j = \begin{pmatrix} 0 & 0 & 1 & 0 \\ 0 & 0 & 0 & 1 \\ 1 & 0 & 0 & 0 \\ 0 & 1 & 0 & 0 \end{pmatrix} ;
$$

$$
k = \begin{pmatrix} 0 & -1 & 0 & 0 \\ 0 & 0 & 1 & 0 \\ 0 & 0 & 0 & -1 \\ 1 & 0 & 0 & 0 \end{pmatrix}.
\tag{A.6.6}
$$

The characteristic determinant can be obtained by means of matrices A and B, according to the property recalled on p. 173. The results is

$$
\begin{aligned}
|q|^4 &= \begin{vmatrix} t+y & -x-z & 0 & 0 \\ x+z & t+y & 0 & 0 \\ 0 & 0 & t-y & z-x \\ 0 & 0 & z-x & t-y \end{vmatrix} \\
&\equiv \left[(t+y)^2 + (x+z)^2 \right] \left[(t-y)^2 - (x-z)^2 \right].
\end{aligned}
\tag{A.6.7}
$$

As it is known that the invariant coincides with the product of the invariants of component systems.

We can also obtain the value of the characteristic determinant by multiplying the number q by the conjugations[2]

$$\bar{q} = t - i\,x + j\,y - k\,z;, \quad \tilde{q} = t + i\,x - j\,y - k\,z; \quad \bar{\tilde{q}} = t - i\,x - j\,y + k\,z\,; \quad \text{(A.6.8)}$$

in fact we have

$$q\,\bar{q} = (t + j\,y)^2 - (i\,x + k\,z)^2 \equiv [t^2 + y^2 + 2\,x\,z + j(x^2 + z^2 + 2\,t\,y)], \quad \text{(A.6.9)}$$
$$\tilde{q}\,\bar{\tilde{q}} = (t - j\,y)^2 - (i\,x - k\,z)^2 \equiv [t^2 + y^2 + 2\,x\,z - j(x^2 + z^2 + 2\,t\,y)], \quad \text{(A.6.10)}$$

and product (A.6.9)·(A.6.10) is equal to (A.6.7).

A.6.1 Generalized Cauchy–Riemann Conditions

From (7.2.1), we have

$$F_{,x} = i\,F_{,t}; \qquad F_{,y} = j\,F_{,t}; \qquad F_{,z} = k\,F_{,t}; \qquad \text{(A.6.11)}$$

and obtain

$$\text{(GCR)} \quad \begin{cases} F_{1,t} = F_{i,x} = F_{j,y} = F_{k,z} & (a) \\ F_{i,t} = -F_{j,x} = F_{k,y} = -F_{1,z} & (b) \\ F_{j,t} = F_{k,x} = F_{1,y} = F_{i,z} & (c) \\ F_{k,t} = -F_{1,x} = F_{i,y} = -F_{j,z} & (d) \end{cases} \quad \text{(A.6.12)}$$

A.6.2 Elementary Functions

The exponential function

For these quaternions we can not use the same method we have used for Segre's quaternions since $(1, i)$ and $(1, k)$ are not subalgebras. Then the series expansions of $\exp[i\,x]$, $\exp[k\,z]$ are composed by terms with all four versors and they do not represent elementary functions. Nevertheless we can obtain the components by applying (A.3.17) to the decomposed system of Table (A.6.1), and going back to the system with unity by means of (A.6.3).

$$F_1 = \frac{1}{2}\exp[t]\,[\exp[y]\cos(x + z) + \exp[-y]\cosh(x - z)]$$

$$F_i = \frac{1}{2}\exp[t]\,[\exp[y]\sin(x + z) + \exp[-y]\sinh(x - z)]$$

$$F_j = \frac{1}{2}\exp[t]\,[\exp[y]\cos(x + z) - \exp[-y]\cosh(x - z)]$$

$$F_1 = \frac{1}{2}\exp[t]\,[\exp[y]\sin(x + z) - \exp[-y]\sinh(x - z)]\,. \qquad \text{(A.6.13)}$$

---

[2]It can be verified that they satisfy the characteristic equation which can be obtained as (A.2.7).

The logarithm function

$$F_1 = \frac{1}{4} \ln \left\{ \left[(t+y)^2 + (x+z)^2 \right] \left[(t-y)^2 - (x-z)^2 \right] \right\}$$

$$F_i = \frac{1}{2} \tan^{-1} \left[\frac{x+z}{t+y} \right]$$

$$F_j = \frac{1}{4} \ln \left[\frac{(t+y)^2 + (x+z)^2}{(t-y)^2 - (x-z)^2} \right] \equiv \frac{1}{2} \tanh^{-1} \left[\frac{x^2 + z^2 + 2\,t\,y}{t^2 + y^2 + 2\,x\,z} \right]$$

$$F_k = \frac{1}{2} \tanh^{-1} \left[\frac{x-z}{t-y} \right]. \tag{A.6.14}$$

The hyperbolic trigonometric functions

Hyperbolic cosine: $\cosh q \equiv (\exp[q] + \exp[-q])/2$

$$F_1 = \cosh (t+y) \cos (x+z) + \cosh (t-y) \cosh (x-z)$$
$$F_i = \sinh (t+y) \sin (x+z) + \sinh (t-y) \sinh (x-z)$$
$$F_j = \cosh (t+y) \cos (x+z) - \cosh (t-y) \cosh (x-z)$$
$$F_k = \sinh (t+y) \sin (x+z) - \sinh (t-y) \sinh (x-z). \tag{A.6.15}$$

Hyperbolic sine: $\sinh q \equiv (\exp[q] - \exp[-q])/2$

$$F_1 = \sinh (t+y) \cos (x+z) + \sinh (t-y) \cosh (x-z)$$
$$F_i = \cosh (t+y) \sin (x+z) + \cosh (t-y) \sinh (x-z)$$
$$F_j = \sinh (t+y) \cos (x+z) - \sinh (t-y) \cosh (x-z)$$
$$F_k = \cosh (t+y) \sin (x+z) - \cosh (t-y) \sinh (x-z). \tag{A.6.16}$$

A.7 Elliptic-Parabolic Generalized Segre's Quaternions

These systems seem, up to now, the most applied in different fields: Sobrero associated their functions to differential equations of elasticity [71], in more recent time Cheng used them for computer analysis of spatial mechanisms [26]. We give this name to the quaternion system which corresponds to the 7th and 8th rows in Table A.1, i.e., a system composed by two-dimensional elliptic numbers coupled by means of the parabolic versor or vice-versa,

$$q = t + i\,x + p\,(y + i\,z) \equiv t + p\,y + i(x + p\,z). \tag{A.7.1}$$

If we put $p\,i = k$, we can write $q = t + i\,x + p\,y + k\,z$ with the following *multiplication table for the versors*

1	i	p	k
i	-1	k	$-p$
p	k	0	0
k	$-p$	0	0

$$\tag{A.7.2}$$

and the characteristic matrix is given by

$$
\mathcal{M} = \begin{pmatrix} t & -x & 0 & 0 \\ x & t & 0 & 0 \\ y & -z & t & -x \\ z & y & x & t \end{pmatrix} \equiv \left(\begin{array}{c|c} A & \emptyset \\ \hline B & A \end{array} \right). \tag{A.7.3}
$$

The A and B matrices are the same as the matrices of complex numbers $t + ix$ and $y + iz$, respectively, composed as the parabolic number $A + pB$. From (A.7.3) we obtain the matrix form of the versors (Section 2.1)

$$
i = \begin{pmatrix} 0 & -1 & 0 & 0 \\ 1 & 0 & 0 & 0 \\ 0 & 0 & 0 & -1 \\ 0 & 0 & 1 & 0 \end{pmatrix}; \; p = \begin{pmatrix} 0 & 0 & 0 & 0 \\ 0 & 0 & 0 & 0 \\ 1 & 0 & 0 & 0 \\ 0 & 1 & 0 & 0 \end{pmatrix};
$$

$$
k = \begin{pmatrix} 0 & 0 & 0 & 0 \\ 0 & 0 & 0 & 0 \\ 0 & -1 & 0 & 0 \\ 1 & 0 & 0 & 0 \end{pmatrix}. \tag{A.7.4}
$$

We can obtain the value of the characteristic determinant by multiplying the number q by the conjugations:[3]

$$
\bar{q} = t - ix + py - kz;, \quad \bar{\bar{q}} = t + ix - py - kz; \quad \tilde{q} = t - ix - py + kz. \tag{A.7.5}
$$

It can be also obtained from the characteristic matrix. The immediate result is

$$
|q|^4 \Rightarrow \begin{vmatrix} t & -x & 0 & 0 \\ x & t & 0 & 0 \\ y & -z & t & -x \\ z & y & x & t \end{vmatrix} \equiv \left[t^2 + x^2 \right]^2. \tag{A.7.6}
$$

A.7.1   Generalized Cauchy–Riemann conditions

From the conditions (7.2.1), we have

$$
F_{,x} = i F_{,t}; \qquad F_{,y} = p F_{,t}; \qquad F_{,z} = k F_{,t}; \tag{A.7.7}
$$

we obtain

$$
\text{(GCR)} \quad \begin{cases} F_{1,t} = F_{i,x} = F_{p,y} = F_{k,z} & (a) \\ F_{i,t} = -F_{1,x} = F_{k,y} = -F_{k,z} & (b) \\ F_{p,t} = F_{k,x}; \; F_{1,y} = F_{i,z} = 0 & (c) \\ F_{k,t} = -F_{p,x}; \; F_{i,y} = F_{1,z} = 0 & (d) \end{cases} \tag{A.7.8}
$$

---

[3]It can be verified that they satisfy the characteristic equation which can be obtained as (A.2.7).

A.7.2   Elementary Functions

We know that the functions of a parabolic variable are given by (C.5.17). This expression can be extended to these quaternions by means of the second side of (A.7.1), and we write

$$f(q) = f(t + i\,x) + p\,(y + i\,z)\,f'(t + i\,x),\qquad\qquad(\text{A.7.9})$$

where $f'(t + i\,x)$ indicates the derivative of the function of the complex variable $f(t + i\,x)$. In particular it follows that for exponential functions as well as for hyperbolic and circular trigonometric functions, we can put in the expressions of elliptic quaternions of Section A.5: $\cosh y \simeq \cosh z \simeq \cos y \simeq \cos z = 1$ and $\sinh y \simeq \sin y \simeq y$; $\sinh z \simeq \sin z \simeq z$, but for the products $\sin\cdot\sinh = \sin\cdot\sin = \sinh\cdot\sinh = 0$.

The exponential function

$$F_1 = \exp[t]\cos x\,; \qquad\qquad F_p = \exp[t]\,(y\cos x - z\sin)\,;$$
$$F_i = \exp[t]\sin x\,; \qquad\qquad F_k = \exp[t]\,(z\cos x + y\sin x)\,. \qquad(\text{A.7.10})$$

The logarithm function

It is obtained as the inverse of the exponential function.

$$F_1 = \frac{1}{2}\ln[t^2 + x^2]\,; \qquad\qquad\qquad F_i = \tan^{-1}\left[\frac{x}{t}\right]$$
$$F_p = \frac{t\,y + x\,z}{t^2 + x^2}\,; \qquad\qquad\qquad F_k = \frac{t\,z - x\,y}{t^2 + x^2}\,. \qquad(\text{A.7.11})$$

The circular trigonometric functions

Cosine: $\cos q \equiv (\exp[i\,q] + \exp[-i\,q])/2$

$$F_1 = \cos t\cosh x\,; \qquad\quad F_p = -y\sin t\cosh x + z\cos t\sinh x\,;$$
$$F_i = -\sin t\sinh x\,; \qquad F_k = -z\sin t\cosh x - y\cos t\sinh x\,. \qquad(\text{A.7.12})$$

Sine: $\sin q \equiv (\exp[i\,q] - \exp[-i\,q])/(2\,i)$

$$F_1 = \sin t\cosh x\,; \qquad\quad F_p = y\cos t\cosh x + z\sin t\sinh x\,;$$
$$F_i = \cos t\sinh x\,; \qquad\quad F_k = z\cos t\cosh x - y\sin t\sinh x\,. \qquad(\text{A.7.13})$$

The hyperbolic trigonometric functions

Hyperbolic cosine: $\cosh q \equiv (\exp[q] + \exp[-q])/2$

$$F_1 = \cosh t\cos x\,; \qquad\quad F_p = y\sinh t\cos x - z\cosh t\sin x\,;$$
$$F_i = \sinh t\sin x\,; \qquad\quad F_k = z\sinh t\cos x + y\cosh t\sin x\,. \qquad(\text{A.7.14})$$

Hyperbolic sine: $\sinh q \equiv (\exp[q] - \exp[-q])/2$

$$F_1 = \sinh t \cos x\,; \qquad F_p = y \cosh t \cos x \; - z \sinh t \sin x\,;$$
$$F_i = \cosh t \sin x\,; \qquad F_k = z \cosh t \cos x \; + y \sinh t \sin x\,. \qquad \text{(A.7.15)}$$

Appendix B

Constant Curvature Segre's Quaternion Spaces

In Chapter 9 we have referred to B. Riemann's work *Ueber die Hypothesen ...* [61] in which he extended Gauss' ideas on surface differential geometry to N-dimensional spaces[1]. The complete development of Riemann's work, which is considered one of his milestones in the development of mathematics, has been accomplished by many mathematicians for more than half a century, leading to generalizations to non-definite differential quadratic forms (semi-Riemannian geometry) [57] and to non-Riemannian geometry [35]. The starting point of Riemann's work is that the infinitesimal distance between two points in N-dimensional space is given by an extension of Gauss' studies on surfaces, i.e., by a differential quadratic form. From these forms all the properties of spaces can be obtained. He introduced what we call today the "Riemann tensor" and the term "line element". He also gave the expressions of line elements for constant curvature spaces, i.e., spaces in which "motions" are the same as in flat (Euclidean) spaces: the roto-translations.

In this appendix we start from the concept introduced in Chapter 3 of geometries associated with N-dimensional hypercomplex numbers, i.e., geometries defined by a metric (distance between two points) given by a form of degree N and "motions" determined by $2N - 1$ parameters. If we start from this definition of distance a completely new differential geometry must be developed in order to study "non-flat hypercomplex spaces" in a complete way.

We begin to study the positive constant curvature elliptic-quaternion space (PCCQS) and make up for the lack of a complete formalized theory by using the mathematics "generated" by quaternions. This approach is an extension of the Gauss and Beltrami application of complex variables for studying surfaces (Chapter 8) and, in particular, constant curvature surfaces, as we have seen in Chapter 9. Actually the two-dimensional hypercomplex numbers provide us with a suitable mathematical tool for studying the geometries they generate, then it seems a natural approach to study the PCCQS by means of the quaternion mathematical apparatus. We shall see that the ideas introduced in this way allow us to formalize a self-consistent work. From a more general point of view, this use of quaternion analysis for studying quaternion geometry, can also be considered as an application of a well-accepted principle, following which a problem is simplified if we use a suitable mathematics with its symmetries.

---

[1]We extend the word "space" to indicate hyperspaces with the number of dimensions we are considering. We use "varieties" as usual and mean by surfaces the two-dimensional varieties.

In particular this appendix is organized in the following way: in Section B.1 we introduce the differential forms of degree 4, and the "conformal representations" in quaternion spaces. In Section B.2 we find the Euler–Lagrange equations for geodesics. These equations, as the usual ones (8.5.6), give a differential system, linear with respect to the second derivatives, which can be reduced to the normal form and, if the appropriate regularity conditions are satisfied, the theorem of existence and uniqueness of solutions holds. Thanks to this property we use an alternative method (an extension of Beltrami's method recalled in Chapter 8, p. 130 and used in Chapter 9) for finding the equations of geodesics and then check that the Euler–Lagrange equations are satisfied. In Section B.3 we introduce the PCCQS, defining with this name the space in which the same motions of flat quaternion space are allowed.

B.1 Introduction of Quaternion Differential Geometry: Differential Forms of Degree 4

It is well known that Riemann introduced the differential geometry in N-dimensional spaces, extending the quadratic form of Euclidean distance to infinitesimal distances. In quaternion geometry we introduce in a similar way *non-flat quaternion spaces*, described by a line element given by a differential form of degree 4,

$$d s^4 = a_{ihkm} d x^i d x^h d x^k d x^m, \tag{B.1.1}$$

where a_{ihkm} (for i, h, k, m running from 1 to 4) is a fourth-order covariant tensor which can be called a *metric tensor* and is supposed symmetric with respect to the four indexes.

We have seen in Chapter 8 that Gauss showed how line elements can be expressed in the isometric orthogonal form

$$d s^2 = F_1(x, y)[d x^2 + d y^2]. \tag{B.1.2}$$

Today x, y are called **conformal coordinates** and (B.1.2) is written, in the language of complex numbers, as ([29], p. 115)

$$d s^2 = F(z, \bar{z})[d z d \bar{z}] \equiv F(z, \bar{z}) \| d z \| \tag{B.1.3}$$

where we have indicated by $\| d z \|$ the characteristic determinant (squared modulus) of the differential of a complex variable: $d z = d x + i d y$. From (B.1.3) it follows at once that the isometric orthogonal form is preserved by any mapping by means of functions of a complex variable $z = f(w)$ [29]. The same result is true for two-dimensional pseudo-Euclidean (space-time) geometry (Chapter 8), and the same expression of (B.1.3) can be obtained as a function of a hyperbolic variable.

It is well known that for Riemannian geometry in more than two dimensions, it is not possible to write all line elements in an isometric form [6] and [34]. The same is true for the general line elements expressed by a differential form of degree 4, since the number of equations for its reduction to an "orthogonal" form is greater than the number of unknowns. Nonetheless Riemann showed that line elements, for constant curvature spaces of any dimension, can be written in isometric-orthogonal form [6], and we start by considering quaternion spaces that can be expressed by an extension of (B.1.3), i.e., a line element that has the structure of quaternion representative space

$$ds^4 = F(q, \bar{q}, \bar{\bar{q}}, \tilde{q}) \, dq \, d\bar{q} \, d\bar{\bar{q}} \, d\tilde{q}. \tag{B.1.4}$$

For Segre's quaternion (p. 172), as a function of coordinates (real variables), (B.1.4) becomes

$$ds^4 = F_1(t, x, y, z) \left[(dt + dy)^2 + (dx + dz)^2 \right] \left[(dt - dy)^2 + (dx - dz)^2 \right]; \tag{B.1.5}$$

in particular, if $F_1(t, x, y, z) = 1$, we call this space *flat quaternion space*.

As it happens in the two-dimensional case, we have

Theorem B.1. *The expression* (B.1.4) *of a line element is preserved from conformal quaternion mapping.*

Proof. Actually let us consider the quaternion conformal mapping (Section A.4.2) $q = G(w)$; we have

$$dq = \left(\frac{dG}{dw} \right) dw, \quad d\bar{q} = \left(\overline{\frac{dG}{dw}} \right) d\bar{w},$$

$$d\bar{\bar{q}} = \left(\overline{\overline{\frac{dG}{dw}}} \right) d\bar{\bar{w}}, \quad d\tilde{q} = \left(\widetilde{\frac{dG}{dw}} \right) d\tilde{w}. \tag{B.1.6}$$

Then (B.1.4) is transformed in

$$ds^4 = \|G'(q)\| \, F_2(w, \bar{w}, \bar{\bar{w}}, \tilde{w}) \, dw \, d\bar{w} \, d\bar{\bar{w}} \, d\tilde{w} \tag{B.1.7}$$

where $\|G'(q)\|$ indicates the characteristic determinant of the derivative of $G(q)$, that is equal to the Jacobian determinant of the mapping (Section 7.2.1). □

Then we call **conformal coordinates** the ones which give to a line element the form (B.1.4), which corresponds to the isometric orthogonal form of two-dimensional varieties (surfaces), and we also extend the name of **isometric form** to line elements in quaternion geometry given by (B.1.4) or (B.1.5). These expressions, as far as this appendix is concerned are adequate for the following developments.

B.2 Euler's Equations for Geodesics

Let us consider a four-dimensional space with a line element given by (B.1.1). As in Riemannian geometry, we define *geodesic* to be the line between two "sufficiently near" points for which the variation of $l = \int_{s_0}^{s_1} ds$ is zero. Here we extend to this line element of degree 4, a method ("calculus of variations") ([49], pp. 128–134) which allows us to obtain the differential equations for the geodesic lines in Riemannian geometry. We have

Theorem B.2. *In a space represented by a differential form of degree 4 (B.1.1), the equations of geodesics are given by*

$$\sum_{i,h,k=1}^{4} a_{ihkm}\ddot{x}^i\,\dot{x}^h\,\dot{x}^k + \frac{1}{3}\sum_{i,h,k,r=1}^{4} \Gamma_{m,ihkr}\dot{x}^i\,\dot{x}^r\,\dot{x}^h\,\dot{x}^k = 0, \qquad (B.2.1)$$

for $m = 1, 2, 3, 4$, and $\Gamma_{m,ihkr}$ is a Christoffel-like five indexes symbol given by

$$\Gamma_{m,ihkr} = \frac{1}{4}\left(\frac{\partial a_{ihkm}}{\partial x^r} + \frac{\partial a_{rhkm}}{\partial x^i} + \frac{\partial a_{irkm}}{\partial x^h} + \frac{\partial a_{ihrm}}{\partial x^k} - \frac{\partial a_{ihkr}}{\partial x^m}\right). \quad (B.2.2)$$

Proof. Actually let us call $x^i(s)$ the parametric equations of a geodesic line between the points represented by the values s_0, s_1 of line coordinates and indicate by δ the variation. If $l = \int_{s_0}^{s_1} ds$ is the length of an arbitrary line between the given points, we must have $\delta l \equiv \int_{s_0}^{s_1} \delta(ds) = 0$. By applying the variation to line element (B.1.1), we have

$$4\,ds^3\,\delta(ds) \quad = \quad \delta(a_{ihkm})\,dx^i\,dx^h\,dx^k\,dx^m \qquad (B.2.3)$$
$$+\ a_{ihkm}\,dx^i\,dx^h\,dx^k\,\delta(dx^m) \quad + \quad a_{ihkm}\,dx^i\,dx^h\,\delta(dx^k)\,dx^m$$
$$+\ a_{ihkm}\,dx^i\,\delta(dx^h)\,dx^k\,dx^m \quad + \quad a_{ihkm}\,\delta(dx^i)\,dx^h\,dx^k\,dx^m\ .$$

Setting $dx^i \equiv (dx^i/ds)\,ds = \dot{x}^i\,ds$ we can divide both sides by $4\,ds^3$, collect the last four terms (which, taking into account the symmetry of a metric tensor, are the same), reverse $\delta \leftrightarrow d$ and express $\delta a_{ihkm} \equiv \frac{\partial a_{ihkm}}{\partial x^r}\,\delta x^r$, then the line integral becomes

$$\delta l \equiv \int_{s_0}^{s_1}\left[\frac{1}{4}\frac{\partial(a_{ihkm})}{\partial x^r}\dot{x}^i\,\dot{x}^h\,\dot{x}^k\,\dot{x}^m\,\delta x^r\,ds + a_{ihkm}\,\dot{x}^i\,\dot{x}^h\,\dot{x}^k\,\delta(dx^m)\right]. \quad (B.2.4)$$

Integrating the second term by parts and taking into account that $\delta x^m = 0$ in s_0 and s_1 and $d\dot{x}^i \equiv \ddot{x}^i\,ds$, we obtain

$$\int_{s_0}^{s_1} a_{ihkm}\,\dot{x}^i\,\dot{x}^h\,\dot{x}^k\,d(\delta x^m) \qquad (B.2.5)$$

$$\equiv \quad a_{ihkm}\,\dot{x}^i\,\dot{x}^h\,\dot{x}^k\,\delta x^m\Big|_{s_0}^{s_1} - \int_{s_0}^{s_1} d(a_{ihkm}\,\dot{x}^i\,\dot{x}^h\,\dot{x}^k)\delta x^m$$

$$\equiv \quad -\int_{s_0}^{s_1}\big[(a_{ihkm}\,\ddot{x}^i\,\dot{x}^h\,\dot{x}^k + a_{ihkm}\,\ddot{x}^h\,\dot{x}^i\,\dot{x}^k + a_{ihkm}\,\ddot{x}^k\,\dot{x}^h\,\dot{x}^i)\,ds$$

$$+\,d(a_{ihkm})\,\dot{x}^h\,\dot{x}^i\,\dot{x}^k\big]\,\delta x^m\ .$$

Since $da_{ihkm} = \frac{\partial a_{ihkm}}{\partial x^r} dx^r \equiv \frac{\partial a_{ihkm}}{\partial x^r} \dot{x}^r ds$ and taking into account the symmetries of multiple sums, i.e., all terms in the following round brackets are equal, we can write

$$\frac{\partial a_{ihkm}}{\partial x^r} = \frac{1}{4} \left(\frac{\partial a_{ihkm}}{\partial x^r} + \frac{\partial a_{rhkm}}{\partial x^i} + \frac{\partial a_{irkm}}{\partial x^h} + \frac{\partial a_{ihrm}}{\partial x^k} \right). \tag{B.2.6}$$

By collecting the five terms with derivative of the metric tensor and by means of the Christoffel-like five indexes symbol (B.2.2), we obtain

$$\delta l \equiv - \int_{s_0}^{s_1} \left[(3 a_{ihkm} \ddot{x}^i + \Gamma_{m,ihkr} \dot{x}^i \dot{x}^r) \dot{x}^h \dot{x}^k \right] \delta x^m ds = 0. \tag{B.2.7}$$

Since (B.2.7) must remain valid for arbitrary variations δx^m (as is stated by the calculus of variations) each term of the sum in m has to be zero. Then, by reinserting the sum symbol, the four differential equations for geodesics (B.2.1) are obtained. □

This method can be extended to differential forms of every degree.

The system of differential equations (B.2.1) can not be solved with respect to second derivatives by using the contravariant form of the metric tensor, as is done in Riemannian geometry [49]. On the other hand, the system is linear with respect to second derivatives and they can be obtained as functions of x^n, $\dot{x}^n$ by solving a linear algebraic system. In particular the second derivatives are obtained as functions of $\dot{x}^n$ which appear explicitly in (B.2.1), and of x^n through the "metric tensor" and its derivative in Christoffel-like symbol. Then system (B.2.1) can be put in normal form and if the coefficients satisfy the necessary regularity conditions, the theorem of existence and uniqueness of solutions keeps valid. It implies that, if we have a solution of (B.2.1) depending on the right number of integration constants, no matter how it has been obtained, it is the only solution. In Section B.4 we obtain the geodesic equations by means of an alternative method.

As a conclusion we observe that in system (B.2.1) the variable s does not explicitly appear and this allows us to reduce it to a first-order differential system. Actually let us put $\dot{x}^i = y^i$, then we have

$$\frac{d^2 x^i}{ds^2} \equiv \frac{d}{ds} \left(\frac{dx^i}{ds} \right) \equiv \frac{dy^i}{ds} \equiv \frac{dy^i}{dx^i} \frac{dx^i}{ds} = y^i \frac{dy^i}{dx^i}$$

and (B.2.1) becomes a first-order differential system with respect to $(dy^i)/(dx^i)$.

B.3 Constant Curvature Quaternion Spaces

Definition

In Riemannian geometry the constant curvature spaces have the following characteristic property [6] and [34]: the motions (roto-translations which correspond to mappings that leave unchanged the line element) are characterized by the same number of parameters as in Euclidean (Riemann-flat) spaces.

We take this property as our **definition**. We call *constant curvature quaternion space* the ones in which *the allowed motions are the same as in a flat quaternion space*. From this definition it follows that *the quaternion line element must remain unchanged for mappings depending on seven parameters*.

Then these mappings may depend on just some constants and, in particular, this condition is satisfied if the mapping depends on two quaternion constants linked by a condition. These conditions are exactly the same which hold for motions in constant curvature Euclidean or pseudo-Euclidean surfaces (Chapter 9). Moreover we know that motions for constant curvature Euclidean and pseudo-Euclidean surfaces are described by bilinear mappings of complex [6] and [25] or hyperbolic [81] variables depending on three parameters obtained by means of two complex or hyperbolic constants, respectively. For these reasons we start from the mapping that represents a straightforward extension of the two-dimensional case (Chapter 9), i.e., a bilinear quaternion mapping with just two constants and their conjugations.

The bilinear quaternion mapping

Among functions of a quaternion variable [53], we recall the linear-fractional mappings and write them in decomposed form. In general these mappings depend on three quaternion constants (12 real parameters). If we call $a^l = e_1 \alpha^l + e_2 \beta^l$ for $l = 1, 2, 3, 4$ with $|a^1 a^4 - a^2 a^3| \neq 0$ and take into account the decomposability property, the linear-fractional mapping is given by

$$q' = \frac{a^1 q + a^2}{a^3 q + a^4} \equiv e_1 \frac{\alpha^1 z^1 + \alpha^2}{\alpha^3 z^1 + \alpha^4} + e_2 \frac{\beta^1 z^2 + \beta^2}{\beta^3 z^2 + \beta^4}. \tag{B.3.1}$$

B.3.1   Line Element for Positive Constant Curvature

We have

Theorem B.3. *The line element of constant curvature Segre's quaternion space is given by the product of two line elements of constant curvature surfaces*

$$ds^4 = R^4 \frac{(d\,\rho^1)^2 + (d\,\phi^1)^2}{\cosh^2 \rho^1} \; \frac{(d\,\rho^2)^2 + (d\,\phi^2)^2}{\cosh^2 \rho^2}. \tag{B.3.2}$$

Proof. We return to (B.1.4) and look for an expression of $F(q, \bar{q}, \tilde{q}, \tilde{\bar{q}})$ for which ds^4 does not change for a bilinear mapping which depends on two quaternion constants (seven parameters) since, as for bilinear mappings in complex analysis, the normalization reduces of one the eight real constants. The mapping is

$$w = \frac{\alpha\, q + \beta}{-\tilde{\bar{\beta}}\, q + \tilde{\bar{\alpha}}}. \tag{B.3.3}$$

By means of elementary calculations it can be verified that mapping (B.3.3) does not change the line element[2]

$$ds^4 = \frac{dq\,d\bar{q}}{(1+q\,\bar{q})^2}\,\frac{d\tilde{\bar{q}}\,d\tilde{q}}{(1+\tilde{\bar{q}}\,\tilde{q})^2}\,.$$ (B.3.4)

We can verify at once that the denominator of (B.3.4) is given by the sum of a real quantity (the characteristic determinant plus 1) plus two number conjugates, in the sub-algebra $(1,\,j)$. Therefore it is real, as it must be.

The line element (B.3.4) can be rewritten by means of (A.2.25) and their extensions to differentials and by introducing, as an extension of the two-dimensional case (Chapter 9), the value of "curvature" by means of a constant R; we obtain

$$
\begin{aligned}
ds^4 &= 16\,R^4 \frac{dq\,d\bar{q}}{(1+q\,\bar{q})^2}\,\frac{d\tilde{\bar{q}}\,d\tilde{q}}{(1+\tilde{\bar{q}}\,\tilde{q})^2} \\[2mm]
&\equiv 16\,R^4 \frac{|dz^1|^2\,|dz^2|^2}{(1+e_1|z^1|^2+e_2|z^2|^2)^2(1+e_1|z^2|^2+e_2|z^1|^2)^2} \\[2mm]
&\equiv 16\,R^4 \frac{|dz^1|^2\,|dz^2|^2}{(1+|z^1|^2)^2(1+|z^2|^2)^2}\,.
\end{aligned}
$$ (B.3.5)

The last passage follows from the coincidence of denominators. In particular by developing the calculations in the denominator of the third expression we obtain a real quantity that can be rearranged, giving the expression of the fourth term. By means of polar transformation (Section A.4.1) we have

$$e_1\,z^1+e_2\,z^2 = e_1\exp[\zeta^1]+e_2\exp[\zeta^2] \equiv e_1\exp[\rho^1+\mathrm{i}\,\phi^1]+e_2\exp[\rho^2+\mathrm{i}\,\phi^2]$$ (B.3.6)

and

$$d\,z^n = \exp[\zeta^n]\,d\,\zeta^n;\quad |\exp[\zeta^n]|^2 \equiv \exp[\zeta^n]\exp[\tilde{\zeta}^n] \equiv \exp[2\,\rho^n];\quad n=1,\,2.$$ (B.3.7)

These substitutions allow us to write

$$
\begin{aligned}
ds^4 &= 16\,R^4 \frac{|\exp[\zeta^1]|^2\,|d\zeta^1|^2\,|\exp[\zeta^2]|^2|d\zeta^2|^2}{(1+|\exp[\zeta^1]|^2)^2(1+|\exp[\zeta^2]|^2)^2} \\[2mm]
&\equiv 16\,R^4 \frac{\exp[2\,\rho^1]\,|d\zeta^1|^2\,\exp[2\,\rho^2]|d\zeta^2|^2}{(1+\exp[2\,\rho^1])^2(1+\exp[2\,\rho^2])^2}\,.
\end{aligned}
$$

The last term corresponds to (B.3.2) which is the product of line elements of two-dimensional constant curvature surfaces (Chapter 9). □

---

[2]For developing the calculations it must be $(\alpha\bar{\alpha}+\beta\bar{\beta})^2\,(\tilde{\bar{\alpha}}\tilde{\alpha}+\tilde{\bar{\beta}}\tilde{\beta})^2 \neq 0$.

B.4   Geodesic Equations in Quaternion Space

Let us consider a four-dimensional line element of degree 4 (B.3.2) in a more general form, i.e., as a product of two two-dimensional line elements of degree 2,

$$
\begin{aligned}
d\,s^4 &\equiv \ [b_{ih}(x^1,\,x^2)d\,x^i\,d\,x^h][c_{lm}(x^3,\,x^4)\,d\,x^l\,d\,x^m] \\
&\equiv \ d\,s^2_{(1)}\,d\,s^2_{(2)} \equiv F_{1-2}\,F_{3-4}\,,
\end{aligned} \tag{B.4.1}
$$

where $i, h = 1, 2;\ \ l, m = 3, 4$ and we have set

$$
\begin{aligned}
d\,s^2_{(1)} &= \ b_{ih}(x^1,\,x^2)d\,x^i\,d\,x^h \equiv F_{1-2}, \\
d\,s^2_{(2)} &= \ c_{lm}(x^3,\,x^4)\,d\,x^l\,d\,x^m \equiv F_{3-4}\,.
\end{aligned} \tag{B.4.2}
$$

We have

Theorem B.4. *The geodesic equations in a quaternion space, characterized by line element* (B.4.1), *can be obtained with two integrating steps, via two differential forms of degree* 2.

For the differential forms of degree 2, geodesics can be calculated by means of classical differential geometry.

Proof. Let us apply to line element (B.4.1) the variational method, by which geodesic equations are obtained,

$$
\delta(d\,s^4) \equiv \delta(F_{1-2}\,F_{3-4}) = F_{1-2}\,\delta(F_{3-4}) + \delta(F_{1-2})\,F_{3-4}\,. \tag{B.4.3}
$$

The total variation is zero if partial variations of (B.4.3) are zero, then geodesic lines on the planes z^1, z^2 are three-dimensional geodesic varieties in quaternion space. Actually the coordinates of points on the geodesic lines on each plane can be expressed as functions of a parameter, in particular of the line parameters which we call θ_1, θ_2, respectively. The intersection of these varieties gives geodesic surfaces in quaternion space. After having found the geodesics on z^1, z^2 planes, we can express $b_{ih}\,d\,x^i\,d\,x^h$ as a function of $\theta_1, d\,\theta_1$ and $c_{lm}\,d\,x^l\,d\,x^m$ as a function of $\theta_2, d\,\theta_2$ and calculate the geodesics in two-dimensional space defined by the variables θ_1, θ_2. $\qquad\square$

By this method, lines in four-dimensional quaternion space are determined. For asserting that these lines are all the geodesics we shall verify, as a first step, that they depend on the right number of integration constants and as a second step that they satisfy Euler's equation (B.2.1). These tests give us confidence that the two-steps integration gives the right equations.

Now we return to the integration problem. In Chapter 9 it is shown that the most suitable method for obtaining the equations of the geodesics for constant curvature surfaces is to use Beltrami's integration method, recalled in Section 8.5.1, p. 130, that will be used in two steps for solving our problem with line element (B.3.2).

Let us write (B.3.2) in the form (B.4.1) with $d\,s_{(1)}$ and $d\,s_{(2)}$ given by

$$d\,s_{(1)}^2 = f^2(x^1)\,[(dx^1)^2 + (dx^2)^2]\,;\ \ d\,s_{(2)}^2 = f^2(x^3)\,[(dx^3)^2 + (dx^4)^2],\quad\text{(B.4.4)}$$

where $f(x^1)$, $f(x^3)$ are two arbitrary functions[3] of the variables x^1 and x^3.

As first step we calculate geodesics equations in the planes with line elements $d\,s_{(1)}^2$ and $d\,s_{(2)}^2$, starting with plane "1". Equation (8.5.4) becomes

$$\Delta_1\theta_1 \equiv \frac{1}{f^2(x^1)}\left[\left(\frac{\partial\theta_1}{\partial x^1}\right)^2 + \left(\frac{\partial\theta_1}{\partial x^2}\right)^2\right] = 1\,.\qquad\text{(B.4.5)}$$

Setting

$$\theta_1(x^1, x^2) = \theta_1'(x^1) + \alpha_1\,x^2\qquad\text{(B.4.6)}$$

and substituting in (B.4.5), we obtain

$$\frac{\partial\theta_1'}{\partial x^1} = \sqrt{f^2(x^1) - \alpha_1^2}\,,\qquad\text{(B.4.7)}$$

and

$$\theta_1(x^1,\ x^2,\ \alpha_1) = \int\sqrt{f^2(x^1) - \alpha_1^2}\ dx^1 + \alpha_1\,x^2\qquad\text{(B.4.8)}$$

and the geodesic equations

$$\Theta_{\alpha_1} \equiv \frac{\partial\theta_1}{\partial\alpha_1} \equiv -\int\frac{\alpha_1}{\sqrt{f^2(x^1) - \alpha_1^2}}\ dx^1 + x^2 = \sigma_1.\qquad\text{(B.4.9)}$$

Equation (B.4.9) links together the variables x^1 and x^2.

By using (B.4.8) and (B.4.9), a relation that links up x^1 and x^2 with θ_1 can be calculated. Actually, let us consider the identity

$$\theta_1 - \alpha_1\,\Theta_{\alpha_1} = \theta_1 - \alpha_1\,\sigma_1\qquad\text{(B.4.10)}$$

and substitute in the left-hand side the values obtained from (B.4.8) and (B.4.9). There results a link between x^1 and θ_1,

$$\theta_1 = \int\frac{f^2(x^1)}{\sqrt{f^2(x^1) - \alpha_1^2}}\ dx^1 + \alpha_1\,\sigma_1\,.\qquad\text{(B.4.11)}$$

By using (B.4.9) and (B.4.11), also x^2 can be written as a function of the parameter θ_1, as will be shown for line element (B.3.2). The obtained result allows one to verify the *coincidence between θ_1 and the line coordinate along the geodesics*, as stated by Theorem 8.2.

---

[3]In the example here considered the function f is the same for both parts of (B.4.4), but the results we obtain are also valid if $f(x^1)$ and $f(x^3)$ are different functions.

Proof. By differentiating (B.4.9) and (B.4.11) and solving the system, we calculate dx^1 and dx^2 as functions of $f(x^1)$, $d\theta_1$,

$$dx^1 = \frac{\sqrt{f^2(x^1) - \alpha_1^2}}{f^2(x^1)} \, d\theta_1 \, ; \quad dx^2 = \frac{\alpha_1}{f^2(x^1)} \, d\theta_1 \, , \qquad (B.4.12)$$

which allows one to verify the relation

$$ds_{(1)}^2 \equiv f^2(x^1) \, [(dx^1)^2 + (dx^2)^2] \equiv d\theta_1^2 \qquad (B.4.13)$$

that demonstrate the assertion. □

Analogous results are obtained for the plane "2", by substituting x^1, $x^2 \Rightarrow x^3$, x^4 and $\alpha_1 \Rightarrow \alpha_2$.

Now we carry out the second step. Thanks to (B.4.13), equation (B.4.1) becomes

$$d s^4 = d\theta_1^2 \, d\theta_2^2 \Rightarrow d s^2 = e \, d\theta_1 \, d\theta_2 \quad \text{with } e = \pm 1. \qquad (B.4.14)$$

The geodesic equations on the surface θ_1, θ_2 are calculated by applying the method of Section 8.5.1 to a quadratic line element of (B.4.14). Referring to (8.5.1) and (8.5.2), in the present case $g_{11} = g_{22} = 0$; $g_{12} = g_{21} = 1/2$, then $g^{11} = g^{22} = 0$; $g^{12} = g^{21} = 2$. By setting $e = 1$, i.e., by considering the same sign for $ds_{(1)}$ and $ds_{(2)}$, (8.5.4) becomes

$$\Delta_1\theta \equiv 2 \, \frac{\partial\theta}{\partial\theta_1} \frac{\partial\theta}{\partial\theta_2} + 2 \, \frac{\partial\theta}{\partial\theta_2} \frac{\partial\theta}{\partial\theta_1} \equiv 4 \, \frac{\partial\theta}{\partial\theta_1} \frac{\partial\theta}{\partial\theta_2} = 1 \, . \qquad (B.4.15)$$

Setting

$$\theta = C_1 \, \theta_1 + C_2 \, \theta_2 \qquad (B.4.16)$$

from (B.4.15) we have

$$C_1 \, C_2 = \frac{1}{4} \quad \Rightarrow \quad C_1 = \frac{\exp[\gamma]}{2} \, , \quad C_2 = \frac{\exp[-\gamma]}{2} \qquad (B.4.17)$$

with the solution

$$\theta = \frac{\exp[\gamma]}{2} \, \theta_1 + \frac{\exp[-\gamma]}{2} \, \theta_2 \, . \qquad (B.4.18)$$

From (B.4.18), we obtain the third equation for geodesics,

$$\Theta_\gamma \equiv \frac{\partial\theta}{\partial\gamma} \equiv \frac{\exp[\gamma]}{2} \, \theta_1 - \frac{\exp[-\gamma]}{2} \, \theta_2 = \sigma_3 \, . \qquad (B.4.19)$$

We have

Theorem B.5. *Equations (B.4.18) and (B.4.19) allow us to express θ_1 and θ_2 as functions of θ.*

Proof. From the identities

$$\theta + \Theta_\gamma = \theta + \sigma_3 \,; \qquad \theta - \Theta_\gamma = \theta - \sigma_3 \,, \qquad\qquad\qquad \text{(B.4.20)}$$

in which the terms on the left-hand sides are evaluated by (B.4.18) and (B.4.19), we obtain

$$\theta_1 = \exp[-\gamma]\,\theta + \exp[-\gamma]\,\sigma_3 \,; \qquad \theta_2 = \exp[\gamma]\,\theta - \exp[\gamma]\,\sigma_3. \qquad \text{(B.4.21)}$$

$\square$

These relations allow one to verify

$$ds^4 \equiv d\theta_1^2 \, d\theta_2^2 = d\theta^4 \qquad\qquad\qquad \text{(B.4.22)}$$

which states *the coincidence between θ and the line coordinate along the geodesics.*

Finally by using (B.4.8) and the equivalent expressions for θ_2, the four variables x^1, x^2, x^3, x^4, can be expressed as functions of the only parameter θ. Equations (B.4.9), the equivalent expressions for θ_2 and (B.4.19) represent three-dimensional varieties which define lines in four-dimensional space. The six arbitrary constants of these lines are determined (in general unambiguously) solving three independent systems, obtained by forcing the lines into passing through two points. With respect to Euler's equations (B.2.1), these lines are in an implicit form; meanwhile we have

Theorem B.6. *Thanks to the coincidence between θ and the line elements along the geodesics, we can express the coordinates as functions of the line parameter θ, and check that Euler's equations are satisfied.*

Proof. Actually the parametric expressions of (x^1, x^2, x^3, x^4) as functions of θ are obtained, in finite or integral form. Anyway in order to verify (B.2.1), we do not need the explicit expression of function f, because only the first and second derivatives of (x^1, x^2, x^3, x^4) with respect to θ are necessary. Actually, from (B.4.12) and (B.4.21), we obtain the first derivative

$$\frac{dx^1}{d\theta} \equiv \frac{dx^1}{d\theta_1}\frac{d\theta_1}{d\theta} = \exp[-\gamma]\,\frac{\sqrt{f^2(x^1) - \alpha_1^2}}{f^2(x^1)}$$

and, in an analogous way,

$$\frac{dx^2}{d\theta} = \exp[-\gamma]\,\frac{\alpha_1}{f^2(x^1)}\,; \quad \frac{dx^3}{d\theta} = \exp[\gamma]\,\frac{\sqrt{f^2(x^3) - \alpha_2^2}}{f^2(x^3)}\,; \quad \frac{dx^4}{d\theta} = \exp[\gamma]\,\frac{\alpha_2}{f^2(x^3)}\,.$$

The second derivatives are calculated by

$$\frac{d}{d\theta}\left(\frac{dx^{1,2}}{d\theta}\right) \equiv \left[\frac{d}{dx^1}\left(\frac{dx^{1,2}}{d\theta}\right)\right]\frac{dx^1}{d\theta}\,; \quad \frac{d}{d\theta}\left(\frac{dx^{3,4}}{d\theta}\right) \equiv \left[\frac{d}{dx^3}\left(\frac{dx^{3,4}}{d\theta}\right)\right]\frac{dx^3}{d\theta}\,.$$

A suitable software allows us to verify Euler's equations (B.2.1). $\square$

B.4.1   Positive Constant Curvature Quaternion Space

For PCCQS, we obtain the function θ and the geodesics equations substituting $f(x^i)$, with the functions given by (B.3.2). Setting (B.4.8), and the equivalent expression for θ_2, into (B.4.18), we get

$$\theta = \frac{\exp[\gamma]}{2} \left[\int \frac{\sqrt{R^2 - \alpha_1^2 \cosh^2 \rho^1}}{\cosh \rho^1} \, d\rho^1 + \alpha_1 \phi^1 \right] \tag{B.4.23}$$

$$+ \frac{\exp[-\gamma]}{2} \left[\int \frac{\sqrt{R^2 - \alpha_2^2 \cosh^2 \rho^2}}{\cosh \rho^2} \, d\rho^2 + \alpha_2 \phi^2 \right]$$

and the geodesics equations

$$\Theta_{\alpha_1} \equiv -\int \frac{\alpha_1 \cosh \rho^1}{\sqrt{R^2 - \alpha_1^2 \cosh^2 \rho^1}} \, d\rho^1 + \phi^1 = \sigma_1 , \tag{B.4.24}$$

$$\Theta_{\alpha_2} \equiv -\int \frac{\alpha_2 \cosh \rho^2}{\sqrt{R^2 - \alpha_2^2 \cosh^2 \rho^2}} \, d\rho^2 + \phi^2 = \sigma_2 , \tag{B.4.25}$$

$$\Theta_{\gamma} \equiv \frac{\exp[\gamma]}{2} \left[\int \frac{\sqrt{R^2 - \alpha_1^2 \cosh^2 \rho^1}}{\cosh \rho^1} \, d\rho^1 + \alpha_1 \phi^1 \right] \tag{B.4.26}$$

$$- \frac{\exp[-\gamma]}{2} \left[\int \frac{\sqrt{R^2 - \alpha_2^2 \cosh^2 \rho^2}}{\cosh \rho^2} \, d\rho^2 + \alpha_2 \phi^2 \right] = \sigma_3 .$$

Performing the integrations and substituting in (B.4.26) the value of ϕ^1 and ϕ^2 obtained from (B.4.24) and (B.4.25), and redefining the integration constants

$$\sin(\epsilon_1) = \alpha_1/R ; \quad \sin(\epsilon_2) = \alpha_2/R ;$$

$$\sigma_3' = \{2\sigma_3 - R \exp[\gamma]\sigma_1 \sin(\epsilon_1) + R \exp[-\gamma]\sigma_2 \sin(\epsilon_2)\}/(2R) ,$$

we obtain

$$\sin(\phi^1 - \sigma_1) = \tan \epsilon_1 \sinh \rho^1 , \tag{B.4.27}$$

$$\sin(\phi^2 - \sigma_2) = \tan \epsilon_2 \sinh \rho^2 , \tag{B.4.28}$$

$$\exp[\gamma] \sin^{-1}\left[\frac{\tanh \rho^1}{\cos \epsilon_1} \right] - \exp[-\gamma] \sin^{-1}\left[\frac{\tanh \rho^2}{\cos \epsilon_2} \right] = \sigma_3' . \tag{B.4.29}$$

By means of (B.4.23), all the coordinates ρ^1, ϕ^1, ρ^2, ϕ^2 can be expressed as functions of the line parameter θ. We obtain

$$
\begin{cases}
\rho^1(\theta) = \tanh^{-1}\left[\cos\epsilon_1 \sin\left(\dfrac{\theta - \theta_0'}{R\,\exp[\gamma]}\right)\right] ; \\[3mm]
\phi^1(\theta) = \sigma_1 + \tan^{-1}\left[\sin\epsilon_1 \tan\left(\dfrac{\theta - \theta_0'}{R\,\exp[\gamma]}\right)\right] ; \\[3mm]
\rho^2(\theta) = \tanh^{-1}\left[\cos\epsilon_2 \sin\left(\dfrac{\theta - \theta_0''}{R\,\exp[-\gamma]}\right)\right] ; \\[3mm]
\phi^2(\theta) = \sigma_2 + \tan^{-1}\left[\sin\epsilon_2 \tan\left(\dfrac{\theta - \theta_0''}{R\,\exp[-\gamma]}\right)\right] .
\end{cases}
\tag{B.4.30}
$$

By (B.3.6) and (A.2.21) we can write geodesics equations (B.4.27)–(B.4.29) as functions of the components of z^1 and z^2,

$$
\tan\epsilon_1\,(\tau^2 + \xi^2 - 1) + 2(\sin\sigma_1\,\tau - \cos\sigma_1\,\xi) = 0 ,
\tag{B.4.31}
$$

$$
\tan\epsilon_2\,(\eta^2 + \zeta^2 - 1) + 2(\sin\sigma_2\,\eta - \cos\sigma_2\,\zeta) = 0 ,
\tag{B.4.32}
$$

$$
\exp[\gamma]\,\sin^{-1}\left[\frac{\tau^2 + \xi^2 - 1}{(\tau^2 + \xi^2 + 1)\cos\epsilon_1}\right]
\tag{B.4.33}
$$

$$
- \exp[-\gamma]\,\sin^{-1}\left[\frac{\eta^2 + \zeta^2 - 1}{(\eta^2 + \zeta^2 + 1)\cos\epsilon_2}\right] = \sigma_3' .
$$

The first two equations are the same obtained, in Chapter 9, for positive constant curvature surfaces. In the planes z^1, z^2 they represent circles and, in the four-dimensional space, "hyper-cylinders" .

Finally geodesics equations can be obtained as functions of original variables (A.2.1). From (A.2.19), we obtain (if $\epsilon_1, \epsilon_2 \neq 0$)

$$
(t + y)^2 + (x + z)^2 + 2\frac{\sin\sigma_1\,(t + y) - \cos\sigma_1\,(x + z)}{\tan\epsilon_1} - 1 = 0 ,
\tag{B.4.34}
$$

$$
(t - y)^2 + (x - z)^2 + 2\frac{\sin\sigma_2\,(t - y) - \cos\sigma_2\,(x - z)}{\tan\epsilon_2} - 1 = 0 ,
\tag{B.4.35}
$$

$$
\exp[\gamma]\,\sin^{-1}\left[\frac{(t + y)^2 + (x + z)^2 - 1}{[(t + y)^2 + (x + z)^2 + 1]\cos\epsilon_1}\right]
\tag{B.4.36}
$$

$$
- \exp[-\gamma]\,\sin^{-1}\left[\frac{(t - y)^2 + (x - z)^2 - 1}{[(t - y)^2 + (x - z)^2 + 1]\cos\epsilon_2}\right] = \sigma_3' .
$$

By using (B.4.34) and (B.4.35) the quadratic terms of (B.4.36) can be substituted by linear terms. This is also true for (B.4.33), by using (B.4.31) and (B.4.32).

Conclusions

A complete development of differential geometry starting from a differential form of degree 4 shall require many important investigations. The approach we have used is an extension of the Gauss work in which he showed the link of complex analysis and functions of a complex variable with surface differential geometry [39]. As has been shown in Section 3.2.1, this application of complex variable analysis is due to the fact that the geometry "generated" by complex numbers is the Euclidean one.

In this appendix the same procedure has been applied to geometries generated by commutative quaternions, i.e., we have applied the quaternion analysis and functions of a quaternion variable which represent *the mathematics with the symmetries of the problem we are dealing with.* This approach also means that we start from the roots of differential geometry, i.e., the works of Gauss (complex variable on surfaces), Riemann (differential geometry in N-dimensional spaces) and Beltrami (complex variable on surfaces and integration of geodesic equations). In this way we have obtained some preliminary, self-consistent results on constant curvature quaternion spaces.

The obtained results can be extended in a straightforward way to negative constant curvature elliptic-quaternion spaces and to constant curvature hyperbolic-quaternion spaces.

Moreover the exposed method can be a starting point for studying the constant curvature spaces associated with other commutative number systems.

Appendix C

A Matrix Formalization for Commutative Hypercomplex Numbers

In Chapter 2 we have introduced hypercomplex numbers following the historical approach and have noted in Section 2.1.3 that, for commutative numbers, a representation by means of matrices is more appropriate than a vector representation. In this appendix we tackle the problem starting from this consideration and requiring that the representative matrices satisfy the properties of hypercomplex numbers, i.e., to be a commutative group with respect to multiplication. This approach allows us *to formalize the algebraic properties of hypercomplex numbers* and, by using the well-established theory of eigenvalues, *to introduce the functions of a hypercomplex variable and the differential calculus* as an analytic continuation from real and complex fields.

This appendix is organized in the following way: in Section C.1, the matrix formalism is associated with hypercomplex numbers. In Section C.2, two-dimensional hypercomplex numbers are discussed in view of the formalism introduced in Section C.1. In Section C.3, the algebraic properties of systems of hypercomplex numbers are outlined. In Sections C.4–C.6, the functions of a hypercomplex variable and the differential calculus are eventually introduced.

C.1 Mathematical Operations

Let us start with some definitions and properties relative to the mathematical operations among hypercomplex numbers. Let us introduce an N-dimensional hypercomplex number as a column vector, whose entries are the components of the number. If $\mathbf{x}$ is the hypercomplex number, then

$$\mathbf{x} = \sum_{k=0}^{N-1} \mathbf{e}_k \, x^k, \qquad (\text{C.1.1})$$

where $x^k \in \mathbf{R}$ are called the *components* of $\mathbf{x}$ and $\mathbf{e}_k \notin \mathbf{R}$ are the *unit vectors* along the N dimensions or *bases*. In matrix representation, we can write $\mathbf{x}$ in its matrix form

$$\mathbf{x} = \begin{pmatrix} x^0 \\ x^1 \\ \vdots \\ x^{N-1} \end{pmatrix} \qquad (\text{C.1.2})$$

and the unit vectors, which we call *natural* unit vectors, as

$$\mathbf{e}_0 = \begin{pmatrix} 1 \\ 0 \\ \vdots \\ 0 \end{pmatrix}, \qquad \mathbf{e}_1 = \begin{pmatrix} 0 \\ 1 \\ \vdots \\ 0 \end{pmatrix}, \qquad \dots, \qquad \mathbf{e}_{N-1} = \begin{pmatrix} 0 \\ 0 \\ \vdots \\ 1 \end{pmatrix}. \qquad \text{(C.1.3)}$$

Although $\mathbf{e}_k$ has a different meaning with respect to e_k introduced in Section 2.1, nevertheless we use the same convention for the positions up and down of indexes x^k and e_k. These positions indicate, in agreement with Section 2.1.4, if by a linear transformation they transform by means of matrix a_α^β or $(a_\beta^\alpha)^{-1}$, as the covariant and contravariant components of a vector, respectively. Equations (C.1.2) and (C.1.3) define the matrix representation of $\mathbf{x}$.

Hereafter, when unnecessary, we do not distinguish between hypercomplex numbers and their matrix representations. Moreover, we label the set of N-dimensional hypercomplex numbers with the symbol $\mathbf{H}_N$ and use the term $\mathbf{H}$-number as a shorthand to hypercomplex number.

C.1.1   Equality, Sum, and Scalar Multiplication

1. **Equality of two Hypercomplex Numbers.** Two $\mathbf{H}$-numbers are equal if all their components are equal, one by one. One can verify that such a definition is an equivalence relationship, since it satisfies the reflexive, symmetric, and transitive properties.

2. **Sum and Difference of Two Hypercomplex Numbers.** As for column vectors, the sum of two $\mathbf{H}$-numbers is defined by summing their components. The result is an $\mathbf{H}$-number. By exploiting the above introduced matrix representation, one can verify that the sum operation is both commutative and associative. The column vector with all zero components, indicated with $\mathbf{0}$, is the null element for the sum, which we name as *zero*. With respect to the sum, the inverse element of $\mathbf{x}$ is $-\mathbf{x}$, which is defined as having all its components changed in sign with respect to the components of $\mathbf{x}$. Therefore, $(\mathbf{H}_N, +)$ is an Abelian group. In analogy with the above definition of sum, the definition of difference between $\mathbf{H}$-numbers can be stated as the difference between column vectors.

3. **Multiplication of a Real Number by a Hypercomplex Number.** In agreement with the multiplication of a scalar by a matrix, the multiplication of $a \in \mathbf{R}$ by $\mathbf{x} \in \mathbf{H}_N$ is an $\mathbf{H}$-number, whose components are the components of $\mathbf{x}$ multiplied by a, one by one. Because of commutativity in the multiplication of two real numbers, the product of a by $\mathbf{x}$ can be indicated with either $a\mathbf{x}$ or $\mathbf{x}a$.

C.1.2 Product and Related Operations

Before defining the multiplication between **H**-numbers, let us introduce the *multiplication* (which we indicate by $\odot$) among *the natural unit vectors* $\mathbf{e}_k$ ($k = 0, 1, \ldots,$ $N - 1$) by

$$\mathbf{e}_i \odot \mathbf{e}_j = \sum_{k=0}^{N-1} \mathcal{C}_{ij}^k \, \mathbf{e}_k. \tag{C.1.4}$$

For commutative systems we must have $\mathbf{e}_i \odot \mathbf{e}_j = \mathbf{e}_j \odot \mathbf{e}_i$, then $\mathcal{C}_{ij}^k = \mathcal{C}_{ji}^k$.

Although the "unit vectors" $\mathbf{e}_i$ have been introduced in a different formalism with respect to versors e_i introduced in Section 2.1, the constants $\mathcal{C}_{ji}^k$ are the same as C_{ji}^k, there introduced (see Section C.8). Then we can write $\mathcal{C}_{ji}^k \Rightarrow C_{ji}^k$.

Now let us define the multiplication of a versor $\mathbf{e}_k$ by a hypercomplex number $\mathbf{y} \in \mathbf{H}_N$ as

$$\mathbf{e}_k \odot \mathbf{y} = \mathcal{A}_k \mathbf{y}, \tag{C.1.5}$$

where $\mathcal{A}_k$ is a proper $N \times N$ real matrix and the operation in the right-hand side of (C.1.5) is the well-known multiplication of a square matrix by a column vector. Note that $\mathcal{A}_k \mathbf{y}$ is a column vector with real components, therefore it is an **H**-number as well. The representation introduced in (C.1.5) of the product of a unit vector by a hypercomplex number gives the same results as the multiplication by means of the structure constants (see Chapter 2.1) if the following relation holds for all elements of the matrices $\mathcal{A}_k$,

$$(\mathcal{A}_k)_j^i = C_{kj}^i, \tag{C.1.6}$$

where, as in Chapter 2, the upper index indicates the rows and the lower one the columns. The multiplication between two **H**-numbers, $\mathbf{x}$ and $\mathbf{y}$, can be conveniently defined by applying (C.1.5) to (C.1.1). It ensues that

$$\mathbf{x} \odot \mathbf{y} = \sum_{k=0}^{N-1} x^k \mathbf{e}_k \odot \mathbf{y} = \sum_{k=0}^{N-1} x^k \left(\mathcal{A}_k \mathbf{y} \right). \tag{C.1.7}$$

From (C.1.6) it follows: if

$$(\mathcal{A}_k)_j^i = (\mathcal{A}_j)_k^i, \forall \, (i, j, k) \tag{C.1.8}$$

is true, then the multiplication between **H**-numbers, defined by (C.1.7), is commutative. Hereafter, *we consider only multiplication-commutative systems of* **H**-*numbers, for which (C.1.8) is true.* Equation (C.1.7) can be written also as

$$\mathbf{x} \odot \mathbf{y} = \mathcal{M}(\mathbf{x})\mathbf{y}, \tag{C.1.9}$$

where the matrix $\mathcal{M}(\mathbf{x})$ — whose dependence on $\mathbf{x}$ has to be understood as an explicit dependence on the components of $\mathbf{x}$, so that we write $\mathcal{M}(\mathbf{x})$ instead of

$\mathcal{M}(x^0, x^1, \ldots, x^{N-1})$ for the sake of brevity — is defined as

$$\mathcal{M}(\mathbf{x}) = \sum_{k=0}^{N-1} x^k \mathcal{A}_k. \tag{C.1.10}$$

As is shown in Section C.8, *the matrix $\mathcal{M}(\mathbf{x})$ matches the characteristic matrix introduced in Section* 2.1.3, *p.* 7. Since (C.1.8) is assumed to be true, instead of (C.1.9), one can alternatively write

$$\mathbf{x} \odot \mathbf{y} = \mathcal{M}(\mathbf{y})\mathbf{x}. \tag{C.1.11}$$

Now, let us state the conditions for the existence of the *neutral element* for the multiplication of **H**-numbers. We name such an element as the *unity* of $\mathbf{H}_N$, and indicate it with the symbol **1**. By definition, $\forall \, \mathbf{x} \in \mathbf{H}_N$,

$$\mathbf{1} \odot \mathbf{x} = \mathbf{x} \tag{C.1.12}$$

must hold. Therefore, if 1^k are the components of **1**, by applying (C.1.7) one finds

$$\sum_{k=0}^{N-1} 1^k \mathcal{A}_k \mathbf{x} = \mathbf{x}. \tag{C.1.13}$$

Since (C.1.13) must hold $\forall \, \mathbf{x}$, the following condition of existence of the $\mathbf{H}_N$ unity is found:

$$\sum_{k=0}^{N-1} 1^k \mathcal{A}_k = \mathcal{I}, \tag{C.1.14}$$

where $\mathcal{I}$ is the identity $N \times N$ matrix. Hereafter, we assume that this condition of existence of the $\mathbf{H}_N$ unity is satisfied — we shall see later, with some examples, that the unity actually exists for the $\mathbf{H}_N$ systems we consider.

Now, let us demonstrate that the unity **1** is unique.

Proof. Let us suppose that another unity, $\mathbf{1}'$, exists in $\mathbf{H}_N$. In such a case, both

$$\mathbf{1} \odot \mathbf{1}' = \mathbf{1}' \text{ and } \mathbf{1}' \odot \mathbf{1} = \mathbf{1}$$

should hold. Because of multiplication commutativity, one concludes that $\mathbf{1}' = \mathbf{1}$. Thus, it is demonstrated that **1**, if it exists, is unique. $\qquad \square$

Inverse Element for Multiplication

Given $\mathbf{x} \in \mathbf{H}_N$, the *inverse element* for multiplication is an **H**-number, $\mathbf{x}^{-1}$, which satisfies

$$\mathbf{x} \odot \mathbf{x}^{-1} = \mathbf{1}. \tag{C.1.15}$$

Since

$$\mathbf{x} \odot \mathbf{x}^{-1} = \mathcal{M}(\mathbf{x})\mathbf{x}^{-1},$$

the inverse element exists only if the determinant of $\mathcal{M}(x)$ (*the characteristic determinant* defined in Section 2.1) is non-zero, and one gets

$$\mathbf{x}^{-1} = \mathcal{M}^{-1}(\mathbf{x})\mathbf{1}, \tag{C.1.16}$$

where $\mathcal{M}^{-1}(\mathbf{x})$ is the inverse matrix of $\mathcal{M}(\mathbf{x})$. Since $\mathbf{1}$ is unique and the inverse matrix is univocally defined, (C.1.16) also states that if the inverse element exists, it is unique.

Equation (C.1.16) implies that, for all $\mathbf{y} \in \mathbf{H}_N$, the following equation holds:

$$\mathbf{x}^{-1} \odot \mathbf{y} = \mathcal{M}^{-1}(\mathbf{x})\mathbf{y}.$$

Therefore,

$$\mathcal{M}(\mathbf{x}^{-1}) = \mathcal{M}^{-1}(\mathbf{x}). \tag{C.1.17}$$

Moreover, since (C.1.15) leads to

$$\left(\mathbf{x}^{-1}\right)^{-1} = \mathbf{x},$$

taking into account (C.1.17), one gets

$$\mathcal{M}^{-1}(\mathbf{x}^{-1}) = \mathcal{M}(\mathbf{x}).$$

A further property is

$$(\mathbf{x} \odot \mathbf{y})^{-1} = \mathbf{x}^{-1} \odot \mathbf{y}^{-1}.$$

Taking into account commutativity, this relation is the same as that which holds for square matrices in matrix algebra. Furthermore, by applying (C.1.15) one can verify that the inverse element of $\mathbf{1}$ is $\mathbf{1}$ itself, i.e.,

$$\mathbf{1}^{-1} = \mathbf{1}.$$

Properties of Multiplication

Theorem C.1. *The multiplication between* **H**-*numbers satisfies the distributive property with respect to the sum, i.e.,*

$$(\mathbf{x} + \mathbf{y}) \odot \mathbf{z} = \mathbf{x} \odot \mathbf{z} + \mathbf{y} \odot \mathbf{z}.$$

Proof. As a matter of fact, being equal to $x^k + y^k$ the kth component of $\mathbf{x} + \mathbf{y}$, one gets from (C.1.7),

$$(\mathbf{x}+\mathbf{y}) \odot \mathbf{z} = \sum_{k=0}^{N-1} \left(x^k + y^k\right) \mathcal{A}_k \mathbf{z} = \sum_{k=0}^{N-1} x^k \mathcal{A}_k \mathbf{z} + \sum_{k=0}^{N-1} y^k \mathcal{A}_k \mathbf{z} = \mathbf{x} \odot \mathbf{z} + \mathbf{y} \odot \mathbf{z}. \qquad \square$$

Now let us demonstrate a *fundamental theorem*.

Theorem C.2. *Commutative* **H**-*numbers are also associative.*

Proof. As a matter of fact, given $\mathbf{x}, \mathbf{y}, \mathbf{z} \in \mathbf{H}_N$, by taking into account commutativity and the above-demonstrated distributive property, it ensues that

$$(\mathbf{x} \odot \mathbf{y}) \odot \mathbf{z} = \mathbf{z} \odot (\mathbf{x} \odot \mathbf{y}) = \mathbf{z} \odot \left(\sum_{k=0}^{N-1} x^k \mathcal{A}_k \mathbf{y} \right) = \sum_{k=0}^{N-1} x^k \mathbf{z} \odot \mathcal{A}_k \mathbf{y} = \sum_{k,n=0}^{N-1} x^k z^n \mathcal{A}_n \mathcal{A}_k \mathbf{y} \, ,$$

$$\mathbf{x} \odot (\mathbf{y} \odot \mathbf{z}) = \mathbf{x} \odot (\mathbf{z} \odot \mathbf{y}) = \mathbf{x} \odot \left(\sum_{n=0}^{N-1} z^n \mathcal{A}_n \mathbf{y} \right) = \sum_{n=0}^{N-1} z^n \mathbf{x} \odot \mathcal{A}_n \mathbf{y} = \sum_{k,n=0}^{N-1} x^k z^n \mathcal{A}_k \mathcal{A}_n \mathbf{y} \, .$$

Therefore, since the matrices $\mathcal{A}_k$ and $\mathcal{A}_n$ satisfy

$$\mathcal{A}_k \mathcal{A}_n = \mathcal{A}_n \mathcal{A}_k \, , \ \forall \, (k, n = 0, 1, \ldots, N-1) \tag{C.1.18}$$

as follows from $\mathbf{e}_k \odot \mathbf{e}_n = \mathbf{e}_n \odot \mathbf{e}_k$, the associative property

$$(\mathbf{x} \odot \mathbf{y}) \odot \mathbf{z} = \mathbf{x} \odot (\mathbf{y} \odot \mathbf{z})$$

is demonstrated. $\square$

Another property which can be easily verified is

$$\mathbf{x} \odot \mathbf{0} = \mathbf{0}.$$

C.1.3   Division Between Hypercomplex Numbers

Given two **H**-numbers, $\mathbf{x}$ and $\mathbf{y}$, with $\det[\mathcal{M}(\mathbf{y})] \neq 0$, let us define the division of $\mathbf{x}$ by $\mathbf{y}$ as the product of $\mathbf{x}$ by the inverse element of $\mathbf{y}$, i.e.,

$$\frac{\mathbf{x}}{\mathbf{y}} \equiv \mathbf{x} \odot \mathbf{y}^{-1}. \tag{C.1.19}$$

It is easy to verify that

$$\frac{\mathbf{x}}{\mathbf{1}} = \mathbf{x}, \quad \frac{\mathbf{1}}{\mathbf{y}} = \mathbf{y}^{-1}.$$

Moreover, as one can verify, many of the rules of fraction algebra of real numbers can be extended to **H**-numbers. Among them, we report

$$\frac{1}{\mathbf{x}/\mathbf{y}} = \frac{\mathbf{y}}{\mathbf{x}} \, ;$$

$$\frac{\mathbf{x}_1}{\mathbf{y}_1} \odot \frac{\mathbf{x}_2}{\mathbf{y}_2} = \frac{\mathbf{x}_1 \odot \mathbf{x}_2}{\mathbf{y}_1 \odot \mathbf{y}_2} \, ;$$

$$\frac{\mathbf{x}_1}{\mathbf{y}_1} + \frac{\mathbf{x}_2}{\mathbf{y}_2} = \frac{\mathbf{x}_1 \odot \mathbf{y}_2 + \mathbf{x}_2 \odot \mathbf{y}_1}{\mathbf{y}_1 \odot \mathbf{y}_2} \, .$$

In the above identities, the non-singularity of the characteristic matrices of **H**-numbers which appear in the denominators is assumed.

Divisors of Zero

We mentioned before that the inverse element of $\mathbf{y}$ for multiplication is not defined if $\det[\mathcal{M}(\mathbf{y})] = 0$ and that, in such a case, the ratio $\mathbf{x}/\mathbf{y}$ cannot be defined. An **H**-number $\mathbf{y} \neq \mathbf{0}$, whose characteristic matrix $\mathcal{M}(\mathbf{y})$ is singular, is called a *divisor of zero*. The reason for such a name is explained in the following. Let us consider, in the domain of **H**-numbers, the equation

$$\mathbf{x} \odot \mathbf{y} = \mathbf{0} \tag{C.1.20}$$

in the unknown $\mathbf{x}$, where $\mathbf{x}, \mathbf{y} \neq \mathbf{0}$ and $\mathbf{y}$ is a divisor of zero. The equation can be rewritten in its equivalent matrix form

$$\mathcal{M}(\mathbf{y})\mathbf{x} = \mathbf{0}. \tag{C.1.21}$$

In matrix language (C.1.21) represents a homogeneous system of N equations with N unknowns, $x^0, x^1, \ldots, x^{N-1}$. Since $\mathcal{M}(\mathbf{y})$ is singular, i.e., its determinant is zero, it is a well-known result that the above system of equations admits infinite non-zero solutions $\mathbf{x}$. Stated in **H**-number terms, since $\mathbf{y}$ is a divisor of zero, a numerable infinity of non-zero **H**-numbers $\mathbf{x}$ exists which satisfy (C.1.20). This means that two (proper) *non*-zero **H**-numbers can have zero product, something which never happens for real or complex numbers. In such a case, we say that the two **H**-numbers, $\mathbf{x}$ and $\mathbf{y}$, are *orthogonal*. This definition can be interpreted by analogy with vector algebra where the scalar product of two non-zero vectors is zero if they are orthogonal. The set of non-zero solutions $\mathbf{x}$ to (C.1.20), indicated as $\perp_{\mathbf{y}}$, is called the *orthogonal set* of $\mathbf{y}$. Equation (C.1.20) can also be formally written as

$$\mathbf{x} = \frac{\mathbf{0}}{\mathbf{y}} \neq \mathbf{0},$$

from which the reason for the name *divisor of zero* is explained for $\mathbf{y}$.

Formally, (C.1.20) could also be written as

$$\mathbf{y} = \frac{\mathbf{0}}{\mathbf{x}}.$$

One can wonder if the above equation makes sense. If it does, a direct implication is that $\mathbf{x}$, which is a non-zero solution of (C.1.20), is a divisor of zero as well. This is indeed true.

Proof. An equivalent matrix form of (C.1.20) is

$$\mathcal{M}(\mathbf{x})\mathbf{y} = \mathbf{0}. \tag{C.1.22}$$

Since $\mathbf{y}$ is a divisor of zero, and therefore $\mathbf{y} \neq \mathbf{0}$, (C.1.22) implies $\det[\mathcal{M}(\boldsymbol{x})] = 0$, thus also $\mathbf{x}$ is a divisor of zero. This demonstrates that all elements of $\perp_{\mathbf{y}}$ are divisors of zero. $\qquad\square$

As a special case, let us demonstrate that $\mathbf{1}$ is *not* a divisor of zero.

Proof. If $\mathbf{1}$ were a divisor of zero, then at least one $\mathbf{x} \neq \mathbf{0}$ would exist such that

$$\mathbf{x} \odot \mathbf{1} = \mathbf{0},$$

something which contradicts the definition of $\mathbf{1}$. □

To conclude, one can verify that if $\mathbf{x}$ is a divisor of zero and $\mathbf{y}$ is any non-zero **H**-number such that $\mathbf{y} \not\subseteq \perp_{\mathbf{x}}$, then also $\mathbf{x} \odot \mathbf{y}$ and $\mathbf{x}/\mathbf{y}$ are divisors of zero (in the latter case, $\mathbf{y}$ is assumed to be not a divisor of zero).

Power of a Hypercomplex Number

Given any integer $n > 0$, the nth power of an **H**-number $\mathbf{x}$ is defined by multiplying $\mathbf{x}$ by itself n times. Being

$$\mathbf{x} = \mathbf{x} \odot \mathbf{1} = \mathcal{M}(\mathbf{x})\mathbf{1},$$

one gets

$$\mathbf{x}^n = \mathcal{M}^n(\mathbf{x})\mathbf{1} \tag{C.1.23}$$

where $\mathcal{M}^n(\mathbf{x})$ indicates the power of a matrix. Conventionally, we set

$$\mathbf{x}^0 = \mathbf{1}.$$

Equation (C.1.23) can be viewed as the actual definition of power. It can be formally extended to negative-exponent powers — only for non-zero **H**-numbers which are not divisors of zero — with

$$\mathbf{x}^{-n} = \left(\mathbf{x}^{-1}\right)^n = \mathcal{M}^{-n}(\mathbf{x})\mathbf{1},$$

where $\mathcal{M}^{-n}(\mathbf{x})$ means

$$\mathcal{M}^{-n}(\mathbf{x}) \equiv \left[\mathcal{M}^{-1}(\mathbf{x})\right]^n.$$

Under proper conditions for the determinants of the involved matrices, one can verify that many of the known properties which hold for real numbers can be extended to **H**-numbers. Among them, we report (m is another integer > 0)

$$\mathbf{0}^n = \mathbf{0}; \qquad \mathbf{1}^n = \mathbf{1};$$

$$(\mathbf{x}^m)^n = \mathbf{x}^{mn}; \quad \mathbf{x}^m \odot \mathbf{x}^n = \mathbf{x}^{m+n}; \quad (\mathbf{x} \odot \mathbf{y})^n = \mathbf{y}^n \odot \mathbf{y}^n;$$

$$\frac{\mathbf{x}^m}{\mathbf{x}^n} = \mathbf{x}^{m-n}; \quad \left(\frac{\mathbf{x}}{\mathbf{y}}\right)^n = \frac{\mathbf{x}^n}{\mathbf{y}^n}.$$

These properties can be demonstrated by taking into account that

$$\mathcal{M}(\mathbf{x} \odot \mathbf{y}) = \mathcal{M}(\mathbf{x})\mathcal{M}(\mathbf{y}), \tag{C.1.24}$$

$$\mathcal{M}\left(\frac{\mathbf{x}}{\mathbf{y}}\right) = \mathcal{M}(\mathbf{x})\mathcal{M}^{-1}(\mathbf{y}). \tag{C.1.25}$$

Proof. As a matter of fact, given another arbitrary **H**-number, **z**, and by exploiting the associativity of **H**-numbers, one gets

$$\mathcal{M}(\mathbf{x} \odot \mathbf{y}) \odot \mathbf{z} = (\mathbf{x} \odot \mathbf{y}) \odot \mathbf{z} = \mathbf{x} \odot (\mathbf{y} \odot \mathbf{z}) = \mathbf{x} \odot \mathcal{M}(\mathbf{y})\mathbf{z} = \mathcal{M}(\mathbf{x})\mathcal{M}(\mathbf{y})\mathbf{z},$$

from which (C.1.24) follows. Equation (C.1.25) can be similarly demonstrated. $\square$

C.2 Two-dimensional Hypercomplex Numbers

As an example, let us consider two-dimensional **H**-numbers. When $N = 2$, the generally defined **H**-number, **z**, is

$$\mathbf{z} = \begin{pmatrix} x \\ y \end{pmatrix}, \tag{C.2.1}$$

and its characteristic matrix (see Section 2.1), is defined as

$$\mathcal{M}\left[\begin{pmatrix} x \\ y \end{pmatrix} \right] = \begin{pmatrix} x & \alpha y \\ y & x + \beta y \end{pmatrix}, \tag{C.2.2}$$

where α and β are real numbers. We have

Theorem C.3. *The two-dimensional **H**-numbers defined by (C.2.2) are*

1. *multiplication-commutative,*

2. *multiplication-associative,*

3. *the multiplicative unity exists.*

Proof.

1. The matrices $\mathcal{A}_0$ and $\mathcal{A}_1$, associated with the natural unity vectors, $\mathbf{e}_0$ and $\mathbf{e}_1$, can be obtained from (C.2.2) by setting in it $(x, y) = (1, 0)$ and $(x, y) = (0, 1)$, respectively. They are

$$\mathcal{A}_0 = \begin{pmatrix} 1 & 0 \\ 0 & 1 \end{pmatrix} \quad \text{and} \quad \mathcal{A}_1 = \begin{pmatrix} 0 & \alpha \\ 1 & \beta \end{pmatrix}. \tag{C.2.3}$$

   It can be easily verified that $\mathcal{A}_0$ and $\mathcal{A}_1$ satisfy (C.1.8), thus the commutativity of multiplication is demonstrated.

2. Since $\mathcal{A}_0 \equiv \mathcal{I}$, where $\mathcal{I}$ is the 2×2 identity matrix, (C.1.18) is satisfied, thus the multiplication is also associative.

3. It can be verified that

$$\mathbf{1} = \mathbf{e}_0 \equiv \begin{pmatrix} 1 \\ 0 \end{pmatrix} \tag{C.2.4}$$

   satisfies both (C.1.12), as $\mathcal{M}(\mathbf{1}) \equiv \mathcal{I}$, and the existence condition, (C.1.14). $\square$

Theorem C.4. *The parameters α and β subdivide the above-introduced two-dimensional* **H**-*numbers into three classes.*

Proof. The characteristic determinant of **z** is:

$$\det[\mathcal{M}(\mathbf{z})] = x^2 + \beta xy - \alpha y^2 . \tag{C.2.5}$$

Setting it equal to zero, gives a second-order algebraic equation in the unknown x, whose solutions are

$$x = \frac{-\beta \pm \sqrt{\Delta}}{2} \, y \, , \tag{C.2.6}$$

where $\Delta = \beta^2 + 4\alpha$. The three main classes of two-dimensional **H**-numbers are distinguished according to the value of Δ (see Section 2.1).

- The first class is that of the *elliptic* **H**-numbers, for which $\Delta < 0$; in this case, (C.2.6) gives no real solution, and therefore there are no divisors of zero.

- The second class is that of *hyperbolic* **H**-numbers, for which $\Delta > 0$; in this case, the two solutions given by (C.2.6) represent two sets — two intersecting straight lines in the (x, y)-plane — of divisors of zero.

- The third class is that of *parabolic* **H**-numbers, for which $\Delta = 0$; in this case, the only set of divisors of zero is described by the equation $x = -\beta y/2$, which is a straight line (the y Cartesian axis for $\beta = 0$).

The names of these three classes come from the conic section described by (C.2.5), as we have already seen in Section 2.2. $\square$

When $\beta = 0$ and $\alpha = \pm 1, 0$, the above three classes are further named as *canonical* (see Section 2.1). Standard complex numbers **Z**, often indicated as $\mathbf{z} \equiv x + \mathbf{i}y \in \mathbf{Z}$, $\mathbf{i} = \sqrt{-1}$ being the imaginary unit, are the canonical elliptic **H**-numbers ($\alpha = -1$ and $\beta = 0$). In this case, the usual imaginary unit, $\mathbf{i}$, is found to be in our formalism

$$\mathbf{i} = \mathbf{e}_1 \equiv \begin{pmatrix} 0 \\ 1 \end{pmatrix} . \tag{C.2.7}$$

This means that in our matrix representation of **z**, see (C.2.1), when applied to standard complex numbers, x and y coincide with the usual real and imaginary parts of the number, respectively.

C.3 Properties of the Characteristic Matrix $\mathcal{M}$

The characteristic matrix $\mathcal{M}(\mathbf{x})$ plays a fundamental role in the definition of **H**-numbers and their operations. Let us explore the main properties of this matrix, before going on with the study of **H**-numbers. Some of these properties allow us to introduce the functions of an **H**-number.

C.3.1 Algebraic Properties

Let us list the main algebraic properties of characteristic matrices. Some of them had been already introduced in the previous section. The rather elementary demonstrations are left to the reader.

$$\mathcal{M}(\mathbf{0}) = \mathcal{O}; \qquad \mathcal{M}(\mathbf{1}) = \mathcal{I}; \tag{C.3.1}$$

$$\mathcal{M}(a\mathbf{x} + b\mathbf{y}) = a\mathcal{M}(\mathbf{x}) + b\mathcal{M}(\mathbf{y}); \tag{C.3.2}$$

$$\mathcal{M}(\mathbf{x} \odot \mathbf{y}) = \mathcal{M}(\mathbf{x})\mathcal{M}(\mathbf{y});$$

$$\mathcal{M}\left((\mathbf{x})^n\right) = [\mathcal{M}(\mathbf{x})]^n. \tag{C.3.3}$$

In (C.3.1), $\mathcal{O}$ and $\mathcal{I}$ are the all-zero $N \times N$ matrix and the identity $N \times N$ matrix, respectively; in (C.3.2), $a, b \in \mathbf{R}$, $\mathbf{x}, \mathbf{y} \in \mathbf{H}_N$. In (C.3.3), n is an integer and when $n < 0$, $\mathbf{x}$ is assumed to be neither $\mathbf{0}$ nor a divisor of zero.

C.3.2 Spectral Properties

Let us consider the eigenvalue problem for the matrix $\mathcal{M}(\mathbf{x})$, that is

$$\mathcal{M}(\mathbf{x})\mathbf{u}_k(\mathbf{x}) = \lambda_k(\mathbf{x})\mathbf{u}_k(\mathbf{x}). \tag{C.3.4}$$

Here, $\mathbf{u}_k(\mathbf{x}) \in \mathbf{H}_N$. As for $\mathcal{M}(\mathbf{x})$, also for the scalar — real or complex — eigenvalues $\lambda_k(\mathbf{x})$ and eigenvectors $\mathbf{u}_k(\mathbf{x})$ the dependence on $\mathbf{x}$ means a dependence on the components of $\mathbf{x}$.

Now let us enumerate some properties of eigenvectors and eigenvalues.

1. *The eigenvectors $\mathbf{u}_k(\mathbf{x})$ are divisors of zero.*

   Proof. As a matter of fact, one can write (C.3.4) also as

$$[\mathbf{x} - \lambda_k(\mathbf{x})\mathbf{1}] \odot \mathbf{u}_k(\mathbf{x}) = \mathbf{0}.$$

   Because

$$\det\left[\mathcal{M}(\mathbf{x}) - \lambda_k(\mathbf{x})\mathcal{I}\right] = \det\left\{\mathcal{M}\left[\mathbf{x} - \lambda_k(\mathbf{x})\mathbf{1}\right]\right\} = 0$$

   must hold, $\mathbf{x} - \lambda_k(\mathbf{x})\mathbf{1}$ is a divisor of zero;[1] moreover, $\mathbf{u}_k(\mathbf{x}) \in \perp_{\mathbf{x}-\lambda_k(\mathbf{x})\mathbf{1}}$. Since all the elements of $\perp_{\mathbf{x}-\lambda_k(\mathbf{x})\mathbf{1}}$ are divisors of zero themselves, it is demonstrated that $\mathbf{u}_k(\mathbf{x})$ is a divisor of zero. □

2. *If (C.3.4) admits N distinct eigenvalues, then the corresponding eigenvectors are mutually orthogonal, i.e.,[2]*

$$\mathbf{u}_k(\mathbf{x}) \odot \mathbf{u}_n(\mathbf{x}) = \mathbf{0}; \qquad \text{for } k \neq n. \tag{C.3.5}$$

---

[1]$\boldsymbol{x} - \lambda_k(\boldsymbol{x})\mathbf{1}$ is assumed to be different from $\mathbf{0}$.

[2]Hereafter, if not otherwise stated, we assume that (C.3.4) admits N distinct eigenvalues.

Proof. As a matter of fact, as (C.3.4) is true for $\mathbf{u}_k(\mathbf{x})$ and

$$\mathcal{M}(\mathbf{x})\mathbf{u}_n(\mathbf{x}) = \lambda_n(\mathbf{x})\mathbf{u}_n(\mathbf{x}) \qquad (C.3.6)$$

is true for $\mathbf{u}_n(\mathbf{x})$, after right-multiplying (C.3.6) on both sides by $\mathbf{u}_k(\mathbf{x})$, one gets

$$\lambda_n(\mathbf{x})\mathbf{u}_n(\mathbf{x}) \odot \mathbf{u}_k(\mathbf{x}) \equiv \mathcal{M}(\mathbf{x})\mathbf{u}_n(\mathbf{x}) \odot \mathbf{u}_k(\mathbf{x}) \equiv \mathbf{x} \odot \mathbf{u}_n(\mathbf{x}) \odot \mathbf{u}_k(\mathbf{x})$$

$$= \mathbf{x} \odot \mathbf{u}_k(\mathbf{x}) \odot \mathbf{u}_n(\mathbf{x}) \equiv \mathcal{M}(\mathbf{x})\mathbf{u}_k(\mathbf{x}) \odot \mathbf{u}_n(\mathbf{x}) \equiv \lambda_k(\mathbf{x})\mathbf{u}_k(\mathbf{x}) \odot \mathbf{u}_n(\mathbf{x}) \,.$$

From the above relationship, since the multiplication is commutative, it ensues that

$$[\lambda_n(\mathbf{x}) - \lambda_k(\mathbf{x})]\,\mathbf{u}_n(\mathbf{x}) \odot \mathbf{u}_k(\mathbf{x}) = \mathbf{0}\,.$$

Since the two eigenvalues are distinct by hypothesis, (C.3.5) is thus demonstrated. □

3. *Given an* **H***-number* $\mathbf{x}$*, such that* $\mathbf{x} \neq \mathbf{0}$*, and its set of eigenvectors* $\mathbf{u}_k(\mathbf{x})$*, if* $\mathbf{x} \not\perp_{\mathbf{u}_k(\mathbf{x})}$*, then the following property holds,*

$$\mathbf{u}_k^2(\mathbf{x}) = \gamma_k(\mathbf{x})\mathbf{u}_k(\mathbf{x})\,, \qquad (C.3.7)$$

where $\gamma_k(\mathbf{x})$ *is a suitable scalar which depends on the components of* $\mathbf{x}$*.*

Proof. $\mathbf{x}$ can be written as a linear combination of the eigenvectors of $\mathcal{M}(\mathbf{x})^3$

$$\mathbf{x} = \sum_{n=0}^{N-1} c_n(\mathbf{x})\mathbf{u}_n(\mathbf{x})\,, \qquad (C.3.8)$$

where the $c_n(\mathbf{x})$ are scalar coefficients which depend on the components of $\mathbf{x}$. By substituting (C.3.8) into (C.3.4), taking into account (C.3.5) and that $\mathcal{M}(\mathbf{x})\mathbf{u}_k(\mathbf{x}) = \mathbf{x} \odot \mathbf{u}_k(\mathbf{x})$, we get

$$c_k(\mathbf{x})\mathbf{u}_k^2(\mathbf{x}) = \lambda_k(\mathbf{x})\mathbf{u}_k(\mathbf{x})\,. \qquad (C.3.9)$$

Since we have assumed $\mathbf{x} \not\perp_{\mathbf{u}_k(\mathbf{x})}$, it must be $c_k(\mathbf{x}) \neq 0$: actually, if $c_k(\mathbf{x})$ were zero, one would get $\mathbf{x} \odot \mathbf{u}_k(\mathbf{x}) = \mathbf{0}$ from (C.3.8), and therefore $\mathbf{x} \in \perp_{\mathbf{x}_k(\mathbf{x})}$ would be true, in contrast with our hypothesis. Therefore, we can divide both sides of (C.3.9) by $c_k(\boldsymbol{x})$ to obtain (C.3.7) with

$$\gamma_k(\mathbf{x}) = \frac{\lambda_k(\mathbf{x})}{c_k(\mathbf{x})}\,. \qquad \square$$

---

[3] As known from linear algebra, if (C.3.4) admits N distinct eigenvalues, then the corresponding N eigenvectors are linearly independent. In such a case, any column vector (i.e., **H**-number) can be written as a linear combination of the eigenvectors.

Equation (C.3.7) can be alternatively written as $\mathcal{M}\left[\mathbf{u}_k(\mathbf{x})\right]\mathbf{u}_k(\mathbf{x}) = \gamma_k(\mathbf{x})\mathbf{u}_k(\mathbf{x})$, which means that $\mathbf{u}_k(\mathbf{x})$ is an eigenvector of the matrix associated with itself.

4. *The eigenvectors of the characteristic matrix $\mathcal{M}(\mathbf{x})$ do not depend on the components of* $\mathbf{x}$.

Proof. As a matter of fact, since the eigenvectors $\mathbf{u}_k(\mathbf{x})$ form a linearly independent system, given any $\mathbf{y} \in \mathbf{H}_N$ one can write it as a linear combination of the eigenvectors

$$\mathbf{y} = \sum_{n=0}^{N-1} c_n(\mathbf{x},\mathbf{y})\mathbf{u}_n(\mathbf{x}), \tag{C.3.10}$$

where the real coefficients $c_n(\mathbf{x},\mathbf{y})$ generally depend on the components of $\mathbf{x}$ and $\mathbf{y}$. Therefore, thanks to (C.3.10), (C.3.2), (C.3.5), and (C.3.7), one can write

$$\mathcal{M}(\mathbf{y})\mathbf{u}_k(\mathbf{x}) = \sum_{n=0}^{N-1} c_n(\mathbf{x},\mathbf{y})\mathcal{M}\left[\mathbf{u}_n(\mathbf{x})\right]\mathbf{u}_k(\mathbf{x}) = \sum_{n=0}^{N-1} c_n(\mathbf{x},\mathbf{y})\mathbf{u}_n(\mathbf{x}) \odot \mathbf{u}_k(\mathbf{x})$$

$$= c_k(\mathbf{x},\mathbf{y})\mathbf{u}_k^2(\mathbf{x}) = c_k(\mathbf{x},\mathbf{y})\gamma_k(\mathbf{x})\mathbf{u}_k(\mathbf{x}). \tag{C.3.11}$$

Equation (C.3.11) shows that $\mathbf{u}_k(\mathbf{x})$ is an eigenvector of $\mathcal{M}(\mathbf{y})$ as well, whatever $\mathbf{y}$ is. □

This result implies that (C.3.7) is true in the following $\mathbf{x}$-independent form $\mathbf{u}_k^2 = \gamma_k\mathbf{u}_k$. Further, it demonstrates that the eigenvectors form a set which only depends on the system of $\mathbf{H}$-numbers under consideration. For this reason, hereafter we will drop the dependence on $\mathbf{x}$ from the eigenvectors.

One can build normalized eigenvectors $\mathbf{v}_k$ which form a set of orthonormal $\mathbf{H}$-numbers. Let us set

$$\mathbf{v}_k = \frac{1}{\gamma_k}\mathbf{u}_k. \tag{C.3.12}$$

It follows from (C.3.5):

$$\mathbf{v}_k \odot \mathbf{v}_n = \mathbf{0} \tag{C.3.13}$$

for $k \neq n$. Moreover, from (C.3.7) one gets

$$\mathbf{v}_k^2 = \mathbf{v}_k. \tag{C.3.14}$$

By virtue of (C.3.13) and (C.3.14), we shall say that the eigenvalue v_k form a set of orthonormal $\mathbf{H}$-numbers. One can verify that the original and normalized eigenvectors, $\mathbf{u}_k$ and $\mathbf{v}_k$, share the same eigenvalues $\lambda_k(\mathbf{x})$ for a chosen $\mathbf{H}$-number $\mathbf{x}$.

5. *The orthonormal eigenvectors $\mathbf{v}_k$ are a complete set [40], therefore any H-number $\mathbf{x}$ can be expressed as a linear superposition of them.*

Proof. As a matter of fact, knowing that

$$\mathcal{M}(\mathbf{x})\mathbf{v}_k \equiv \mathbf{x} \odot \mathbf{v}_k = \lambda_k(\mathbf{x})\mathbf{v}_k\,, \qquad (C.3.15)$$

one can verify — by recalling completeness of the eigenvectors of a matrix — that

$$\mathbf{x} = \sum_{k=0}^{N-1} \lambda_k(\mathbf{x})\mathbf{v}_k\,. \qquad (C.3.16)$$

□

Equation (C.3.16), together with (C.3.13) and (C.3.14), plays a fundamental role for the **H**-numbers: it states that the eigenvectors $\mathbf{v}_k$ can be used as a new set of unit vectors (i.e., a basis) instead of the natural unit vectors $\mathbf{e}_k$, and that the components of any **H**-number for them are the eigenvalues of the associated matrix. Also, if the components $\lambda_k(\mathbf{x})$ of $\mathbf{x}$ happen to be complex in this new $\{\mathbf{v}_k\}$ basis, this fact does not constitute a problem for the developments that follow. Equation (C.3.16) will come in handy when functions of **H**-numbers will be dealt with.

6. *As a consequence of (C.3.16), it follows that the unity $\mathbf{1}$ can be written as*

$$\mathbf{1} = \sum_{k=0}^{N-1} \mathbf{v}_k\,. \qquad (C.3.17)$$

Proof. As a matter of fact, since $\mathcal{M}(\mathbf{1}) = \mathcal{I}$, the eigenvalues of the associated matrix are all equal to 1, so that (C.3.17) follows. □

In the theory of hypercomplex numbers (Section 2.2.2), the basis satisfying relations (C.3.13), (C.3.14) and (C.3.17) is called an *idempotent basis*.

7. *The eigenvalues of the associated matrix satisfy the following properties:*

$$\lambda_k(\mathbf{0}) = 0\,; \qquad \lambda_k(\mathbf{1}) = 1\,; \qquad (C.3.18)$$

$$\lambda_k(a\mathbf{x} + b\mathbf{y}) = a\lambda_k(\mathbf{x}) + b\lambda_k(\mathbf{y})\,; \qquad (C.3.19)$$

$$\lambda_k(\mathbf{x} \odot \mathbf{y}) = \lambda_k(\mathbf{x})\lambda_k(\mathbf{y})\,; \qquad (C.3.20)$$

$$\lambda_k(\mathbf{x}^n) = [\lambda_k(\mathbf{x})]^n\,. \qquad (C.3.21)$$

In the above equations, $a, b \in \mathbf{R}$; $\mathbf{x}, \mathbf{y} \notin \mathbf{H}$, and n is an integer. When $n < 0$ in (C.3.21), $\mathbf{x}$ is assumed to be neither zero nor a divisor of zero. The proof is straightforward.

C.3.3 More About Divisors of Zero

If $\mathbf{x} \in \mathbf{H}_N$ is a divisor of zero, then

$$\mathbf{x} \not\in \perp_{\mathbf{x}} . \tag{C.3.22}$$

Proof. As a matter of fact, if $\mathbf{x} \in \perp_{\mathbf{x}}$ were true, one would get $\mathbf{x}^2 = \mathbf{0}$. On the other hand, (C.3.21) allows writing

$$\mathbf{x}^2 = \sum_{k=0}^{N-1} [\lambda_k(\mathbf{x})]^2 \mathbf{v}_k . \tag{C.3.23}$$

Since $\mathbf{x} \neq \mathbf{0}$ for its being a divisor of zero, at least one eigenvalue $\lambda_k(\mathbf{x})$ must be non-zero, so that also $\mathbf{x}^2 \equiv x \odot x \neq \mathbf{0}$. This demonstrates (C.3.22) to be true. □

It follows that the only solution to equation $\mathbf{x}^2 = \mathbf{0}$, is $\mathbf{x} = \mathbf{0}$.

C.3.4 Modulus of a Hypercomplex Number

Let us consider $\mathbf{x} \in \mathbf{H}_N$ for which $\det[\mathcal{M}(\mathbf{x})] \geq 0$. The *modulus* of $\mathbf{x}$, indicated as $|\mathbf{x}|$, is defined as

$$|\mathbf{x}| = \{\det[\mathcal{M}(\mathbf{x})]\}^{1/N} . \tag{C.3.24}$$

Demonstrating that the modulus of a divisor of zero is zero is straightforward.

C.3.5 Conjugations of a Hypercomplex Number

Let us consider $\mathbf{x} \in \mathbf{H}_N$. We define $N-1$ conjugations of $\mathbf{x}$, indicated with $^1\overline{\mathbf{x}}, {}^2\overline{\mathbf{x}}, \ldots, {}^{N-1}\overline{\mathbf{x}}$, as

$$^s\overline{\mathbf{x}} = \sum_{k=0}^{N-1} \lambda_{[k+s]}(\mathbf{x})\mathbf{v}_k \qquad (s = 1, 2, \ldots, N-1), \tag{C.3.25}$$

where $[k + s]$ is the congruent[4] of $k + s$ module N.

Equation (C.3.25) states that the sth conjugation of $\mathbf{x}$ is obtained by scaling in a cyclic way the subscript of the eigenvalues by s places with respect to those of the eigenvectors $\mathbf{v}_k$.

The hypercomplex conjugations satisfy the following properties:

1. Their product is equal to the characteristic determinant

$$\mathbf{x} \odot {}^1\overline{\mathbf{x}} \odot {}^2\overline{\mathbf{x}} \cdots {}^{N-1}\overline{\mathbf{x}} = \det[\mathcal{M}(\mathbf{x})]\mathbf{1} . \tag{C.3.26}$$

Proof. It follows from the property that the product of the eigenvalues of a matrix is equal, but for the sign, to the matrix determinant. □

---

[4]We recall that, given a positive integer q, the congruent $[q]$ of q module N is equal to $q - mN$, where m is a suitable integer number such that $0 \leq q - mN \leq N - 1$.

2. It can be demonstrated that

$$^s\overline{\mathbf{0}} = \mathbf{0}; \qquad ^s\overline{\mathbf{1}} = \mathbf{1}; \tag{C.3.27}$$

$$^r\overline{(^s\overline{\mathbf{x}})} = \,^{[r+s]}\overline{\mathbf{x}}; \tag{C.3.28}$$

$$^s\overline{(\mathbf{x} \odot \mathbf{y})} = \,^s\overline{\mathbf{x}} \odot \,^s\overline{\mathbf{y}}; \tag{C.3.29}$$

$$^s\overline{(a\mathbf{x} + b\mathbf{y})} = a\,^s\overline{\mathbf{x}} + b\,^s\overline{\mathbf{y}}; \tag{C.3.30}$$

$$^s\overline{(\mathbf{x}^n)} = (^s\overline{\mathbf{x}})^n . \tag{C.3.31}$$

In the above equations, $a, b \in \mathbf{R}$; $\mathbf{x}, \mathbf{y} \in \mathbf{H}_N$, and n is an integer. In (C.3.31), $\det[\mathcal{M}(\boldsymbol{x})] \neq 0$ is assumed when $n < 0$.

3. Since $^s\overline{\mathbf{x}}$ is an $\mathbf{H}$-number, it can be written with an expression similar to (C.3.16), that is

$$^s\overline{\mathbf{x}} = \sum_{k=0}^{N-1} \lambda_k(^s\overline{\mathbf{x}})\mathbf{v}_k . \tag{C.3.32}$$

By comparing (C.3.25) and (C.3.32), one finds that

$$\lambda_k(^s\overline{\mathbf{x}}) = \lambda_{[k+s]}(\mathbf{x}) . \tag{C.3.33}$$

C.4   Functions of a Hypercomplex Variable

C.4.1   Analytic Continuation

Let us consider a real function of a real variable, $f : \mathbf{R} \to \mathbf{R}$. Let us assume that f can be written, in its domain, as an absolutely convergent power series. That is

$$f(x) = \sum_{n=0}^{\infty} a_n x^n , \tag{C.4.1}$$

where $a_n \in \mathbf{R}$ are suitable coefficients. Similarly to the case of complex numbers, the *analytic continuation* $\mathbf{f}$ of f is defined as

$$\mathbf{f}(\mathbf{x}) = \sum_{n=0}^{\infty} a_n \mathbf{x}^n . \tag{C.4.2}$$

Clearly, $\mathbf{f} : \mathbf{H}_N \to \mathbf{H}_N$. Since we are dealing with algebraic operations (sums and products), as final result $\mathbf{f}(\mathbf{x})$ is given by real functions of the components x^n, multiplied by the unit vectors, as for functions of a complex variable.

We are going to see that it is possible to establish a link between the functions defined by means of (C.4.2) and matrix functions. For what we reported in Section C.1.3, (C.4.2) can also be written as

$$\mathbf{f}(\mathbf{x}) = \sum_{n=0}^{\infty} a_n [\mathcal{M}(\mathbf{x})]^n \mathbf{1} = \left[\sum_{n=0}^{\infty} a_n [\mathcal{M}(\mathbf{x})]^n \right] \mathbf{1} . \tag{C.4.3}$$

Note that the series within square brackets is the power series of a matrix function [40] that will be indicated with $\mathcal{F}[\mathcal{M}(\mathbf{x})]$. Therefore, if this series converges, also the series in (C.4.2) converges. Moreover *The following fundamental relationship can be stated between the analytic continuation* $\mathbf{f}$ *of* f *and the matrix function* $\mathcal{F}$ *associated with* f,

$$\mathbf{f}(\mathbf{x}) = \mathcal{F}[\mathcal{M}(\mathbf{x})]\,\mathbf{1}, \tag{C.4.4}$$

which can also be written

$$\mathcal{F}[\mathcal{M}(\mathbf{x})] = \mathcal{M}[\mathbf{f}(\mathbf{x})]. \tag{C.4.5}$$

As previously anticipated, (C.4.4) enables one to study the properties of hypercomplex functions by exploiting the results known for matrix functions. A first important result is given by the following

Theorem C.5. *If* $\mathcal{M}(\mathbf{x})$ *admits* N *normalized eigenvectors* $\mathbf{v}_k$ *which form a basis* [40], *then*

$$\mathcal{F}[\mathcal{M}(\mathbf{x})]\,\mathbf{v}_k = f[\lambda_k(\mathbf{x})]\,\mathbf{v}_k. \tag{C.4.6}$$

Proof. It follows from the theory of matrix functions since, by virtue of (C.3.17), we can write the unity $\mathbf{1}$ as a sum of the $\mathbf{v}_k$ and (C.4.4) becomes

$$\mathbf{f}(\mathbf{x}) = \mathcal{F}[\mathcal{M}(\mathbf{x})] \sum_{k=0}^{N-1} \mathbf{v}_k = \sum_{k=0}^{N-1} \mathcal{F}[\mathcal{M}(\mathbf{x})]\,\mathbf{v}_k = \sum_{k=0}^{N-1} f[\lambda_k(\mathbf{x})]\,\mathbf{v}_k. \tag{C.4.7}$$

$\square$

Equation (C.4.7) establishes an important link between f and its analytic continuation $\mathbf{f}$. Generally, it holds even when the eigenvalues of $\mathcal{M}(\mathbf{x})$ are complex numbers. In such a case, one has to consider, in the right-hand side of (C.4.7), the *complex* analytic continuation of f. The resulting hypercomplex function has, nonetheless, pure real components in the $\{\mathbf{e}_k\}$ basis, since (C.4.3) contains just real coefficients and matrices with real entries.

C.4.2 Properties of Hypercomplex Functions

We list below, leaving the straightforward demonstrations to the reader, some properties of the functions of a hypercomplex variable:

$$\text{if} \quad f(ax+by) = af(x) + bf(y), \quad \text{then} \quad \mathbf{f}(a\mathbf{x}+b\mathbf{y}) = a\mathbf{f}(\mathbf{x}) + b\mathbf{f}(\mathbf{y}); \tag{C.4.8}$$

$$\text{if} \quad f(xy) = f(x) + f(y), \quad \text{then} \quad \mathbf{f}(\mathbf{x}\odot\mathbf{y}) = \mathbf{f}(\mathbf{x}) + \mathbf{f}(\mathbf{y}); \tag{C.4.9}$$

$$\text{if} \quad f(x+y) = f(x)f(y), \quad \text{then} \quad \mathbf{f}(\mathbf{x}+\mathbf{y}) = \mathbf{f}(\mathbf{x})\odot\mathbf{f}(\mathbf{y}). \tag{C.4.10}$$

In the above equations, $a, b \in \mathbf{R}$ and $\mathbf{x}, \mathbf{y} \in \mathbf{H}_N$. By using (C.4.7), other interesting properties can be investigated. By means of it, for instance, the analytic continuations of the exponential and logarithmic functions can be defined. Consistent with the adopted notation, we indicate these analytic continuations in bold characters, that is **exp** and **ln**. Regarding these two functions, from (C.4.9) and (C.4.10) it follows that they possess the same properties of the corresponding real and complex functions

$$\ln(\mathbf{x} \odot \mathbf{y}) = \ln(\mathbf{x}) + \ln(\mathbf{y});$$
$$\exp(\mathbf{x} + \mathbf{y}) = \exp(\mathbf{x}) \odot \exp(\mathbf{y}).$$

C.5 Functions of a Two-dimensional Hypercomplex Variable

As one can verify, for two-dimensional **H**-numbers the eigenvalues of the associated matrix are

$$\lambda_{\pm}(x, y) = x + \frac{1}{2} \left(\beta \pm \sqrt{\Delta} \right) y. \tag{C.5.1}$$

To them, the following orthonormal eigenvectors correspond,

$$\mathbf{v}_{\pm} = \frac{1}{\sqrt{\Delta}} \begin{pmatrix} \frac{\sqrt{\Delta} \mp \beta}{2} \\ \pm 1 \end{pmatrix}, \tag{C.5.2}$$

where we have assumed, for the moment, $\Delta \neq 0$ (*elliptic* and *hyperbolic* numbers). The case of *parabolic* numbers, for which $\Delta = 0$, will be tackled by applying $\Delta \to 0$ to the results obtained with $\Delta \neq 0$. By applying (C.4.7), one gets

$$\mathbf{f} \left[\begin{pmatrix} x \\ y \end{pmatrix} \right] = \begin{pmatrix} f_{\Re}(x, y) \\ f_{\Im}(x, y) \end{pmatrix}, \tag{C.5.3}$$

with

$$f_{\Re}(x, y) = \frac{1}{2} \left[f \left(x + \frac{\beta}{2}y + \frac{\sqrt{\Delta}}{2}y \right) + f \left(x + \frac{\beta}{2}y - \frac{\sqrt{\Delta}}{2}y \right) \right] \tag{C.5.4}$$
$$- \frac{\beta}{2\sqrt{\Delta}} \left[f \left(x + \frac{\beta}{2}y + \frac{\sqrt{\Delta}}{2}y \right) - f \left(x + \frac{\beta}{2}y - \frac{\sqrt{\Delta}}{2}y \right) \right],$$

$$f_{\Im}(x, y) = \frac{1}{\sqrt{\Delta}} \left[f \left(x + \frac{\beta}{2}y + \frac{\sqrt{\Delta}}{2}y \right) - f \left(x + \frac{\beta}{2}y - \frac{\sqrt{\Delta}}{2}y \right) \right]. \tag{C.5.5}$$

In the limit $\Delta \to 0$, the above equations give the following expression for *parabolic* numbers,

$$\mathbf{f} \left[\begin{pmatrix} x \\ y \end{pmatrix} \right]_{\Delta \to 0} = \begin{pmatrix} f \left(x + \frac{\beta}{2}y \right) - \frac{\beta}{2}y f' \left(x + \frac{\beta}{2}y \right) \\ y f' \left(x + \frac{\beta}{2}y \right) \end{pmatrix}. \tag{C.5.6}$$

In Section C.5.2 we show that the definition of functions of a parabolic variable allows us to demonstrate, by means of plain algebra, the fundamental theorems concerning the derivative of functions of a real variable. Now we show a first application of the functions of two-dimensional hypercomplex numbers.

C.5.1 Function of 2×2 Matrices by Two-dimensional Hypercomplex Numbers

If $f(x)$ is a real function of a real variable and $\mathcal{F}$ is the corresponding function of a matrix, an analytic expression of $\mathcal{F}(\mathcal{A})$ can be obtained by means of the developed theory of two-dimensional hypercomplex number. Let us consider the 2×2 matrix

$$\mathcal{A} = \begin{pmatrix} A & B \\ C & D \end{pmatrix}, \tag{C.5.7}$$

and the characteristic matrix of the two-dimensional hypercomplex number $\mathbf{z}$ as given by (C.2.1),

$$\begin{pmatrix} x & \alpha y \\ y & x + \beta y \end{pmatrix}, \tag{C.5.8}$$

where α and β are real numbers and x, y are the "real" and "imaginary" parts of $\mathbf{z}$. Let us observe that the number of unknown elements in matrix (C.5.7) is the same as in (C.5.8). Therefore we can associate $\mathbf{z}$ with matrix (C.5.7) by setting

$$\begin{cases} \alpha = B/C; & x = A; \\ \beta = (D - A)/C; & y = C. \end{cases} \tag{C.5.9}$$

It follows that

$$\Delta \equiv \beta^2 + 4\alpha = [(D - A)^2 + 4BC]/C^2. \tag{C.5.10}$$

By using (C.4.5), one gets

$$\begin{pmatrix} f_\Re(\mathbf{z}) & \alpha f_\Im(\mathbf{z}) \\ f_\Im(\mathbf{z}) & f_\Re(\mathbf{z}) + \beta f_\Im(\mathbf{z}) \end{pmatrix} = \mathcal{F}\left[\begin{pmatrix} x & \alpha y \\ y & x + \beta y \end{pmatrix}\right], \tag{C.5.11}$$

where $f_\Re(\mathbf{z})$ and $f_\Im(\mathbf{z})$, are the "real" and "imaginary" parts of $\mathbf{f}(\mathbf{z})$, given by (C.5.4) and (C.5.5), respectively, if $\Delta \equiv \beta^2 + 4\alpha \neq 0$, or by (C.5.6) if $\Delta = 0$. Let us start by considering $C \equiv y \neq 0$ in (C.5.9) and (C.5.10).

Discriminant $\Delta \neq 0$

By means of (C.5.9) and (C.5.10), taking into account that $\mathbf{z} = \begin{pmatrix} A \\ C \end{pmatrix}$, (C.5.4) and (C.5.5) become

$$f_{\Re}\left[\left(\begin{array}{cc} A \\ C \end{array}\right)\right] = \frac{1}{2}\left[f\left(\frac{A+D}{2} + \text{sgn}(C)\frac{\sqrt{(D-A)^2+4BC}}{2}\right)\right.$$

$$\left. +f\left(\frac{A+D}{2} - \text{sgn}(C)\frac{\sqrt{(D-A)^2+4BC}}{2}\right)\right]$$

$$-\frac{\text{sgn}(C)\,(D-A)}{2\sqrt{(D-A)^2+4BC}}\left[f\left(\frac{A+D}{2} + \text{sgn}(C)\frac{\sqrt{(D-A)^2+4BC}}{2}\right)\right.$$

$$\left. -f\left(\frac{A+D}{2} - \text{sgn}(C)\frac{\sqrt{(D-A)^2+4BC}}{2}\right)\right], \quad (C.5.12)$$

$$f_{\Im}\left[\left(\begin{array}{cc} A \\ C \end{array}\right)\right] = \frac{|C|}{\sqrt{(D-A)^2+4BC}}\left[f\left(\frac{A+D}{2} + \text{sgn}(C)\frac{\sqrt{(D-A)^2+4BC}}{2}\right)\right.$$

$$\left. -f\left(\frac{A+D}{2} - \text{sgn}(C)\frac{\sqrt{(D-A)^2+4BC}}{2}\right)\right]. \quad (C.5.13)$$

Inserting $(C.5.12)$ and $(C.5.13)$ into $(C.5.11)$, we obtain $\mathcal{F}(\mathcal{A})$.

Discriminant $\Delta = 0$

If the elements of matrix $\mathcal{A}$ are such that $(D-A)^2 + 4BC = 0$, we can calculate $\mathcal{F}(\mathcal{A})$ by applying (C.5.6). We get

$$f_{\Re}\left[\left(\begin{array}{cc} A \\ C \end{array}\right)\right] = f\left(\frac{A+D}{2}\right) - \frac{D-A}{2}f'\left(\frac{A+D}{2}\right), \quad (C.5.14)$$

$$f_{\Im}\left[\left(\begin{array}{cc} A \\ C \end{array}\right)\right] = Cf'\left(\frac{A+D}{2}\right). \quad (C.5.15)$$

Again, by inserting $(C.5.14)$ and $(C.5.15)$ into $(C.5.11)$, we obtain $\mathcal{F}(\mathcal{A})$. Equations (C.5.9) and (C.5.10) seem not to hold for $C = 0$ since C appears alone in some denominators. Anyway, in (C.5.12) and (C.5.15) C does not appear as a single term in the denominators. This means that (C.5.12) and (C.5.15) hold also for the limiting case $y \equiv C \to 0$, on the condition that in the function $\text{sgn}(C)$ one uses the left or right limit for $C \to 0$, i.e., $\text{sgn}(0) = 1$ (or -1), instead of the usual convention $\text{sgn}(0) = 0$.

Some Considerations

It is known that the functions of a matrix are usually obtained by calculating the eigenvalues of the matrix, transforming the vectors represented by the matrix lines, then calculating the functions and coming back by means of an inverse transformation. The association here proposed gives, of course, the same result but with one step only.

C.5.2 Functions of a Parabolic Variable and the Derivatives of Functions of a Real Variable

Functions of a parabolic variable feature the property that their "imaginary" part is proportional to the derivative of their "real" part. We have

Theorem C.6. *The functions of a parabolic variable allow demonstrating the rules of the differential calculus for the real functions of a real variable in an algebraic way.*

Proof. From the expressions of (C.5.6), one can obtain the functions of the canonical parabolic variable ($\beta = 0$, $\alpha = 0$) which, as usual for the functions of a complex variable, can be written as

$$f(z) = U(x,\, y) + p\, V(x,\, y),\tag{C.5.16}$$

where p is the "imaginary" versor (p represents the first letter of *parabolic*). From (C.5.6), we get

$$f(z) = f(x) + p\, y\, f'(x),\tag{C.5.17}$$

where, as usual, we have indicated with $f'(x)$ the derivative of the real function of a real variable $f(x)$. Thanks to (C.5.17), we can write

$$f'(x) = \frac{\Im\,[f(z)]}{\Im(z)},$$

$\Im(\cdot)$ representing the "imaginary" part of the parabolic quantity within parentheses.

The above expression allows one to demonstrate, by elementary algebra, the fundamental differentiation rules for the functions of a real variable.

Let us indicate with $f(x)$, $g(x)$ and $h(x)$ three functions of a real variable and $f(z)$, $g(z)$ and $h(z)$ their analytic parabolic continuations, as defined by (C.5.17). We get

1. *The Derivative of the Sum Functions*

   Proof. Since

   $$f(z) + g(z) = [f(x) + g(x)] + p\, y\, [f'(x) + g'(x)]\,,$$

   it follows at once that

   $$\frac{\mathrm{d}}{\mathrm{d}x}\,[f(x) + g(x)] = f'(x) + g'(x).\tag{C.5.18}$$

   $\square$

2. *The Derivative of the Product of Two Functions*

Proof. Taking into account the property $p^2 = 0$, we have

$$f(z) \cdot g(z) = [f(x) + p\,y\,f'(x)][g(x) + p\,y\,g'(x)]$$

$$= f(x)g(x) + p\,y\,[f'(x)g(x) + f(x)g'(x)]\,.$$

This demonstrates that

$$\frac{d}{dx}[f(x)g(x)] = f'(x)g(x) + f(x)g'(x). \qquad\qquad (C.5.19)$$

$\square$

3. *The Derivative of the Quotient Function*

Proof. Let us consider the quotient of the analytic continuations of the functions $f(x)$ and $g(x)$,

$$\frac{f(z)}{g(z)} = \frac{f(x) + p\,y\,f'(x)}{g(x) + p\,y\,g'(x)}.$$

If we multiply the numerator and the denominator of the right-hand side by the conjugate parabolic function of $g(z)$, $(\overline{g(z)} = g(x) - p\,yg'(x))$, we obtain

$$\frac{f(z)}{g(z)} = \frac{[f(x) + p\,yf'(x)]\,[g(x) - p\,yg'(x)]}{[g(x) + p\,yg'(x)][g(x) - p\,yg'(x)]}$$

$$\equiv \frac{f(x)}{g(x)} + p\,y\frac{f'(x)g(x) - f(x)g'(x)}{g^2(x)}.$$

The following relation is thus demonstrated,

$$\frac{d}{dx}\left[\frac{f(x)}{g(x)}\right] = \frac{f'(x)g(x) - f(x)g'(x)}{g^2(x)}. \qquad\qquad (C.5.20)$$

$\square$

4. *The Derivative of the Inverse Function*

Proof. Let us indicate

$$f(z) = f(x) + p\,yf'(x) \equiv X + pY,$$

and let $f^{-1}(x)$ and its analytic continuations exist. Because of the analyticity of the inverse function, we know that

$$f^{-1}(z) = f^{-1}(x) + p\,y(f^{-1})'(x)$$

and the definition of the inverse function gives the identity

$$f^{-1}\left[f(z)\right] \equiv z.$$

Then, it follows that

$$
\begin{aligned}
f^{-1}\left[f(z)\right] &= f^{-1}(X+pY) = f^{-1}(X) + pY\,(f^{-1})'(X) \\
&= x + pyf'(x)\,(f^{-1})'[f(x)] \equiv x + py.
\end{aligned}
$$

From the previous expressions, it ensues that

$$f'(x)\,(f^{-1})'[f(x)] = 1.$$

Since $x \equiv f^{-1}(X)$, the relation

$$\frac{\mathrm{d}}{\mathrm{d}X}\left[f^{-1}(X)\right] = \left.\frac{1}{f'(x)}\right|_{x \equiv f^{-1}(X)} \tag{C.5.21}$$

is demonstrated. □

5. *The Derivative of the Function of a Function*

Proof. Let us consider the composite function

$$h(z) = f[g(z)]$$

and set

$$g(z) = g(x) + py\,g'(x) \equiv X + pY.$$

Because of the analyticity of f, we have

$$h(z) = f(X + pY) = f(X) + pY\,f'(X).$$

Since $X \equiv g(x)$ and $Y \equiv y\,g'(x)$, we obtain

$$h(z) = f[g(x)] + py\,g'(x)f'[g(x)].$$

The following result is thus demonstrated

$$\frac{\mathrm{d}}{\mathrm{d}x}f[g(x)] = f'[g(x)]\,g'(x). \tag{C.5.22}$$

□

Some Considerations. By means of simple algebraic calculations, we have obtained the fundamental rules of differential calculus in the real domain. This has been possible thanks to the property of the versor p of parabolic numbers which acts as a first-order infinitesimal quantity, with the following difference: the square of p is exactly zero, whereas the square of an infinitesimal quantity is considered smaller than the quantity itself.

C.6 Derivatives of a Hypercomplex Function

Before introducing the derivative, it is useful to introduce the limit of a function of a hypercomplex variable. Given $\mathbf{x} \in \mathbf{H}_N$ and $\mathbf{f} : \mathbf{H}_N \to \mathbf{H}_N$, we define the limit of $\mathbf{f}(\mathbf{x})$ for $\mathbf{x} \to \mathbf{y}$ to be equal to $\boldsymbol{\ell} \equiv (\ell_0, \ell_1, \dots, \ell_{N-1})$ if

$$\lim_{x^0 \to y^0} \lim_{x^1 \to y^1} \cdots \lim_{x^{N-1} \to y^{N-1}} f^k(\mathbf{x}) = \ell^k \quad \text{for } k = 0, 1, \dots, N-1, \tag{C.6.1}$$

under the assumption that all the limits of the components do exist.

C.6.1 Derivative with Respect to a Hypercomplex Variable

Assuming that $f(x)$ is a differentiable function, let us define the derivative of $\mathbf{f}(\mathbf{x})$ with respect to the hypercomplex variable $\mathbf{x}$. The *incremental ratio* of this function is defined as

$$\frac{\Delta \mathbf{f}}{\Delta \mathbf{x}}(\mathbf{x}, \mathbf{t}) = \frac{\mathbf{f}(\mathbf{x} + \mathbf{t}) - \mathbf{f}(\mathbf{x})}{\mathbf{t}}, \tag{C.6.2}$$

with $\mathbf{t} \in \mathbf{H}_N$. Obviously, the division in the previous definition is defined only if $\det[\mathcal{M}(\mathbf{t})] \neq 0$. Under such an assumption, we define the *derivative* of $\mathbf{f}(\mathbf{x})$ with respect to $\mathbf{x}$ as the limit of the incremental ratio for $\mathbf{t} \to \mathbf{0}$, that is

$$\frac{d\mathbf{f}}{d\mathbf{x}}(\mathbf{x}) = \lim_{\mathbf{t} \to \mathbf{0}} \frac{\Delta \mathbf{f}}{\Delta \mathbf{x}}(\mathbf{x}, \mathbf{t}). \tag{C.6.3}$$

We say that a derivative of $\mathbf{x}$ exists if the limit (C.6.3) exists and is independent of the direction with respect to which the incremental ratio is taken, see (C.6.1). Alternatively, we could introduce the derivative by applying (C.4.7) to the real function $f'(x)$ (the derivative of $f(x)$), that is

$$\frac{d\mathbf{f}}{d\mathbf{x}}(\mathbf{x}) = \sum_{k=0}^{N-1} f'[\lambda_k(\mathbf{x})] \mathbf{v}_k. \tag{C.6.4}$$

We have

Theorem C.7. *The definitions, (C.6.3) and (C.6.4), are equivalent if $\mathbf{t}$ is not a divisor of zero.*

Proof. As a matter of fact, (C.6.3), together with (C.4.7), gives

$$\begin{aligned}
\frac{d\mathbf{f}}{d\mathbf{x}}(\mathbf{x}) &= \lim_{\mathbf{t} \to \mathbf{0}} \sum_{k=0}^{N-1} \{f[\lambda_k(\mathbf{x} + \mathbf{t})] - f[\lambda_k(\mathbf{x})]\} \, \mathcal{M}^{-1}(\mathbf{t}) \mathbf{v}_k \\
&= \lim_{\mathbf{t} \to \mathbf{0}} \sum_{k=0}^{N-1} \frac{f[\lambda_k(\mathbf{x}) + \lambda_k(\mathbf{t})] - f[\lambda_k(\mathbf{x})]}{\lambda_k(\mathbf{t})} \mathbf{v}_k = \sum_{k=0}^{N-1} f'[\lambda_k(\mathbf{x})] \mathbf{v}_k,
\end{aligned}$$

since $\lim_{\mathbf{t} \to \mathbf{0}} \lambda_k(\mathbf{t}) = 0$. $\qquad \square$

Equation (C.6.4) does not need the introduction of an increment **t** for the definition of the derivative of $\mathbf{f}(\mathbf{x})$, and therefore it is preferable with respect to (C.6.3). As far as the *higher-order derivatives* are concerned, one can demonstrate that a proper definition is

$$\frac{\mathrm{d}^n \mathbf{f}}{\mathrm{d}\mathbf{x}^n}(\mathbf{x}) = \sum_{k=0}^{N-1} f^{(n)}[\lambda_k(\mathbf{x})]\mathbf{v}_k \,,$$

where n is any positive integer.

With (C.6.4), one can demonstrate that properties of the derivative of real functions (see p. 233) can be straightforwardly extended to the derivative of functions of a hypercomplex variable.

This result can be formulated as

Theorem C.8. *For commutative and associative systems, (Theorem C.2) differential calculus exists.*

Then it represents the sufficient condition for Scheffers Theorem 2.1

C.6.2 Partial Derivatives

Equation (C.4.7) allows us to introduce the partial derivative of $\mathbf{f}(\mathbf{x})$ with respect to one component of $\mathbf{x}$, say x^l. The resulting definition is

$$\frac{\partial \mathbf{f}}{\partial x^l}(\mathbf{x}) = \frac{\partial}{\partial x^l} \sum_{k=0}^{N-1} f[\lambda_k(\mathbf{x})]\mathbf{v}_k = \sum_{k=0}^{N-1} f'[\lambda_k(\mathbf{x})]\frac{\partial \lambda_k}{\partial x^l}(\mathbf{x})\mathbf{v}_k \,. \tag{C.6.5}$$

Thanks to (C.3.20), according to which the product of two **H**-numbers can be evaluated, in the orthonormal basis $\{\mathbf{v}_k\}$, by multiplying their components one by one (internal product), the above equation gives

$$\frac{\partial \mathbf{f}}{\partial x^l}(\mathbf{x}) = \frac{\mathrm{d}\mathbf{f}}{\mathrm{d}\mathbf{x}}(\mathbf{x}) \odot \frac{\partial}{\partial x^l} \sum_{k=0}^{N-1} \lambda_k(\mathbf{x})\mathbf{v}_k = \frac{\mathrm{d}\mathbf{f}}{\mathrm{d}\mathbf{x}}(\mathbf{x}) \odot \frac{\partial \mathbf{x}}{\partial x^l} \,. \tag{C.6.6}$$

Noticing that $\partial \mathbf{x}/\partial x^l \equiv \mathbf{e}_l$, one gets the following fundamental result, which relates the full derivative with the partial derivatives of a function of a hypercomplex variable:

$$\frac{\partial \mathbf{f}}{\partial x^l}(\mathbf{x}) = \frac{\mathrm{d}\mathbf{f}}{\mathrm{d}\mathbf{x}}(\mathbf{x}) \odot \mathbf{e}_l \,. \tag{C.6.7}$$

In a similar way, one can verify that

$$\frac{\partial^2 \mathbf{f}}{\partial x^m \partial x^l}(\mathbf{x}) = \frac{\mathrm{d}^2 \mathbf{f}}{\mathrm{d}\mathbf{x}^2} \odot \mathbf{e}_m \odot \mathbf{e}_l$$

holds, and that analogous relationships hold for higher-order derivatives.

From (C.6.7), the following relationship among the partial derivatives of $\mathbf{f}(\mathbf{x})$ follows,

$$\mathbf{e}_m \odot \frac{\partial \mathbf{f}}{\partial x^l}(\mathbf{x}) = \mathbf{e}_l \odot \frac{\partial \mathbf{f}}{\partial x^m}(\mathbf{x}) \tag{C.6.8}$$

and is true $\forall \, l, m = 0, 1, \ldots, N - 1$. Equation (C.6.8) can be considered as the vectorial form of the generalized Cauchy–Riemann (GCR) conditions (see Section 7.3.2). A function of a hypercomplex variable which satisfies the GCR conditions — i.e., (C.6.8) — is called *holomorphic* (or holomorphic analytic). Clearly, any analytic continuation of a real function is holomorphic, since (C.6.8) has been derived for functions which are analytic continuations.

C.6.3 Components of the Derivative Operator

One can define a *derivative operator*, d/d$\mathbf{x}$, by *formally* right-multiplying it by the function subjected to the derivative operation

$$\frac{\mathrm{d}}{\mathrm{d}\mathbf{x}} \odot \mathbf{f}(\mathbf{x}) \equiv \frac{\mathrm{d}\mathbf{f}}{\mathrm{d}\mathbf{x}}(\mathbf{x}) \,. \tag{C.6.9}$$

In such a case, by formally dealing with the derivative operator as if it were an **H**-number, one can decompose it in the orthonormal basis $\{\mathbf{v}_k\}$

$$\frac{\mathrm{d}}{\mathrm{d}\mathbf{x}} = \sum_{k=0}^{N-1} \lambda_k \left(\frac{\mathrm{d}}{\mathrm{d}\mathbf{x}} \right) \mathbf{v}_k \,. \tag{C.6.10}$$

Since

$$\mathbf{f}(\mathbf{x}) = \sum_{k=0}^{N-1} f\left[\lambda_k(\mathbf{x})\right] \mathbf{v}_k \,, \tag{C.6.11}$$

by formally multiplying (C.6.10) by (C.6.11) and recalling that the $\mathbf{v}_k$ are orthonormal, one gets

$$\frac{\mathrm{d}}{\mathrm{d}\mathbf{x}} \odot \mathbf{f}(\mathbf{x}) \equiv \frac{\mathrm{d}\mathbf{f}}{\mathrm{d}\mathbf{x}}(\mathbf{x}) = \sum_{k=0}^{N-1} \lambda_k \left(\frac{\mathrm{d}}{\mathrm{d}\mathbf{x}} \right) f\left[\lambda_k(\mathbf{x})\right] \mathbf{v}_k \,. \tag{C.6.12}$$

By comparing (C.6.12) with (C.6.4), one gets the following expression for the eigenvalues of the derivative operator

$$\lambda_k \left(\frac{\mathrm{d}}{\mathrm{d}\mathbf{x}} \right) = \frac{\mathrm{d}}{\mathrm{d}\lambda_k(\mathbf{x})} \,. \tag{C.6.13}$$

Thus, these eigenvalues are the derivative operators applied to the single components of $\mathbf{f}(\mathbf{x})$ in the $\{\mathbf{v}_k\}$ basis.

C.6.4 Derivative with Respect to the Conjugated Variables

By exploiting the above results, we demonstrate a generalization of a well-known property of holomorphic functions of a complex variable:

Theorem C.9. *The derivatives of a holomorphic function with respect to the conjugated variable are null.*

Proof. Let us decompose, within the $\{\mathbf{v}_k\}$ basis, the derivative operator with respect to ${}^s\overline{\mathbf{x}}$, which is the s-conjugate of $\mathbf{x}$. Equation (C.6.13) formally gives

$$\lambda_k\left(\frac{\mathrm{d}}{\mathrm{d}\,{}^s\overline{\mathbf{x}}}\right) = \frac{\mathrm{d}}{\mathrm{d}\lambda_k({}^s\overline{\mathbf{x}})}. \tag{C.6.14}$$

Since (C.3.33) holds, (C.6.14) can also be written as

$$\lambda_k\left(\frac{\mathrm{d}}{\mathrm{d}\,{}^s\overline{\mathbf{x}}}\right) = \frac{\mathrm{d}}{\mathrm{d}\lambda_{[k+s]}(\mathbf{x})}, \tag{C.6.15}$$

therefore

$$\frac{\mathrm{d}}{\mathrm{d}\,{}^s\overline{\mathbf{x}}} = \sum_{k=0}^{N-1} \frac{\mathrm{d}}{\mathrm{d}\lambda_{[k+s]}(\mathbf{x})}\mathbf{v}_k. \tag{C.6.16}$$

Moreover, the latter two expressions give

$$\lambda_k\left(\frac{\mathrm{d}}{\mathrm{d}\,{}^s\overline{\mathbf{x}}}\right) = \lambda_{[k+s]}\left(\frac{\mathrm{d}}{\mathrm{d}\mathbf{x}}\right). \tag{C.6.17}$$

By using the above results, it should be noticed that, if $\mathbf{f}(\mathbf{x})$ is a holomorphic function, one gets for $s \neq 0$,

$$\frac{\mathrm{d}\mathbf{f}}{\mathrm{d}\,{}^s\overline{\mathbf{x}}}(\mathbf{x}) \equiv \frac{\mathrm{d}}{\mathrm{d}\,{}^s\overline{\mathbf{x}}} \odot \mathbf{f}(\mathbf{x}) \equiv \sum_{k=0}^{N-1} \frac{\mathrm{d}f[\lambda_k(\mathbf{x})]}{\mathrm{d}\lambda_{[k+s]}(\mathbf{x})}\mathbf{v}_k = \mathbf{0}. \tag{C.6.18}$$

This last equation states that the derivatives of a holomorphic function $\mathbf{f}(\mathbf{x})$ with respect to the conjugated variables ${}^s\overline{\mathbf{x}}$ ($s \neq 0$) are null. □

C.7 Characteristic Differential Equation

Theorem C.10. *Equation (C.6.18) gives rise to systems of differential equations of order lower than or equal to N. The components of any holomorphic function must satisfy such differential equations. In particular, all the components of $\mathbf{f}(\mathbf{x})$ must satisfy the same differential equation of order N, which we call the characteristic* *equation of the **H**-numbers,*

$$\left(\prod_{s=0}^{N-1} \frac{\mathrm{d}}{\mathrm{d}\,^s\overline{\mathbf{x}}}\right) \odot \mathbf{f}(\mathbf{x}) = \left[\prod_{s=0}^{N-1} \left(\sum_{k=0}^{N-1} \frac{\mathrm{d}}{\mathrm{d}\lambda_{[k+s]}(\mathbf{x})} \mathbf{v}_k\right)\right] \odot \mathbf{f}(\mathbf{x})$$

$$= \left(\prod_{s=0}^{N-1} \frac{\mathrm{d}}{\mathrm{d}\lambda_s(\mathbf{x})}\right) \odot \left(\sum_{k=0}^{N-1} \mathbf{v}_k\right) \odot \mathbf{f}(\mathbf{x}) = \left[\prod_{s=0}^{N-1} \lambda_s\left(\frac{\mathrm{d}}{\mathrm{d}\mathbf{x}}\right)\right] \odot \mathbf{1} \odot \mathbf{f}(\mathbf{x})$$

$$= \left|\frac{\mathrm{d}}{\mathrm{d}\mathbf{x}}\right|^N \odot \mathbf{f}(\mathbf{x}) = \mathbf{0}. \tag{C.7.1}$$

Proof. We begin by considering the composite operator $(\mathrm{d}/\mathrm{d}\mathbf{x}) \odot (\mathrm{d}/\mathrm{d}\,^s\overline{\mathbf{x}})$, with $s \neq 0$. Since

$$\frac{\mathrm{d}}{\mathrm{d}\,^s\overline{\mathbf{x}}} \odot \mathbf{f}(\mathbf{x}) = \mathbf{0},$$

also

$$\frac{\mathrm{d}}{\mathrm{d}\mathbf{x}} \odot \frac{\mathrm{d}}{\mathrm{d}\,^s\overline{\mathbf{x}}} \odot \mathbf{f}(\mathbf{x}) = \mathbf{0} \tag{C.7.2}$$

holds. By setting equal to zero the N components of the left-hand side of (C.7.2), one finds a system of differential equations of order 2 for the N components of $\mathbf{f}(\mathbf{x})$.

Now we apply the operator

$$\prod_{s=0}^{N-1} \frac{\mathrm{d}}{\mathrm{d}\,^s\overline{\mathbf{x}}} \equiv \frac{\mathrm{d}}{\mathrm{d}\mathbf{x}} \odot \frac{\mathrm{d}}{\mathrm{d}\,^1\overline{\mathbf{x}}} \cdots \frac{\mathrm{d}}{\mathrm{d}\,^{N-1}\overline{\mathbf{x}}}$$

to a holomorphic function $\mathbf{f}(\mathbf{x})$. By recalling orthonormality of the unit vectors $\mathbf{v}_k$ and (C.6.16), one gets (C.7.1), and by setting as equal to zero the N components of (C.7.1), one finds the asserttion. $\qquad\square$

We also have

Theorem C.11. *An explicit form of the characteristic equation in the basis $\mathbf{e}_k$ can be obtained by evaluating the Nth power of the symbolic modulus of the derivative operator that appears in the last row of (C.7.1).*

Proof. After recalling (C.3.24) and (C.6.13) and the known property that the product of the eigenvalues of a matrix is equal to the matrix determinant, one has

$$\left|\frac{\mathrm{d}}{\mathrm{d}\mathbf{x}}\right|^N = \prod_{s=0}^{N-1} \frac{\mathrm{d}}{\mathrm{d}\lambda_s(\mathbf{x})} = \prod_{s=0}^{N-1} \left[\sum_{k=0}^{N-1} \frac{\partial x^k}{\partial\lambda_s(\mathbf{x})} \frac{\partial}{\partial x^k}\right]. \tag{C.7.3}$$

Since

$$\mathbf{x} = \sum_{s=0}^{N-1} \lambda_s(\mathbf{x})\mathbf{v}_s = \sum_{k=0}^{N-1} \left[\sum_{s=0}^{N-1} \lambda_s(\mathbf{x})v_s^k\right] \mathbf{e}_k,$$

where the v_s^k are the component of $\{\mathbf{v}_s\}$ in the $\{\mathbf{e}_k\}$ basis,

$$\mathbf{v}_s = \sum_{k=0}^{N-1} v_s^k \mathbf{e}_k .$$

From (C.3.16) it follows that the kth component of $\mathbf{x}$ is

$$x^l = \sum_{s=0}^{N-1} \lambda_s(\boldsymbol{x}) v_s^l ,$$

from which one finds

$$\frac{\partial x^k}{\partial \lambda_s(\mathbf{x})} = v_s^k . \tag{C.7.4}$$

By substituting (C.7.4) into (C.7.3) and recalling (C.7.1), the explicit form of the characteristic equation is

$$\prod_{s=0}^{N-1} \left(\sum_{k=0}^{N-1} v_s^k \frac{\partial}{\partial x^k} \right) \mathbf{f}(\mathbf{x}) = \mathbf{0} . \tag{C.7.5}$$

Since in (C.7.5) the operator applied to $\mathbf{f}(\mathbf{x})$ is real, (C.7.5) must hold for all the components of $\mathbf{f}(\mathbf{x})$ in the $\{\mathbf{e}_k\}$ basis, that is

$$\prod_{s=0}^{N-1} \left(\sum_{k=0}^{N-1} v_s^k \frac{\partial}{\partial x^k} \right) f^k(x^0, x^1, \ldots, x^{N-1}) = 0 \tag{C.7.6}$$

$\forall\,(k = 0, 1, \ldots, N-1)$. □

C.7.1 Characteristic Equation for Two-dimensional Hypercomplex Numbers

After recalling the form of the orthonormal unit vectors for two-dimensional **H**-numbers, see (C.5.2), the characteristic equation resulting from (C.7.5) is

$$\left(\frac{\sqrt{\Delta} - \beta}{2\sqrt{\Delta}} \frac{\partial}{\partial x} + \frac{1}{\sqrt{\Delta}} \frac{\partial}{\partial y} \right) \left(\frac{\sqrt{\Delta} + \beta}{2\sqrt{\Delta}} \frac{\partial}{\partial x} - \frac{1}{\sqrt{\Delta}} \frac{\partial}{\partial y} \right) \left(\begin{array}{c} f_\Re(x, y) \\ f_\Im(x, y) \end{array} \right) = \left(\begin{array}{c} 0 \\ 0 \end{array} \right).$$

A few mathematical manipulations transform this equation into a more compact form, that is

$$\left(\alpha \frac{\partial^2}{\partial x^2} + \beta \frac{\partial^2}{\partial x\,\partial y} - \frac{\partial^2}{\partial y^2} \right) \left(\begin{array}{c} f_\Re(x, y) \\ f_\Im(x, y) \end{array} \right) = \left(\begin{array}{c} 0 \\ 0 \end{array} \right). \tag{C.7.7}$$

Equation (C.7.7) is a partial differential equation of elliptic, parabolic or hyperbolic type [72], depending on the values of α and β. In particular, (C.7.7) becomes the well-known Laplace equation for *canonical elliptic* two-dimensional **H**-numbers ($\alpha = -1$, $\beta = 0$) and the wave equation for *canonical hyperbolic* two-dimensional **H**-numbers ($\alpha = 1$, $\beta = 0$).

C.8   Equivalence Between the Matrix Formalization of Hypercomplex Numbers and the one in Section 2.1

In this section we show that, for commutative hypercomplex numbers, the classical approach of Section 2.1 and the matrix formalization of the present appendix are equivalent.

Let us start from the matrix representation (C.1.2) and (C.1.3) and indicate by $(\mathbf{e}_k)^l$ the lth component of the kth unit vector. From (C.1.3), we get

$$(\mathbf{e}_k)^l = \delta_k^l. \tag{C.8.1}$$

Now, let us consider the multiplication of a versor $\mathbf{e}_k$ for the hypercomplex number $\mathbf{y}$, and call it $\mathbf{t}_k$. We have

$$\mathbf{t}_k = \mathbf{e}_k \cdot \mathbf{y} \equiv \mathbf{e}_k \cdot \left(\sum_{j=0}^{N-1} y^j \mathbf{e}_j \right) \equiv \sum_{j=0}^{N-1} y^j (\mathbf{e}_k \cdot \mathbf{e}_j) \tag{C.8.2}$$

$$\equiv \sum_{j=0}^{N-1} y^j \left(\sum_{m=0}^{N-1} C_{kj}^m \mathbf{e}_m \right) \equiv \sum_{m=0}^{N-1} \left(\sum_{j=0}^{N-1} C_{kj}^m y^j \right) \mathbf{e}_m \equiv \sum_{m=0}^{N-1} \mathcal{Y}_k^m \mathbf{e}_m .$$

The fourth passage is obtained from the third one by applying the distributive property of multiplication with respect to the sum. This property is easily demonstrated in theorem C.2 within the matrix formalization and it is assumed as an axiom in linear algebra ([81], p. 243).

In the penultimate passage, the term in round brackets is a two-indexes element which we have indicated, in the last passage, by $\mathcal{Y}_k^m$. These elements, for commutative numbers, correspond to the ones of the characteristic matrix (2.1.4), as we shall see in the following. Let us demonstrate some preliminary theorems.

Theorem C.12. *The elements $\mathcal{Y}_k^m$ can be considered as the components of $\mathbf{t}_k$ in the basis $\mathbf{e}_m$.*

Proof. Let us consider the lth component of $\mathbf{t}_k$. From the first and last members of (C.8.2), we have

$$(\mathbf{t}_k)^l = \sum_{m=0}^{N-1} \mathcal{Y}_k^m (\mathbf{e}_m)^l \equiv \sum_{m=0}^{N-1} \mathcal{Y}_k^m \delta_m^l \equiv \mathcal{Y}_k^l . \tag{C.8.3}$$

$\square$

Theorem C.13. *The elements $(\mathbf{t}_k)^l$ can be interpreted as the product between a matrix $\mathcal{A}_k$ and a vector $\mathbf{y}$,*

$$(\mathbf{t}_k)^l = \sum_{j=0}^{N-1} (\mathcal{A}_k)_j^l \, y^j \qquad \text{with} \qquad (\mathcal{A}_k)_j^l \equiv C_{kj}^l . \tag{C.8.4}$$

Proof. By considering the lth component of $\mathbf{t}_k$, from the first and fifth terms of (C.8.2), we get

$$
\begin{aligned}
(\mathbf{t}_k)^l &= \sum_{j=0}^{N-1} y^j \left(\sum_{m=0}^{N-1} C_{kj}^m \, \mathbf{e}_m \right)^l \equiv \sum_{j=0}^{N-1} y^j \sum_{m=0}^{N-1} C_{kj}^m \, (\mathbf{e}_m)^l \\
&\equiv \sum_{j=0}^{N-1} y^j \sum_{m=0}^{N-1} C_{kj}^m \, \delta_m^l \equiv \sum_{j=0}^{N-1} y^j \, C_{kj}^l \,.
\end{aligned}
\tag{C.8.5}
$$

In the last member of this equation, we can group the elements C_{kj}^l in N real $N \times N$ matrices with elements $_j^l$ and call these matrices $\mathcal{A}_k$. $\square$

In conclusion, for any $\mathbf{y} \in \mathbf{H}_N$ we can write

$$
\mathbf{t}_k \equiv \mathbf{e}_k \cdot \mathbf{y} = \mathcal{A}_k \mathbf{y}.
\tag{C.8.6}
$$

Note that $\mathcal{A}_k \mathbf{y}$ is a column vector with real components, therefore $\mathbf{t}_k$ is an **H**-number. By equating the expressions of $(\mathbf{t}_k)^l$ in (C.8.3) and (C.8.4) we obtain

$$
\mathcal{Y}_k^l = \sum_{j=0}^{N-1} (\mathcal{A}_k)_j^l \, y^j \,.
\tag{C.8.7}
$$

From this expression follows

Theorem C.14. *For commutative numbers, $\mathcal{Y}_k^m$ can be considered as the elements of a matrix. This matrix is a linear combination of the matrices $\mathcal{A}_k$ by means of the components y^k*

Proof. By using in the passage from the third to fourth member, the commutativity of the numbers, we have

$$
\begin{aligned}
\mathcal{Y}_k^m &\equiv \sum_{j=0}^{N-1} C_{kj}^m y^j \equiv \sum_{j=0}^{N-1} y^j \, C_{kj}^m \equiv \sum_{j=0}^{N-1} y^j \, C_{jk}^m \\
&= \sum_{j=0}^{N-1} y^j \, (\mathcal{A}_j)_k^m \equiv \sum_{j=0}^{N-1} (y^j \, \mathcal{A}_j)_k^m \equiv \left(\sum_{j=0}^{N-1} y^j \, \mathcal{A}_j \right)_k^m \\
&\Rightarrow \mathcal{Y} = \sum_{j=0}^{N-1} y^j \, \mathcal{A}_j \,.
\end{aligned}
\tag{C.8.8}
$$

$\square$

Theorem C.15. *The matrix $\mathcal{Y}$ matches the characteristic matrix (2.1.4) as well as the matrix $\mathcal{M}(\mathbf{y})$ (C.1.10). Therefore, the elements of $\mathcal{M}(\mathbf{y})$ can be obtained by means of the practical rule (2.1.6), which is equivalent to (C.8.2).*

Proof. By means of (C.8.6) and (C.1.1), it ensues that

$$\sum_{k=0}^{N-1} z^k \, \mathbf{e}_k \;\;\equiv\;\; \mathbf{z} = \mathbf{x} \cdot \mathbf{y} \equiv \sum_{j=0}^{N-1} x^j \, (\mathbf{e}_j \cdot \mathbf{y}) \equiv \sum_{j=0}^{N-1} x^j \, \mathbf{t}_j \qquad (C.8.9)$$

$$\equiv \;\; \sum_{j=0}^{N-1} x^j \sum_{k=0}^{N-1} \mathcal{Y}_j^k \, \mathbf{e}_k \equiv \sum_{k=0}^{N-1} \left(\sum_{j=0}^{N-1} \mathcal{Y}_j^k \, x^j \right) \mathbf{e}_k.$$

Then, from the first and last members, we have

$$z^k = \sum_{j=0}^{N-1} \mathcal{Y}_j^k \, x^j \Rightarrow \mathbf{z} = \mathcal{Y} \, \mathbf{x}. \qquad (C.8.10)$$

In a similar way, we can develop (C.8.9) by means of the matrix $\mathcal{A}$ (C.8.4), and obtain

$$\mathbf{z} = \mathbf{x} \cdot \mathbf{y} = \sum_{k=0}^{N-1} x^k \mathbf{e}_k \cdot \mathbf{y} = \sum_{k=0}^{N-1} x^k \, (\mathcal{A}_k \mathbf{y}) \equiv \mathbf{y} \cdot \mathbf{x} = \sum_{k=0}^{N-1} y^k \, (\mathcal{A}_k \mathbf{x}) \, . \qquad (C.8.11)$$

Considering the multiplication between two **H**-numbers, $\mathbf{x}$ and $\mathbf{y}$ in a new way with respect to Section 2.1, has allowed us the developments and formalizations exposed in the present appendix. $\square$

Bibliography

[1] J. Bailey, et al., *Precise Measuremen of the Anomalous Magnetic Moment of the Muon*, Nuovo Cimento A, **9**, 369 (1972). And in: Nature **268**, 301 (1977)

[2] E. Beltrami, *Saggio di interpretazione della geometria non-euclidea (1868) (Interpretation of Non-Euclidean Geometry)* in Opere Matematiche, Vol. **I**, 374, Hoepli, Milano, 1903; French translation in Ann. Scient. Éc. Norm. Sup. **VI**, 251, (1869)

[3] E. Beltrami, *Delle variabili complesse sopra una superficie qualunque (The Complex Variable on a Surface)*, in Opere Matematiche, Vol. **I**, p. 318, Hoepli, Milano, (1904).

[4] E. Beltrami, *Sulla teorica generale dei parametri differenziali (A General Theory on Differential Parameters)*, published in Memorie dell'Accademia delle Scienze dell'Istituto di Bologna, serie II, tomo VIII, pp. 551–590 (1868), and reprinted in Opere Matematiche, Vol. **II**, pp. 74–118, Hoepli, Milano, (1904). A unabridged English translation is reported in [7].

[5] L. Bianchi, *Lezioni sui gruppi continui finiti di trasformazioni (The Continuous Lie Groups)* , Spoerri, Pisa 1918

[6] L. Bianchi, *Lezioni di geometria differenziale (Treatise on Differential Geometry)*, 2nd edition (Spoerri, Pisa, 1902); German translation by M. Lukat (B. G. Teubner, Leipzig 1910); 3rd Italian edition (Spoerri, Pisa, 1922)

[7] D. Boccaletti, F. Catoni, R. Cannata and P. Zampetti, *Integrating the Geodesic Equations in the Schwarzschild and Kerr Space-times Using Beltrami's "Geometrical" Method,* General Relativity and Gravitation, **37** (12), (2005)

[8] D. Boccaletti, F. Catoni, V. Catoni, *The "Twins Paradox" for Uniform and Accelerated Motions*, Advances in Applied Clifford Algebras, **17** (1), 1 (2007); *The "Twins Paradox" for non-Uniformly Accelerated Motions*, Advances in Applied Clifford Algebras, **17** (4), 611 (2007)

[9] L. Bonolis, *From the Rise of the Group Concept to the Stormy Onset of Group Theory in the New Quantum Mechanics.* Rivista del Nuovo Cimento, **27** (4–5), 1 (2005)

[10] G. H. Bryan, *Commemoration of Beltrami (his life and works)* Proceeding of The London Mathematical Society, Vol. XXXII, pp. 436–8 (1900)

[11] W. K. Bühler, *Gauss — A Biographical Study* (Springer, 1981), Chap. 9

[12] J. J. Callahan: *The Geometry of Spacetime*, Springer, Berlin (2000)

[13] E. Cartan, *"Oeuvres Complètes"*, Gauthier Villars, Paris, (1953)

[14] G. Casanova, *L'algebre de Clifford et ses applications,* Advances in Applied Clifford Algebras, **12** (S1), 1 (2002)

[15] F. Catoni and P. Zampetti, *Two-dimensional Space-time Symmetry in Hyperbolic Functions,* Nuovo Cimento. B, **115** (12), 1433 (2000)

[16] F. Catoni, R. Cannata, E. Nichelatti and P. Zampetti, *Hyperbolic Variables on Surfaces with Non-definite Metric (Extension of Beltrami Equation),* (in Italian), ENEA Tecnical Report, RT/ERG/2001/12 (2001)

[17] F. Catoni, R. Cannata, V. Catoni and P Zampetti, *Introduction to Function of Hyperbolic Variables and h-conformal Mapping,* (in Italian), ENEA Tecnical Report, RT/2003/23/INFO (2003)

[18] F. Catoni, R. Cannata, V. Catoni and P Zampetti, *Hyperbolic Trigonometry in Two-dimensional Space-time Geometry,* Nuovo Cimento B, **118** (5), 475 (2003);
Two-dimensional Hypercomplex Numbers and Related Trigonometries and Geometries, Advances in Applied Clifford Algebras, **14** (1), 48 (2004)

[19] F. Catoni, R. Cannata and E. Nichelatti, *The Parabolic Analytic Functions and the Derivative of Real Functions,* Advances in Applied Clifford Algebras, **14** (2), 185 (2004)

[20] F. Catoni, R. Cannata, V. Catoni and P Zampetti, *N-dimensional Geometries Generated by Hypercomplex Numbers,* Advances in Applied Clifford Algebras, **15** (1), 1 (2005)

[21] F. Catoni, R. Cannata, V. Catoni and P. Zampetti, *Lorentz Surfaces with Constant Curvature and their Physical Interpretation,* Nuovo Cimento B, **120** (1), 37 (2005)

[22] F. Catoni, R. Cannata, E. Nichelatti and P Zampetti, *Hypercomplex Numbers and Functions of Hypercomplex Variable: a Matrix Study,* Advances in Applied Clifford Algebras, **15** (2), 183 (2005)

[23] F. Catoni, R. Cannata, and P. Zampetti, *An Introduction to Commutative Quaternions,* Advances in Applied Clifford Algebras, **16** (1), 1 (2006);
An Introduction to Constant Curvature Segre's Quaternion spaces, Advances in Applied Clifford Algebras, **16** (2), 85 (2006)

[24] F. Cecioni, *Sopra una relazione tra certe forme differenziali quadratiche e le algebre commutative (A Relation Between the Differential Quadratic Forms and the Commutative Algebras),* Annali di Matematica, **XXX**, 275 (1923)

[25] B. Chabat, *Introduction à l 'Analyse Complexe,* Vol. **I**, Mir, Moscou (1990).

[26] H.H. Chen, S. Thompson, *Computer-aided Displacement Analysis of Spatial Mechanisms Using the C^{II} Programming Language*. Advances in Engineering Software, Vol. 23 (3), pp. 163–172 (1995);
Dual polinomial and Complex Dual Numbers for Analysis of Spatial Mechanisms, Proc. 1996 ASME design engineering technical conference and computers in engineering conference. August 18–22, 1996, Irvine, California

[27] C. Davenport, *Commutative Hypercomplex Mathematics*
http://home.usit.net/ cmdaven/cmdaven1.htm

[28] M.P. do Carmo, *Differential Geometry of Curves and Surfaces*, Prentice-Hall, Inc., New Jersey (1976)

[29] B. Doubrovine, S. Novikov and A. Fomenko, *"Géométrie contemporaine"*, **I**, Mir, Moscou (1985)

[30] N. Éfimov, *Géométrie supérieure*, Mir, Moscou (1985)

[31] A. Einstein, *On the electrodynamics of moving bodies*, English translation published as an appendix of: Arthur I. Miller, *Albert Einstein's special theory of relativity: Emergence (1905) and Early Interpretation (1905–1911)*, Springer (1998)

[32] A. Einstein, *Die Grundlagen der allgemeinen Relativitäts theorie*, Annalen der Physik, (4), Vol. **49** pp. 769–822 (1916). We refer to the English translation in *The Collected Papers of Albert Einstein* Vol. **6** p. 146, Princeton University Press, Princeton (1997)

[33] A. Einstein, *On the Method of Theoretical Physics*, Clarendon Press, Oxford (1934)

[34] L.P. Eisenhart, *Riemannian Geometry*, Princeton University Press, Princeton (1949)

[35] L.P. Eisenhart, *Non-Riemannian Geometry*, Am. Math. Society, New York, (1927)

[36] G. Fano *Geometria non Euclidea (Non-Euclidean geometry)*, Zanichelli, Bologna, (1935)

[37] P. Fjelstad, *Extending Relativity Via the Perplex Numbers*, Am. J. Phys., **54**, 416 (1986)

[38] P. Fjelstad and S. G. Gal, *Two-dimensional Geometries, Topologies, Trigonometries and Physics Generated by Complex-type Numbers*, Advances in Applied Clifford Algebras, **11**, 81 (2001)

[39] C.F. Gauss, *Astronomischen Abhandeungen* **III** (1825)

[40] G. Goertzel and N. Tralli, *Some Mathematical Methods of Physics*, Mc Graw-Hill Paperbacks, New York (1960)

[41] W.R. Hamilton, *Elements of Quaternions*, first edition (1866), reprinted from the second edition (1899) by Chelsea Pub. Com., New York (1969)

[42] D. Hestenes and G. Sobczyk, *Clifford Algebra to Geometric Calculus*, Reidel, Dordrecht (1984)

[43] D. Hestenes, P. Reany and G. Sobczyk, *Unipodal Algebra and Roots of Polynomials*, Advances in Applied Clifford Algebras, **1**, (1) 51 (1991)

[44] J. Keller, *Complex, Duplex and Real Clifford Algebras*, Advances in Applied Clifford Algebras, **4**, 1 (1994)

[45] J. Keller, *Theory of Electron*, Kluver Academic Publishers, Dordrecht (2001)

[46] L. Koulikov, *Algèbre et théorie des nombres*, Mir, Moscou (1982)

[47] M. Lavrentiev and B. Chabat, *Effets Hydrodynamiques et modèles mathématiques*, Mir, Moscou (1980)

[48] M. Lavrentiev and B. Chabat, *Méthodes de la théorie des fonctions d'une variable complexe*, Mir, Moscou 1972

[49] T. Levi-Civita, *The Absolute Differential Calculus*, Blackie & son limited, London, (1954)

[50] S. Lie and M. G. Scheffers, *Vorlesungen über continuerliche Gruppen*, Kap. **21**, Teubner, Leipzig (1893)

[51] E. Minguzzi, *Differential Aging from Acceleration, an Explicit Formula*, Am. J. Phys., **73** (9), 876 (2005) and the references therein.

[52] C.M. Misner, K.S. Thorne and J.C. Wheeler, *Gravitation*, W.H. Freeman and Company, S. Francisco (1970)

[53] U. Morin, *Ricerche sull'algebra bicomplessa (Researches on Bicomplex Algebra)*, Memorie Accademia d'Italia, **VI**, 1241 (1935)

[54] P.M. Morse and H. Feshbach, *Methods of Theoretical Physics*, Mc Graw-Hill, New York (1953)

[55] G.L. Naber, *The Geometry of Minkowski Spacetime. An Introduction to the Mathematics of the Special Theory of Relativity*, Springer-Verlag, New York (1992)

[56] H.C. Ohanian, *Gravitation and Spacetime*, W.W. Norton e Company Inc., New York (1976)

[57] B. O'Neill, *Semi-Riemannian Geometry*, Academic Press, New York (1983)

[58] A. Pais, *"Subtle is the Lord..." The Science and the Life of A. Einstein*, Oxford University Press, (1986)

[59] H. Poincareè, *Sur les hypothèses fondamentales de la géométrie*, Bull. Soc. Math. de France, **XV**, 203, (1887)

[60] G.B. Price, *An Introduction to Multicomplex Spaces and Functions*, Marcel Dekker Inc., New York (1991)

[61] B. Riemann, *Gesammelte Mathematische Werke (Collected Papers)*, Springer-Verlag, Berlin (1990)

[62] W. Rindler, *Essential Relativity*, Van Nostrand Reinhold Company (1969), pp. 53–54.

[63] S. Roberts, *King of Infinite Space: Donald Coxeter, the Man Who Saved Geometry*, Walker and Company (2006)

[64] M.G. Scheffers, *Sur la généralisation des fonctions analytiques*, C. R. Acad. Sc., **116**, 1114 (1893); and Berichte Verhandl. Sächsischen Akad. Wiss. Leipzig Mat.-Phis. Kl **45** (1893) and **46** (1894)

[65] G. Scorza, *Corpi numerici ed Algebre (Numerical corps and algebras)* Principato, Messina (1921)

[66] G. Scorza-Dragoni, *Sulle funzioni olomorfe di una variabile bicomplesa (The Analytic Functions of a Bicomplex Variable)* Memorie Accademia d'Italia, **V**, 597 (1934)

[67] C. Segre, *Le rappresentazioni reali della forme complesse e gli enti iperalgebrici (The Real Representations of Complex Elements and the hyperalgebric entities)* Math. Ann., **40**, 413 (1892)

[68] F. Severi, *Opere Matematiche (Collected Papers)*, Vol. III, pp. 353–461, Acc. Naz. Lincei, Roma (1977)

[69] Y.V. Sidorov, M.V. Fednyuk and M.I. Shabunin, *Lectures on the Theory of Functions of a Complex Variable*, Mir, Moscou (1985)

[70] G. Sobczyk, *Hyperbolic Number Plane,* The College Mathematical Journal, **26** (4), 268 (1995)

[71] L. Sobrero, *The Functions of Hypercomplex Variable and Application to Mathematical Theory of Elasticity* (in German), Theorie der ebenen Elastizität, Teubner, Leipzig (1934); Italian translation in Memorie Accademia d'Italia, **VI**, 1 (1935)
J. Edenhofer, *Function Theoretic Methods for Partial Differential Equations*, Lecture notes in math., Springer, Berlin, **561**, 192 (1976)

[72] A. Sommerfeld, *Partial Differential Equations in Physics*, Academic Press, New York (1949)

[73] N. Spampinato, *Sulle funzioni totalmente derivabili in un'algebra reale o complessa dotata di modulo (The Analytic Functions in Real or Complex Algebra with Unity)*, Rendiconti Accademia Lincei, **XXI**, 621 and 683 (1935)

[74] A. Sveshnikov and A. Tikhonov, *The Theory of Functions of a Complex Variable*, Mir, Moscou (1978)

[75] I.N. Vekua, *Generalized Analytic Functions*, Pergamon Press, London (1963)

[76] J.H. Wedderburn, *On Hypercomplex Numbers*, Proc. London Math. Soc. **6**, 77 (1908)

[77] T. Weinstein , *An Introduction to Lorentz Surfaces*, Walter de Gruyter, Berlin, (1996)

[78] H. Weyl , *Space, Time, Matter*, Dover, New York (1950)

[79] S. Wolfram, *Mathematica: a System for Doing Mathematics by Computer*, Addison Wesley, Reading (1992)

[80] I.M. Yaglom, *Complex Numbers in Geometry*, Academic Press, New York (1968)

[81] I.M. Yaglom, *A Simple Non-euclidean Geometry and its Physical Basis*, Springer-Verlag, New York (1979)

Index

Frontiers in Mathematics

This series is designed to be a repository for up-to-date research results which have been prepared for a wider audience. Graduates and post-graduates as well as scientists will benefit from the latest developments at the research frontiers in mathematics and at the "frontiers" between mathematics and other fields like computer science, physics, biology, economics, finance, etc.

Advisory Board

Leonid Bunimovich (Atlanta), Benoît Perthame (Paris), Laurent Saloff-Coste (Rhodes Hall), Igor Shparlinski (Sydney), Wolfgang Sprössig (Freiberg), Cédric Villani (Lyon)